CONTEMPLATING CONCEPTS(41)

Exploring the Conceptual Landscape

David Alan Brown

ISBN: 978-0-615-35363-0

Produced in the United States of America

SUMMARY OUTLINE

CONTEMPLATING CONCEPTS: Exploring the Conceptual Landscape

Table of Contents

PREFACE

Self-discovery is an adventure like no other. It can take many forms, but the process is always engaging and surprising. The collected essays and notes included here are a record of one such adventure through inner space.

The exploration of mind has been compared to exploring the last great frontier. The trouble with this metaphor is that it is the mind that, in some sense, creates the landscape. As a result you can never really be sure of what it is you are exploring! Is the landscape real or imagined? But, that just makes it all the more sublime. Something remarkable is going on inside our heads, but you have to look carefully to begin to understand or appreciate it. We take our deepest concepts of ourselves and the world for granted, but they are anything but obvious.

I have attempted to organize my notes in a more-or-less logical progression, but research follows meandering trails. These notes are ultimately directed toward the problem of how best to understand and represent conceptual knowledge for analysis, but the precursor investigations cover a wide swath from perception to conception. As alluded to, the studies seem as much construction as discovery.

The book covers three main topics. Part I is a fairly wide-ranging examination of the general nature of concepts—what they might be, how they might work, etc. Part II develops an explicit method for representing concepts to facilitate analysis and knowledge organization. The third major topic, perception, is covered in a rather large Appendix. Perception is the grist for our conceptual mill, and an attempt has been made to build a reasonable, if speculative, narrative for perception as a basis for conception, although with radically different forms of representation.

To give this adventure significance beyond mere enjoyment (although that seems enough), I often contemplate how things might be different if we could readily clarify the content of our thoughts and test the consequences of our opinions. Lately, I find I can only listen to debating political pundits for about three minutes, before switching off the set. Clear thinking may turn out to be a Sisyphean goal, but few would deny a need for it. Much frustration in our everyday dealings with each other may lie in our inability to clearly understand the content of our own or others' convictions. If you happen to be one of those adventurers who enjoy the frolic of ideas and wonder about why you think the way you do, read on and join in!

Now a cautionary note on terminology: Philosophers engaged in highly technical discussions over the years have developed their own technical terminology as a means of avoiding mere verbal disputes about substantive issues. Properly understanding much of this technical terminology presumes a deep familiarity with the issues, and relies on shared meanings. For example, certain terms such as "universal" and "property" often entail a metaphysical commitment to their existence in reality. But in the present discussion of concept networks, no such metaphysical commitment or prior technical understanding is implied. Such terms are only used as labels for concepts. Even the term "concept" should be regarded carefully, since its meaning in this text differs subtly from one of its common philosophical meanings. For example, a concept (as a *representation* in someone's brain) is distinguished from the content of a concept (as an *interpretation* of a representation that can be shared with others). While the concept net terminology is broadly suggestive of colloquial meaning, technical meaning is still latent in many contexts.

Any major project with a wide scope, like this book, necessarily builds on the contributions of many people. In addition to the scores of dedicated researchers mentioned throughout, I would personally like to thank Carol Sue Brown for her excellent rendering of the graphic materials, and Jeffrey Robert Gunn for our many insightful discussions about every concept under the sun. My artist wife, Lynne, is also a major contributor, providing a stable base for creative thinking and offering comments which have nothing to do with the topic under discussion, at first glance! Finally, thank you John Scott McVicker for the 3D painting appearing on the book cover. Entitled *One Tear Only*, I like to think of it as a philosopher's tear, shed when facing the frustratingly strange comprehensibility and incomprehensibility of our own minds. If you like this kind of "deep" art, you can contact him at mcvickerstudios@sbcglobal.net. Comments on the book are, of course, welcome too at contemplatingconcepts@gmail.com.

PART I: THE NATURE OF CONCEPTS

Part I examines the idea of a concept from a variety of viewpoints—namely, concepts as an elaboration of underlying perceptual schemas, the interpretation of conceptual representations, and concept structure and function. These initial studies form the basis for the explicit conjectures and representations developed in Part II.

1. PERCEPTUAL-CONCEPTUAL-LINGUISTIC CONNECTIONS

An Information Processing View of the Brain

Information is one of those eminently useful, but frustratingly nebulous, terms that seems to turn up everywhere. It figures prominently in our best theories about the evolution of the whole universe, as well as in the tiny strange world of quantum effects in atoms. In between these extremes, it governs the unfolding of many natural processes, including most notably life itself. It is also the basic idea behind the invention of the computer as a man-made information processing device.[1]

Some physicists and philosophers (von Baeyer, 2003; Wheeler, 1990) are beginning to think of information as one of the fundamental building blocks of the universe, on a par with other fundamental physical properties like space, time, mass, and energy. In fact, a recent nanoscale experiment (Toyabe, Sagawa, Muneyuki, & Sano, 2010) has demonstrated that information can be converted into energy, so information does appear to have a physical basis.[2] Yet while the 20th century saw great progress in understanding such properties, especially Einstein's famous unifications of space-time, mass-energy, and geometry-gravitation, a comprehensive theory of information remains elusive. The ideas put forward here hint that information and *meaning* may be coupled similarly in some fundamental way, as yet only dimly considered. Perhaps an insightful unification of the two concepts, which at the same time explains their mutual significance, awaits a 21st century breakthrough.

In applying the concept of information more locally to ourselves, few will doubt that our brains process information. The brain receives sensory signals from the environment, analyzes those signals, and then produces changes for our benefit. This *input>process>output* pattern is the core of the information processing paradigm. Information processing, as usually understood, is inextricably linked with the computer, but the brain is nothing like a computer.[3]

To put it mildly, the flow of information in the brain is convoluted compared to a computer. The flows between different areas are highly recurrent and dynamically adaptive; the flows within structured

---

[1] From a cosmic perspective, expansion of the universe apparently started from an ordered, nonhomogeneous state, whose minimal description expressed the information content of the universe at that time. Since the big bang, the universe has been gradually increasing entropy in accordance with the 2nd law of thermodynamics, apparently increasing its overall information content. In the quantum world, information seems to exist in a state of superposition, where several distinct pieces of information become entangled with each other. For life processes, information is encoded in DNA. For a computer, information is produced by programs operating on data.

[2] One bit of information converts to a miniscule 3×10^{-21} joules of energy at room temperature.

[3] A computer basically executes an explicit program of instructions. The brain, on the other hand, seems to manipulate models of various situations. Both process information, however.

nuclei seem to involve statistical population coding, rather than direct representation; and the firing of the same neurons often means different things at different times. Perhaps this should not be surprising, since evolution weaves a patchwork quilt based on trial and error.

Nevertheless, there are some significant regularities to the flow of information. One straightforward way of visualizing the flow is to distinguish two major information pathways: one leading from sensory-related inputs on up to the rear half of the cortex and finally to the prefrontal cortex; the other from the prefrontal cortex back toward the action-related outputs of the motor cortex and down to the effectors. These kinds of pathways embody the famous Stimulus-Response loops investigated behaviorally in the middle of the last century. Although there are important exceptions to this simplistic information processing model, most modern models seem to have reinterpreted rather than rewritten its basic precepts.[4]

The following figure illustrates the central ideas. Environmental stimuli (light, sound, pressure, etc) serve as input to the various receptor and low-level sensory processing functions. Following data filtering and transformation, the information is often consolidated in topographic maps, where a map depicts a particular perceptual attribute over a spatial or temporal field—eg, contour lines where luminance changes dramatically in the visual field, highly correlated with the edges of objects. Higher-level perceptual processes ranging over multiple maps then appear to lead to perceptual schemas, templates, and script representations of environmental objects, relations, and situations—eg, a schema of basic car parts, or a script about how to change a tire. At the highest level of processing, core domain models are believed to analyze object occlusions/contact/support, spatiotemporal relationships like cause/effect, others' actions, push/pull force dynamics, and to provide intuitive estimates of quantity (among many other functions). Such core domain models likely rely on geometric and dynamic simulations of object behavior. Executive functions evaluate the global state-of-affairs, make decisions, and direct attention; while the goal and planning processes develop elaborate conceptual models, involving beliefs and inference about what is and can happen. The conceptual models may then trigger the formulation of ideomotor[5] actions, which are subsequently passed on to motor programs for execution by low-level motor control processes.[6]

Bridging the input and output pathways are a variety of mental processes, the most salient of which are the reflex arcs, the automatic (eg, habitual) processes, and the executive processes. The reflexes are instinctive responses to stimuli, such as the well-known knee-jerk reaction or the grasping action when something is placed in an infant's hand. Reflex responses are normally smoothly integrated with responses coming from the higher-level automatic and executive processes, but they are difficult to inhibit. Automatic processes mediate more complex, but still nonvoluntary, responses to stimuli. These processes include learned skills and habits, and draw on implicit procedural memories. Such

[4] One major exception is the discovery of massive back-directed connections between processing stages, in addition to the well-documented forward connections. Such back connections form complex processing loops that severely impact any straightforward understanding of the processing stages.

[5] Having an idea about performing an action can automatically trigger the action itself, unless it is inhibited by higher-level processes.

[6] As usual, take any computer-related metaphors with a grain of salt.

automatisms (eg, walking, vocalization, stereotypical behaviors) are generally unconscious and effortless, although well-integrated and sometimes hard to dissect from higher-level processes. The executive processes include conscious awareness and thinking, along with many unconscious creative problem-solving and choice/selection activities.

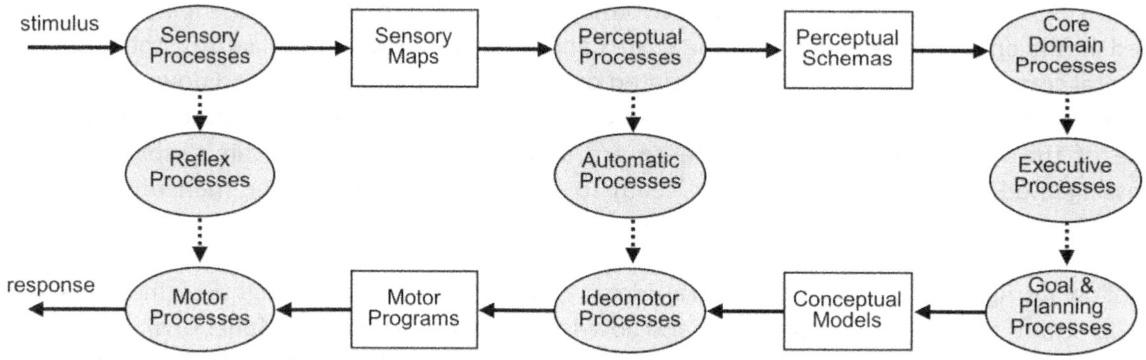

Figure 1: Information Processing Loops in the Brain

In 1990 MacLean (MacLean, 1990) proposed his influential triune model of brain structure. In this model the newer primate brain is viewed as an evolutionary elaboration on an older reptilian brain. Starting with the instinctual behaviors provided by a spinal cord and brainstem (reptile), a limbic system (mammal) subsequently evolved on top of the brainstem to enable the expression of complex emotions and motivations, like foraging, sex rituals, fear. The neocortex (primate) is likewise an addition on top of the limbic system that enables cognitive behaviors like problem-solving and socialization. The three bridging functional loops (reflex, automatic, executive) mentioned above would seem to dovetail with MacLean's structural proposal. If so, each major functional loop corresponds to a qualitatively more sophisticated elaboration of a previous information processing capacity (as opposed to simply the addition of more-of-the-same functions). Qualitatively higher levels of functionality (eg, language and thinking) evidently depend on qualitatively higher levels of organization (eg, neocortex). It makes one wonder what potential qualitative advances for organizing and using information beyond today's executive faculties might await our posthuman descendants—advances which we do not have the capacity to grasp.[7]

[7] It is tempting to propose "conscious processes" as the latest qualitative advance, but many formal studies—eg, see (Wegner, 2002; Dennett, 1991)—currently support the hypothesis that consciousness is an epiphenomenon, as opposed to a causal process affecting our actions. We probably only believe our conscious thoughts cause our actions. This conclusion, however, does not mean we are not agents acting on the world. It's just that our actions are probably the result of unconscious, rather than conscious, processes.

Nascent Meaning

The previous diagram spotlights the difference between sensory and perceptual processes, and their associated representations (labeled above as sensory maps and perceptual schemas). Although the structural distinction is commonly accepted in psychology, here the emphasis is on a subtle, but important, difference in what each system is doing for us: a difference that traces the lineage of meaning to perception. Sensation and perception are viewed as performing qualitatively different information processing functions for animate beings, functions which have evolved more-or-less in chronological sequence to increase overall environmental responsiveness. But it is the evolution of perception which launched the first glimmers of meaning and true awareness.

A basic sensory map is an area of the brain organized to represent a stimulus feature (eg, contact, temperature, vibration) over a receptive or stimulus field, oftentimes an external body surface of some kind. Sensory maps essentially decompose complex stimulus fields into separate feature fields. In other words, a sensory map represents separable features of its stimulus field, with different maps representing different features of the same field. There are sensory maps that correspond to the sensation of touch, vibration, and temperature on the body's skin; to colors and level of brightness derived from sensing light on the eye's retina; and to sound frequency (pitch) and volume from the vibrations of the cochlea in the inner ear, among many others.

A sensory map is said to register *proximal* information (Mack, 1978). The information in a sensory map is called proximal (with respect to the body) because it typically reflects the distribution of environmental energy impinging on body surfaces. The distribution of neural activation within the map more-or-less corresponds (or "maps") to the distribution of a feature in the stimulus field. A description of the content of a sensory map might include features like various edges, color shades, and intensities (of regions), or directions (of motion). Basic sensory maps are common in virtually all animate organisms, but less sophisticated organisms like worms (with contact sensing fields) and jellyfish (with light sensing fields) seem to operate without the benefit of perception. Such maps are typically tied directly to effector muscles which respond automatically when noticeable patterns (such as stroking or shadows) are present in the maps. The behavior of these organisms is probably best described in terms of the standard stimulus-response paradigm.[8]

Perceptual schemas, on the other hand, are structured representations of *distal* (Palmer, 1999) figures located out in the environment (away from the body). These perceived figures are quite varied, but prototypically include various kinds of environmental conditions, bounded figures and their backgrounds, and object perceptual properties and relationships. The schemas derive from individual sensory maps via binding information about multiple sensory features into potentially useful perceptual figures. While sensory feature processing remains mostly unconscious, perceptual feature processing can become conscious and forms the basis for our experience of perceptual figures and their properties. But the perceptual bindings can theoretically be composed in myriad different ways, potentially yielding

[8] An even simpler stimulus-response ties a sensory receptor directly to a motor effector (eg, through an interneuron for reflex responses), but this configuration does not make use of representations, like maps, at all.

many different perceptual figures and arrangements from the same underlying features.[9] Built-in perceptual biases (ie, innate expectations about likely and unlikely perceptual figures) ultimately and automatically select just one global binding—the single figure or scene we end up perceiving. Thus, a perceptual schema necessarily serves to represent a particular *interpretation* of the information available in the feature maps. That one interpretation of the raw sensory data then becomes privileged over all others, and may become conscious. In a wider sense, perception provides the organism with a potentially useful interpretation of chronically ambiguous sensory data, within a framework established by the organism's cognitive capacities and current goals.

The perceptual interpretation of sensory data entails much more than the random or arbitrary selection of one schema from among other competing schemas. The selection is informed by the way perception is internally organized to prefer some kinds of interpretation over others, based on a multitude of innate perceptual expectations about the environment (eg, that shadows normally appear below a body, because sunlight normally comes from above). Such expectations guide perception every step of the way, providing basic criteria for progressively filling in and grouping initially dissociated featural elements into one coherent figure.[10] A perceived figure is necessarily *meaningful* to the organism because perception only delivers figures that have been constructed according to adaptive criteria selected by evolution for that organism. Since the perceptual interpretive process was specifically fashioned to benefit the organism in some way, it will always at least be meaningful (in the sense of being comprehensible), even if it turns out that the interpretation is counterproductive (or illusory) in the present circumstances.

Uninterpreted sensory maps illustrate the general principle, to be expanded later, that *meaning* is not contained in a representation, any more than meaning is contained in an inert photograph. We can, however, properly say that sensory maps and photographs do convey *information* that the organism uses to produce meaning.[11] For any particular organism, the information conveyed by a representation is a measure of the maximal amount of meaningful *content* the organism is capable of extracting from the representation (assuming that the organism's interpretive processes are fully employed and operating under ideal conditions). In other words, from the organism's point-of-view, the total available information content is a quantitative property of the representation, reflecting an upper limit on exploitable content, typically measured in bits. But the amount of information actually acquired on any occasion will vary with the circumstances, subject to the competence and performance of the

---

[9] Technically, the organization of any environmental setting is *underdetermined*. Perceptual underdetermination is perhaps most clearly operative in our extraordinary construction of a 3D perceptual object from 2D sensory features, but it also figures in many perceptual illusions, where we mistakenly perceive an incorrect binding.

[10] As discussed later, interpretive processes operate in conception as well as perception. While perceptual biases are based mostly on innate expectations about the way things normally appear in the world, conceptual biases are based on innate and acquired knowledge about the way the world normally is.

[11] Certain objects and events, regarded as signs or representations, are often said to "contain" or "carry" information. But this is potentially misleading, since it is the agent who actually produces the information from environmental stimuli, frequently by convention. My preference for "convey" reflects the more abstract nature of information.

organism's interpretive processes.[12] In this understanding, meaning is constructed by interpretive processes from information conveyed by the representation. And oftentimes the meaning goes considerably beyond the literal information derived from the representation. In a wider cognitive sense, sensory maps are *used by perception* to develop perceptual meaning, while perceptual schemas are used by conception to develop conceptual meaning.

It is oftentimes plausibly said that the primary function of perception is to develop a veridical (ie, accurate or objective) representation of the environment.[13] And indeed, perception does seem to represent reliably the vast majority of the time, barring unusual external or internal conditions. This is because perception is based on sensation, which directly responds to the environment. However representational accuracy in perception is a dubious claim, because the only interpretations it can produce are ones that it has evolved to produce, within the framework of the organism's perceptual capacities. Furthermore, perception does not deliver reality to us on a plate. What we call external perceptual reality is better described as a sophisticated internal perceptual model derived from piecemeal perceptual clues. Perception produces model-based interpretations consistent with historically useful, *good enough* solutions to problems of perceptual interpretation, ones that have previously generated a differential benefit for the organism. And good enough is not quite the same thing as veridical. Many good ideas are pragmatically useful without being true. (However, good enough interpretations can still be *objective*, in the sense of being preferred among those organisms with roughly the same perceptual faculties.) Perception does not have the scope to represent reality as it is—only as it appears from the current point-of-view. Conception, on the other hand, does aim to represent veridically, although even here, our best theories of reality are fundamentally limited by our local perspective and by inherent limitations on conception itself.

Perceptual & Conceptual Objects

This is a book about concepts. But concepts are just one part of the versatile system of conceptual representations and processes we call *conception*, in which concepts are created, combined, evaluated, and otherwise used. Conception, in turn, is only one component of a wider information processing capacity (often labeled *cognition*), broadly encompassing perception, conception, and language. In their respective roles, perception delivers rich sense impressions of the world, conception develops coherent beliefs from those impressions, and language expresses our beliefs in ways that can be logically analyzed and shared with others. However, such subdivisions are somewhat controversial in psychology (Gershkoff-Stowe & Rakison, 2005), because it is invariably difficult to say precisely where one cognitive

---

[12] One could also talk about the amount of *raw* information inherent in the representation itself, independent of the organism, as is common in the communications information theory pioneered by Shannon. In this case, the maximal amount of raw information simply depends on how many possible distinct patterns the representation can hold, independent of interpretation. Such raw information considerations are mainly important in analyzing noisy communications, rather than in interpreting meaningful communications.

[13] For one example of this position see (Burge, Origins of Objectivity, 2010), who has developed a comprehensive theory of perception incorporating tacit conditions of veridicality.

component ends and another begins. The structural (anatomic) and functional (behavioral) boundaries are blurred.

In particular, finding a clear boundary between perception and conception is problematic, when compared with the more noticeable boundary between conception and language. Few would confuse the concept of a dog (having four legs) with the word "dog" (having three letters), even though both seem to involve closely interacting discrete mental representations (namely, concepts and words). On the other hand, perception is likely grounded in fuzzy continuous judgments about perceptual objects (Gardenfors, 2000), subsequently sliced, diced, and cooked by analog-to-digital conversions of information to support conceptual representations (Dretske, 1981). The distinction between one's sensory-based percept of a particular dog and the concept of that dog intuitively seems more nuanced, without a sharp intervening boundary, probably due to an overlap of information derived from perceptual and conceptual representations. We readily comprehend that the individual dog before us has both a tail (observable) and a heart (unobservable), without fully appreciating that these pieces of information are necessarily derived from different sources. Further dissociations between perception and conception are evident in conceptual abstractions, which often exhibit little or no perceptual content.

Therefore, it seems worthwhile in this study of conceptual representation to investigate the perceptual/conceptual interface in greater depth, to see if a useful theoretical boundary can be justified. The objective is to distinguish in a methodical way between our intuitions of *perceptual objects,* as contrasted with *conceptual objects,* each serving as a representational paradigm in perception and conception, respectively. The following analysis suggests that there is a relatively clear dividing line between them, when looked at from an uncommon perspective, in spite of their high degree of integration.

If we are asked to reflect on our simplest possible concepts, we are likely to think about the concrete objects of everyday experience. We seem to create concepts first and foremost from the things we perceive. We merge all of the salient perceptual data coming to us from a localized region of space and time into the impression of a unified *perceptual object* (Palmer, 1999). The perceptual object is what we are paying attention to when we pick out an individual thing in our environment. But the impression of a unified perceptual object masks a complex, behind-the-scenes, construction process. Clinical observations suggest that the perceptual object is fashioned from an underlying sequence of movie-like "frames" that perception is automatically cutting and splicing together during the encounter (Zihl, Von Cramon, & Mai, 1983). As the focal object moves or otherwise changes its appearance, perception segments the experience into *percepts* and *episodes*, which are routinely stored in long-term memory as the objects and events comprising the experience (Coren, Ward, & Enns, 2004; Baars & Gage, 2007). Percepts and episodes are the relatively fixed "frames" characterizing the stable constancies of the objects and events between significant changes. You can reexperience an individual percept or episode by recalling a prior experience, and focusing in your mind's eye on one of its components—eg, how the toast looked after it landed jelly-side down on the floor (percept), or the way it flipped out of your hand (episode).

Percepts

Although the idea of persisting perceptual objects as being made up of snapshot-like frames is suggestive of the perceptual segmentation process, it is a rather poor analogy for describing its operation. Percepts (and episodes, as discussed in a subsequent section) are flexibly individuated according to our current goals and the current environment, rather than a fixed clipping process. The term "percept", as used above, is adopted to specifically denote this most basic unit of perceptual representation. It includes visual, auditory, olfactory/gustatory, pain, and probably emotional/feeling perceptual registrations, among others, whether conscious or unconscious, temporary or permanent. It is also more than the phenomenal experience of something, since it is capable of being recorded in a physical memory trace.

A percept is viewed as an elementary record of a perceptual object's perceptual attributes, normally acquired over a short period of time, during which the attribute values are more-or-less unchanging. A significant change in one or more attributes triggers a new percept for that perceptual object.[14] A percept, therefore, will be said to represent one *aspect* of the complete perceptual object, where each aspect reflects a different set of attribute values, and possibly even different attributes in case of major changes in the perceptual object. As discussed further in the Appendix, the idea of a percept is based on the binding of numerous figural attributes (eg, for vision: shape, color, size, etc), located at roughly the same time and place, and organized according to built-in perceptual grouping, tracking, and attentional processes. A percept can be thought of more concretely as a neural node with a number of connections to particular locations in various perceptual spaces, where each terminal location exemplifies some value of an attribute.[15] Different perceptual attributes are important for characterizing different objects, and the binding process selects relevant attributes depending on both built-in and learned selection preferences. From a conceptual viewpoint, the perceptual bindings can deliver up a rather eclectic assortment of phenomenally experienced percepts including bounded and dispersed figures, shadows, backgrounds, environmental conditions, settings, relationships, and parts, in addition to the whole concrete material objects of everyday experience. Any of these perceptual bindings can become a percept, ultimately attaining a degree of permanence in long-term perceptual memory.

Perceptual Objects

We normally experience the objects in our environment as fully three dimensional, in spite of the fact that our eyes deliver only two dimensional images projected onto our retinas. To better appreciate this constructive achievement, consider the picture of a vase below. Even though conceptually we realize it is a 2D flat image of a vase, by default our initial impression is of a 3D object, possessing a volume with handles arranged in 3D space. The *perceptual object* is the 3D model we intuit, in this case based on the

---

[14] Some percepts (eg, one perhaps formed while watching it snow) appear to record smoothly changing situations as constant rate units, rather than constant value units.

[15] Many percepts will also possess pointers to perceptual imagery (including visual, aural, haptic, pain, etc) representations. Imagery is thought to support our ability to run simulations in our mind's eye (Barsalou, 2003).

single aspectual *percept* derived directly from the 2D image. But typically, multiple percepts, representing different views, will become associated with a single perceptual object.

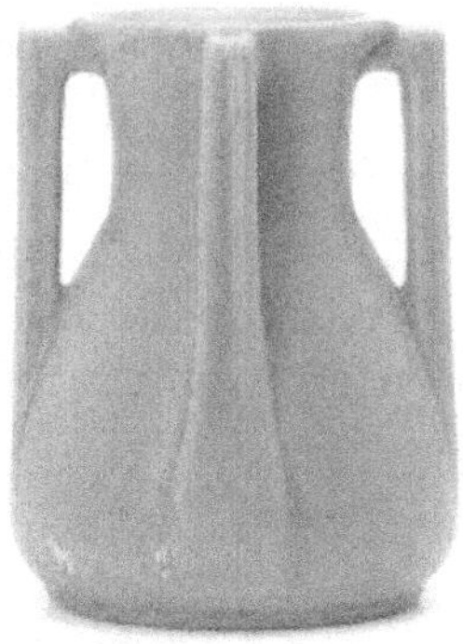

Figure 2: Example of 3D Perceptual Object Construction from 2D Visual Imagery

Constructing a coherent perceptual object from a sequence of percepts probably rests on our innate ability to track a localized perceptual binding over time, even as the binding attributes change. If each more-or-less unchanging set of bound attributes constitutes a single percept (eg, such that the heads and tails of a particular coin, as it is turned over, become registered as two percepts in sequence), then as a general rule, the linking of distinct percepts acquired during extended, mostly uninterrupted, tracking constitutes the perceptual object. A coherent perceptual object grouping is like an album of pictures organized chronologically of different poses and costumes you might get from a single "photo shoot" session. It reflects a simple clustering together of individual percepts, rather than an integration of those percepts into a single summary object, as may happen in conception. Moreover, each separate occasion of presentation or encounter with the same individual (from a conceptual viewpoint) may produce a new perceptual object. Each separate occasion initiates a new tracking event. These distinct perceptual objects, acquired on different occasions, may or may not subsequently be merged with prior perceptual objects or recognized as the same individual, depending on many perceptual and conceptual factors. The morning star (Phosphorus) and the evening star (Hesperus) are two famous examples of recurring perceptual objects that the ancient Greeks did not have the technology to unify as the planet Venus. The left side of the following figure illustrates the way individual percepts cluster together during a tracking event to produce the representation of a perceptual object.

The figure also shows a correspondence between perceptual representations diagrammed on the left side, and conceptual representations diagrammed on the right. Representations near the top of the perceptual hierarchy (ie, percepts and perceptual objects) are here interpreted as overlapping with representations at the bottom of the conceptual hierarchy (ie, modes and conceptual objects). The corresponding perceptual and conceptual representations are not identical, but have comparable content. While the perceptual representations are limited to sensory-related information, the conceptual representations include elaborations beyond sensation. The representational overlap facilitates the smooth transition between lower-level perceptual and higher-level conceptual operations, a transition so subtle that we are rarely aware of it. In fact, the construction of conceptual beliefs from perceptual impressions is normally automatic, only requiring that we actively pay attention to the impression, whether it is occurrent or remembered.

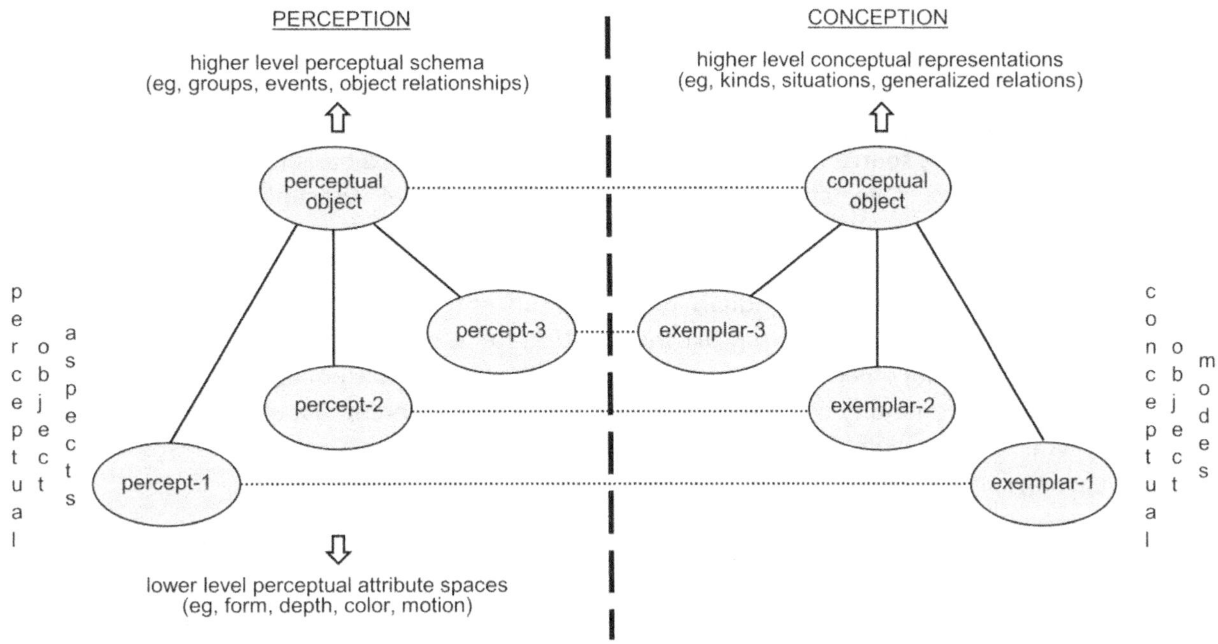

Figure 3: Perceptual/Conceptual Representation Interfaces

Conceptual Object Modes

The previous example of the vase lends itself to talking about the transition from a percept to what will be called a corresponding *conceptual mode* of an object. Under certain conditions of attention, an initial percept of the vase may trigger a conceptualization. Like a percept, previously described as a perceptual aspect of an object, the corresponding conceptual mode loosely represents a more-or-less fixed "snapshot" of the object over an interval of minimal change. But the interpretation associated with a conceptual mode is generally more objective, more quantized, and more elaborated than its underlying percept. A perceptual aspect is a way of appearing, but the corresponding conceptual mode is a way of being (related to an object's identity). Inference is used to draw conclusions about latent characteristics and possible origin, fill in gaps requiring background conceptual knowledge, and convert analog-based perceptual features into discrete conceptual characteristics.[16] For example, a conceptual mode of the

[16] To help limit semantic confusion, the term "feature" will be used to talk about a descriptive component of a perceptual object, while "characteristic" will apply to a descriptive component of a conceptual object. The semantic

vase might include an estimate of its current weight (say, 1 lb), value ($60), number of handles (4), and the average color (light green) based on active regions in the underlying perceptual color space. A different conceptual mode of the vase on a subsequent visit might characterize it as damaged and having only three handles.[17]

The need for recognizing the conceptual mode of an object, in addition to the conceptual object itself (see next section), may seem confusing at first. However, this characterization of percepts and corresponding conceptual modes aims to interface with the well-established research line from a certain *exemplar* model of concepts (Nosofsky, 1988; Barsalou, Huttenlocher, & Lamberts, 1998) in cognitive science, as discussed in more detail in Chapter 2: *Three Basic Varieties of Concept Representation*. If conceptual modes are the source of exemplars, then they can serve as a basis for the generalization and abstraction to conceptual objects generally. In this cognitive model, conception first individually re-represents, and perhaps consolidates, the multiple percepts in a perceptual object cluster, such that the percepts become reinterpreted as exemplar concepts. The multiple exemplars then serve as different modes of a complete, unique, individual, enduring conceptual object. Other perceptual object clusterings, from further encounters, may likewise participate in characterizing other significant modes of the enduring conceptual object. Such conceptual assignments are based largely on logical inference, identity, and panoptic continuity considerations, rather than mere perceptual tracking. As Quine notes, determining whether the penny in my pocket today is the same one that was there yesterday normally involves weighing some evidence (Quine, 1987).

Note that a conceptual mode (ie, exemplar), under this proposal, is not to be construed as an *instance* of the enduring conceptual object, as is common in the cognitive science literature. That is, Betsy the cow sleeping and Betsy the cow grazing are not two instances of Betsy the cow—particular things cannot have instances, as explained later. But particular things can have constituents of various kinds. A mode of an object is viewed as one of its generic constituents, along with other constituents like spatial and temporal parts (eg, stages in its life), compositional ingredients, and many others, depending on the object.

The enduring conceptual object is then simply the aggregate of all of its modes. In other words, it represents the mereological sum of its modes, including all potential modes for that object, whether those modes have actually been perceived or not. It does not represent a summary or integration of modes into a single prototype, although such higher-level concepts for some conceptual objects are proposed in Chapter 5: *Particulars as Unique Universals*. When we talk about the intuitive *concept* of an ordinary object, we seem to be talking about the representation of one concrete particular thing, termed here the conceptual object. But that unitary thing is really a composite idealization, depicting the single object in terms of the multiple ways it can be identified and understood.

---

distinction is based principally on the fundamental difference between perceptual and conceptual representational format—namely, analog vs discrete.

[17] The notion of an object having various modes of presentation is later generalized to the notion of a situation. From this broader perspective, an object participates in various situations (expressed in sentences), which describe relations between it and other objects of interest.

Conceptual Objects

The idea of a conceptual object, in distinction from the perceptual object, suggests an entity persisting beyond the perceptual occasion. It is also objective, and potentially shareable with others, in a way the private experiential perceptual object is not. And it seems to possess an intrinsic unity over an extended lifetime, no matter how long or short, through small or sometimes large changes. It also constitutes an extension of the sensory-limited content of the perceptual object to embrace generalized entities, ranging from reified perceptual figures to all manner of nonperceptual phenomena and abstractions. If something can be perceptually individuated in any way, it seems it can be elaborated into a conceptual object; and even if it can only be imagined, the conceptual object representation can serve as a useful locus for all manner of thoughts about it.

Conceptual objects (more accurately, the interpreted content of our object concepts) are individuated according to a specific set of strategies built-in to our conceptual faculties. These strategies use different representations and operations from those employed in perception. First, conceptual objects are understood as belonging to some *kind* (or type). Establishing a kind for an object is a basic organizing principle for concepts, and allows us to classify the object, giving it a stable place in a wider context vis-à-vis other objects. Second, a conceptual object is expected to possess discrete *characteristics*. An object's characteristics provide useful information about its relations with other objects as well as discrete attribute value assessments. Third, a conceptual object is expected to participate in occurrent *situations* along with other objects. The kinds of situations an object can participate in advises us in using the object for our specific goals. As an example, I conceptualize my pet Angus as a cat (kind), as weighing 15 lb (characteristic), and as tending to sleep in the garage during the summer (situation). Having a description in terms of kinds, characteristics, and situations amounts to fundamental information about a conceptual object: no description = no concept[18].

The foregoing analysis of the difference between the perceptual and conceptual object can also be viewed in terms of the controversy in cognitive science between the view-dependent and view-independent theories of object representation, in general. View-dependent theories (Ullman, 1998; Edelman & Duvdevani-Bar, 1997) argue that objects are mentally represented in an egocentric frame of reference relative to the perceiver (analogous to the perspective inherent in a photograph, based on the location of the camera). View-independent theories (Marr & Nishihara, 1978; Biederman, 1987) instead hypothesize that objects are stored in an allocentric frame of reference centered on the object itself (analogous to a generic model of the object, abstracted away from any particular point-of-view). Evidence favoring one of these two theories of object representation over the other has proven difficult to obtain experimentally, with each being roundly criticized for fundamental weaknesses (Biederman & Gerhardstein, 1993; Tarr & Bulthoff, 1995). However, the previously outlined framework suggests that both theories may nevertheless be correct. If perceptual aspects and conceptual modes can both be seen as representing objects in a view-dependent manner (based on individual encounters), and if

[18] But there may be active percepts and perceptual objects, not based on a description. For example, humans can reliably track four points moving randomly on a computer screen (called subitizing) evidently without forming any concepts (Pylyshyn, 2003).

perceptual object and conceptual object composite representations are seen as portraying objects in view-independent ways (based on multiple encounters), both theories can be accommodated. View-dependent and view-independent representations in both perception and conception may then team up to provide a more effective perceptual/conceptual interface than either could provide alone.

The foregoing analysis is also relevant to the controversy over whether nonhuman animals possess some rudimentary form of concepts. Animals obviously interact with their environment in sophisticated ways, but does this mean they possess primitive concepts in the same sense that young children possess immature concepts? The above model suggests that the divide is more fundamental, reflecting the distinction between perception and conception, rather than between immaturity and maturity of conceptual structure. We then share many high-level perceptual capacities with other animals, but conceptual capacities are, for the most part, uniquely human.[19] If we look at high-level perception and low-level conception as overlapping capacities, it becomes apparent how the two could be intuitively confused. Both are representing the same environmental stimuli in different ways. In the simplest cases, both represent the same observable object in terms of the different ways the object appears. Moreover, high-level perception need not culminate with representations of the perceptual object, as discussed further in Appendix C, *Late Perception*. There is now substantial evidence for nonhuman primates and many species of birds that novel perceptual objects are readily grouped into perceptual categories, and that such classifications support expectations of unobserved properties (Mareschal, Quinn, & Lea, 2010). Higher-order perceptual category *schemas* based on perceptual similarity in some nonhuman animals, therefore, probably correspond to conceptual *kinds* based on rule-like criteria in humans. Among the more advanced perceptual schemas, over and above those serving object classification, include: representations of spatial and temporal relations between objects and their parts, straightforward statistical generalizations of perceptual object features, and perceptual object functional affordances and settings. Taken together, such advanced perceptual capacities seem more than enough to account for nonhuman animal behavior, without positing a distinct conceptual capacity involving abstraction.[20]

---

[19] In a series of well-controlled experiments, Kanzi the chimp effectively demonstrated an understanding of English comparable to a two-and-a-half-year-old toddler (Savage-Rumbaugh, Murphy, & al., 1993). Such results provide evidence that the great apes may have already laid the foundation for the bridge between perception and conception. However, if it turns out other large-brained mammals like dolphins, whales, and elephants exhibit certain aspects of conceptualization, these will undoubtedly be cases of convergent evolution with significant differences in neurological structure, function, and probably content.

[20] The difference between *abstraction* and *generalization* is crucial to this argument. Generalization groups object instances together based on similarity of given attributes, while abstraction groups instances together based on conformance to assigned attributes (sometimes called rules or criteria). Dog, mammal, animal are successive generalizations of collie; but dog, species, taxon are successive abstractions. Generalization operates in both perception and conception, but abstraction only in conception. Thus dogs, spoons, and bombings are all amenable to various classifications based on attributes provided by perception (as well as conception for humans); but abstractions like species, utensils, and terrorist acts are not, as explained further in Chapter 4. Note also that particular immaterial things, like sets or patterns, are *immaterialities* and not abstractions, as commonly thought (discussed in Chapter 5).

From Perceptual to Conceptual Information Processing

This discussion of how information processing is organized in the brain talks about a fundamental separation of mental labor between perception and conception. However, evidence from early childhood development (Mandler, 2004; Carey, 2009) and statistical learning simulations (Rogers & McClellan, 2008) are interpreted by different researchers in different ways, leading some to favor a continuum of representational modes (Fodor, 1980; Macnamara, 1986; Schyns, Goldstone, & Thibaut, 1998) extending from sensory reception to executive processes, rather than two more-or-less distinct systems. The approach here reconciles the two views by distinguishing between information flow vs knowledge organization in the brain.[21]

Assuming the primary flow of information in the brain follows more-or-less established structural pathways laid down by evolution, information is usefully thought of as being processed in successive stages, with radical branching into many interconnected functional units occurring at the higher stages of processing.[22] The succession of stages is apparent anatomically in both the ascending sensory and descending motor pathways, while our flexibility to mix-and-match strategies depending on the situation suggests a wide array of interacting executive functions at the top. The processing of information is continuous from stage to stage and between high level functional units in the sense that information from one stage or unit feeds smoothly into other connected stages or units that add value to the information. The flow of information is continuous, without significant discontinuities, and is experienced as such. However, normal maturation of the structural stages and functional units also proceeds in *developmental* stages. And the basis for organizing the information in a unit may change between developmental stages. Most perceptual representations are relatively fixed[23] in the way they organize information throughout development (assuming more-or-less normal development conditions), but conceptual representations are more flexible, and undergo periodic reorganizations of knowledge. Some major reorganizations of content are tied to physical development (eg, those underlying the fundamental differences between childhood and adult conceptions of the self and reality, in general), while others are based on learning (eg, education and discovered inconsistencies in reasoning). In an organizational sense, then, conceptual knowledge can go through discontinuous jumps when existing knowledge is suddenly reorganized under a new paradigm.

In addition to the relatively new results on early childhood development indicating major developmental transitions in knowledge organization, the argument for two systems here is based primarily on hypothesizing a general mechanism or explanation for how perceptual organization in the child may evolve into conceptual organization in the adult. That the acquisition of useful hierarchies of categories, both concrete (eg, Fido) and abstract (eg, dog, animal, living thing), can be simulated in the laboratory

[21] Note, however, there is widespread agreement that conceptual knowledge and expectation can have a pronounced effect on perception—eg, even before the stage of figure/ground segmentation (Peterson & Gibson, 1994) or afterimage processing (Moscovici & Personnaz, 1991). But perception and conception serve different functional roles in spite of being highly interconnected.

[22] However, a functional unit is not necessarily localized to only one brain region. Many appear to be distributed over several regions.

[23] See *Appendices* for more detail on perceptual representations.

from purely statistical correlations among purely perceptual qualities, would seem to allow a purely perceptual starting point for acquiring all knowledge (Rogers & McClellan, 2008). Our capacity for detecting various perceptual qualities is largely innate, and innate statistical correlation processes among qualities can get higher-level categorization off and running.[24] Perception thus gives rise to conception. But adult categories go beyond perceptual considerations to reflect more sophisticated characteristics related to the nature or function or origin of things, rather than the mere appearance of things. And representing the world the way it "is" or "works" rather than how it "appears" requires a major shift in perspective during development. Perceptual qualities are not discarded in adulthood, but they are superseded by inferred nonperceptual characteristics, which then become the new basis for organizing knowledge. Such reorganizing processes are uniquely conceptual—they do not elaborate perceptual qualities; they add to, redefine, or override them. The transition process is distinctive, even though we often cling to pockets of early organization and thinking, as when we become superstitious, fall back on folk explanations of psychology, or simply forget that dolphins are not fish.

Although perception and conception are often thought of as constituting the two broadest categories of mental functioning in the ascending neural pathways, a number of other related information processing systems have also emerged in recent research, only some of which were included in Figure 1. These intermediate systems are strongly connected to perception and conception, but have distinctive representations and functions. For example, mental simulation programs (eg, imagining riding a horse) are fed by both perception and conception, and involve the dynamic integration of perceptual and conceptual representations over time. Mental spatial maps (eg, imagining the route from home to office) exhibit both types of information, as well. Other domain-specific innate processing systems, ranging from analog quantity estimations to cause/effect predictions to social mirroring projections, have more recently been classified as preconceptual core knowledge (Spelke & Kinzler, 2007) or core cognition (Carey, 2009). Likewise, intuitive feelings (eg, having a sense of a horse's weight) are part of a different operational system, normally identified with force dynamics (Talmy, 2003). A horse's weight, when represented conceptually, is an abstraction, but is grounded in the underlying force dynamics representations. Although perception and conception will continue to serve as the broad covering terms for mental information processing in subsequent discussions, it is important to remember that these global systems are composed of a number of interacting subsystems, interactions mediated in many cases by separate representational systems.

Perceptual vs Conceptual Representations

One would be hard pressed to find a structural boundary between perceptual and conceptual processing in the brain. But the various stages of information processing, extending from sensory reception, to the higher cognitive functions, and on to motor output, apparently use different kinds of

[24] The term "perceptual qualities" applies to anything observed or noticed in the environment. It includes not only simple properties like color and size of objects; but also richer qualities like numerical estimates of quantity, detecting the psychological state of a conspecific, and noticing cause/effect relationships.

representations of meaningful content, which can at least be functionally distinguished. The framework below highlights the major functions (Baars & Gage, 2007).

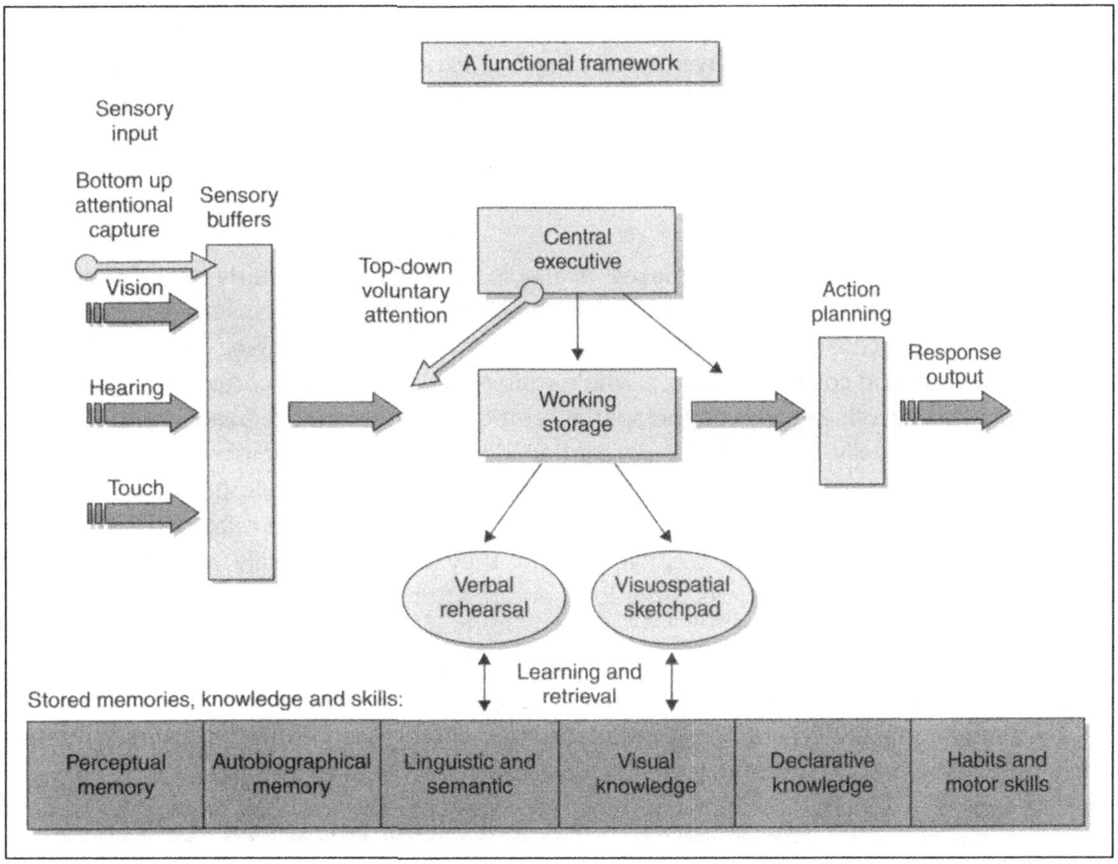

Figure 4: Functional Framework for Information Processing in the Brain (Baars & Gage, 2007)

Embedded in this framework, but not explicitly shown, are the multiple stages of information processing linking perception to conception and executive processes. As previously remarked, the various perceptual representations and operations are generally thought to culminate in *percepts* or *perceptual schemas*—eg, (Mandler, 2004)—encoding bundles of correlated multisensory qualities and their values.[25] Perceptual schemas are viewed here as akin to so-called *object files* (Kahneman & Treisman,

[25] The term "percept" is used in many conflicting ways in the literature on perception, but here it refers to the potential activation pattern that correlates with a coherent perceptual recognition or memory (eg, the imagery and feeling of petting one's dog). Physically, it is the tendency of a delimited interrelated part of the sensory cortex to become active when appropriately stimulated. Since many percepts probably share the same underlying cortical structure, which one becomes active would depend on how the structure is stimulated (and on the activity in neighboring structures). We experience the activation of some percepts consciously, others not; but all can have behavioral effects. It is a useful perceptual analogy to the term "concept" which refers to a similar nexus in the conceptual system.

1984), which are built up over the course of attending to objects in the environment, and which are thought to contain structured entries summarizing perceptual object *qualities*. An object's quality values are typically represented as points or regions in a variety of continuous, integral value spaces (eg, for color, size, tactile pressure). We seem to experience the activation of different regions in these multidimensional value spaces as different *qualia* (eg, the experienced redness of a rose or the texture of a tennis ball).[26] On the other hand, conceptual representations and operations functioning at the highest stages of mental processing encode bundles of model-like and proposition-like associations between more-or-less discrete entities, otherwise known as *concepts*. Concepts frequently arise as symbolic abstractions from percepts.

As already noted, perception usually makes available to conception fairly well-formed percepts, consisting of perceptual objects along with their perceptual qualities, parts, settings, affordances, etc. Any of these percepts is capable of evoking a concept. For example, on seeing a rabbit hop by, we are capable of focusing on and conceptualizing a whole rabbit object, its white cotton tail, the characteristic hopping motion, the bucolic setting, or the spatial relationship between the rabbit and a nearby hole. Which concepts we actually extract from perception depends on our mostly unconscious goals and expectations acting in accord with perceptual saliences. Furthermore, while percepts are constructed according to built-in perceptual faculties for processing color, shape, spatial relationships, episodes, etc; conceptual representations are truly creative in that they can additionally represent an infinity of nonperceptual objects, properties, kinds, situations, and relations of all sorts.

The primary evolutionary function of conception, as envisioned here, is to more effectively model the world than it could be modeled merely through perception. Conception accomplishes this wider function by augmenting underlying perceptual schemas constructed from sensorimotor interactions with the world. Perception can only model the superficial aspects of the world as it is experienced through the senses. Conception builds on the appearances delivered by perception, and though inference-like operations, corrects illusory first impressions and derives additional hidden aspects of experience. Conception is a useful evolutionary tool because it enables knowledge about things far removed from direct experience.

Perceptual Episodes and Impressions

When a coherent set of percepts occurs at roughly the same time and place, they often become bound to each other, forming a unitary perceptual *episode,* which then is recorded in long-term memory. Segmented episodes are organized and stored in episodic (or autobiographical) memory, describing what has previously happened to us. They are generally formed unconsciously, probably in the

[26] However, successfully separating out the individual qualia composing a percept is not straightforward, and requires developmental maturity. Children, somewhat surprisingly, have trouble analyzing the qualities of individual things, instead tending to perceive things as integral wholes. For example, learning the meanings of adjectives (eg, red) functioning independently of nouns seems to begin a couple of years after the meanings of nouns (eg, ball) are well established (Mintz & Gleitman, 2002).

hippocampus (Kumaran, Summerfield, Hassabis, & Maguire, 2009), and are roughly indexed by time and place. For example, an episode might record the percepts related to an encounter with an object or the unfolding of an event which we witnessed. Episodes are thus static perceptual representations in long-term memory, encoding the salient connections among concurrent percepts, subsequently interpreted as a delimited dynamic event.

The presence of similar or repeating episodes in episodic memory subsequently enables an unconscious automatic induction process to create a concept generalizing the individual episodes. The concept joins together the cluster of common features extracted from the episodes, and thus comes to represent a class of similar episodes, as characterized by similar features. This concept represents the holistic permanent particular object, place, event, or circumstance as constructed from the multiple views and sensory modalities recorded in each episode. Also transferred to the permanent object concept is discretized information derived from the multiple encounters about the relative distribution of perceptual features (eg, that Mama is sometimes smiling and sometimes frowning, sometimes talking and sometimes singing, with different likelihoods, but without explicit linguistic labels), as well as correlations among features (eg, that a smiling Mama is likely to be singing). The conceptual object, as an individual, is abstract, because it typically encapsulates multiple viewpoints and contexts, some of which may be incompatible with each other, over multiple encounters.

There are perhaps two fundamentally distinct processes of generalization at work, both using episodes as raw input. The first object generalization process consolidates episodes by extracting common perceptual qualities. These generalizations become reifications of entities and (samples of) substances, as characterized by their attributes (Millikan, 2000; Quine, 1969). In the infant, one cluster of qualities might eventually come to represent Mama, another milk, another animal, each serving to collect similar episodes under its purview, where overgeneralization is common. Once the concept is in place, a novel episode can then be classified by detecting its qualities. This generalization process gives rise to *object* (more technically, *continuant*) *and kind concepts*.

The second event generalization process consolidates episodes by event action type (eg, feeding, diaper changing, experiencing red).[27] The bottle feeding experiences thus become distinguished from the diaper changing experiences since their event action types are dissimilar enough to discriminate. Events also seem to be distinguished by various qualities, but the relevant qualities are very few (eg, static/dynamic, conclusive/nonconclusive, punctual/durative) and more sophisticated, when compared with object qualities. The consolidation of perceptual episodes into conceptual events gives rise to *situation* (more technically, *occurrent*) *and relation concepts*. It is primarily the ability to represent object, kind, situation, and relation concepts that enable and organize higher-level conceptual thought.

When a given percept or episode becomes temporarily active (eg, when looking at or remembering a scene), it will be said to produce a perceptual *impression*, a term first used by Hume to distinguish vivid

[27] The exposure to an object (as in object generalization) can also be viewed as an event, but we seem predisposed to discriminate between objects and events, perhaps reflecting a tangible/intangible distinction, or that objects typically participate in events, and so are different kinds of things.

perceptual experiences from the wispier experiences associated with thinking. The impression correlates with the spatiotemporal activity of the included percepts, and is available to both conscious experience (depending on attentiveness) and to unconscious cognition in general. An impression of some percept or some episode is a dynamic activation pattern in the perceptual neural substrate, corresponding to a partial or full activation of the underlying percept or episode. The dynamic pattern of a percept/episode activation therefore constitutes a temporary perceptual representation, just like the static pattern of connectivity among percepts in the episode constitutes a permanent perceptual representation. As representations, both support content when interpreted.

Encoding

A major issue in perception and conception is how we decide what objects, kinds, situations, and relations are contextually relevant in our diverse encounters with the world. James' "blooming, buzzing confusion" of sense data can be sliced and diced in many different ways, some ways being more useful than others to our purposes. The term *encoding* refers to the ability to focus on and register relevant aspects of experience in different contexts. The encoding process imposes order on stimuli that leads to conscious and unconscious interpretations of experience by creating useful percepts and concepts.

Experiments with early-grade-level children reveal that they usually cannot learn how a balance beam works of their own accord. They do not naturally encode the distance from the fulcrum to objects on the beam. Once they are taught the importance of this relation, they are then able to figure out balancing problems almost as well as older children (Siegler, 1983). Similarly, in learning to read handwriting we unconsciously come to encode numerous aspects of letter strokes, spacing, transitions, etc, based on repeated exposure to examples. A handwriting expert will additionally encode, either consciously or unconsciously, diagnostic characteristics thought to be correlated with psychological or emotional states. In all these examples what is encoded in a given circumstance depends partly on preparation, partly on presentation, partly on attention, and partly on goals.

It would appear that we come ready-made to explicitly encode object,[28] kind, situation, and relation elements from everyday experiences. These given elements are structured differently, and carry different assumptions about what they represent. An object typically has parts (which are other objects), while a kind has instances (which typically are objects). The core of a situation is typically an action of some sort, along with some participants in the action (normally, objects and kinds). A relation is a high-level abstraction (like a kind) which has situations as instances (instead of objects). Given the confines of our innate encoding tendencies, we are still free to encode the world in a wide variety of different ways. The selection of an encoding determines how things are subsequently perceived and conceptualized, in much the same way that choice of a problem representation determines how a problem is subsequently solved.

[28] "Object" is used in the generic sense of an entity or thing here. "Continuant" is often used in philosophy as a term of art for this generic concept.

As noted previously, some experiences are easier to encode than others. Encounters that are formalized like sports events, academic subjects, or occupations are relatively easy to encode. But first-time encounters like a ritual in a foreign culture may be difficult, since there are few default expectations to help organize the experience. But once an experience is classified or characterized, the classification can suggest other latent percepts and concepts that were not initially apparent. Encodings can also vary from very general to very specific (eg, red vs crimson, animal vs collie). The best encoding level of generality for an experience usually depends on context and goals.

Different domains and fields of knowledge (physical domain, social domain, mathematical domain, etc) encode and organize elements quite differently. Mature domains, like chemistry, tend to have clearly specified (whether or not clearly understood) percepts and concepts, while newer domains, like ecology, are still searching for useful encodings. Moreover, some domains are just inherently harder to encode than others—compare relatively concrete domains like phonetics with abstract domains like semantics.

To be encodable, the elements must be distinguishable from background noise, or capable of being enhanced in some way—eg, by attention. In other words, they must have sufficient salience, and be individuable from others of their kind and other similar things. Encoding is facilitated when the stimuli are presented together, rather than temporally or spatially separated; generalized associations are normally not learned if the stimuli are widely separated. Encoding is also easier if there is high correlation between stimuli (eg, one reliably follows the other). And properties that reflect an objective or quantitative attribute (eg, temperature, length) are in general easier to encode than subjective attributes (eg, beauty, wisdom).

Although the previous examples emphasize the encoding of conceptual information, at least two other major sources of information that we regularly encode into mental representations are noteworthy; namely, perceptual information and linguistic information. While conceptual information is encoded into concepts, perceptual information normally becomes encoded into percepts and episodes, and linguistic information into propositions. Any experience can become encoded into all three representational forms, and the memory of the experience (as measured by recall ability) increases as more forms are employed (Craik, 2007). For example, if you witness an automobile accident, your visual and auditory perceptual encodings of the occurrence will probably record the salient features of the situation in visual and auditory imagery. At the same time, you may initially organize the experience into perceptual schemas and episodes. Subsequently, as you review and reimagine the perceptual representations, some of your perceptions may undergo further encoding to classify certain conceptual objects (eg, the truck and the car), properties (eg, red car), and events (eg, force of the impact). Finally, you may be called upon to give a verbal description of the accident, further encoding the occurrence in language. Several months later, when you are asked to remember details of the accident, your ability to recall will depend both on how many forms of encoding you employed, and on how deeply you processed each (eg, based on intensity and degree of attention).[29]

[29] Craik argues that it is the level of processing (LOP) that determines the strength of the encoding, where perceptual, structural, and semantic processing signify increasing depth of processing (Craik, 2007). However, in the

Reference, Abstraction & Conceptual Organization

That perception and conception usefully model the world does not mean our percepts and concepts necessarily correspond to entities actually existing in the world. We say some representations, like my concept of my cat Rosie, do correspond to separable counterparts existing in the world, but the existential status of the content of many other percepts and concepts, like her skittish personality, though demonstrable, is more questionable. It is simply useful to model her personality as a distinct entity. The question of what exists in the world is philosophically vexed, but our mental faculties are powerful enough to represent a range of possibilities, and sometimes even impossibilities. Contemporary philosophy would say that a perceptual or conceptual object formally exists only as part of a coherent explanatory theory, where "theory" includes folk as well as scientific musings. (See Chapter 3: Concepts & Existence for a deeper analysis.) Accordingly in this text, percepts and concepts are elements of our mental models of the world—they do not directly refer to preexisting objects and properties in the world; they merely correlate with ways the world presents itself to us.

Concepts themselves are normally not thought of as part of the world. They are part of our mental system of conceptualization used to model the world. It is their subject matter (ie, what they represent or correlate with) which we take to be part of the world. Somewhat confusingly however, this does not mean concepts cannot be used to model concepts themselves—they can, or we would not be having this discussion about concepts. Concepts can hoist themselves by their own bootstraps, so to speak. It is no accident that our capacity for open-ended thinking is much like our linguistic capacity for universal expression, where we can talk about elements of language as topics in their own right (or any other abstractions at any level), just like we talk about concrete objects in the world. We represent the objects in our perceived world in our conceptual models. But if we want to think about concepts themselves, we need to explicitly represent them as objects alongside our worldly objects.

Conceptual representations are usefully thought of as being organized into levels of *abstraction*. At a basic level concepts partition into two basic categories, depending on whether they represent individual things or collective things. We use *particular* concepts, first and foremost, to represent individual entities, like my cat Rosie or her personality. Particular concepts likely evolved initially to enhance the analysis of perceptual encounters at the first level of abstraction. Many of these particular concepts we deem as correlating with persisting objects and states of the world, and these particulars we label as existing.[30] But many other conceptual particulars, like Pegasus, have no such correlation (as understood today). The capacity for representing nonexisting particulars frees us from direct involvement with the perceived world, and enables us to mentally "play" with scenarios and other conceptual possibilities. The downside is that we become vulnerable to mistakes in judging what is actual vs possible.

---

current proposal these correspond to distinct mental subsystems, each of which can be processed to varying depths. The strength of an encoding then depends both on the number of subsystems employed and on the depth in each. For example, a conceptual encoding can be 1) relatively superficial, as when we offhandedly comment on someone's bizarre behavior, 2) relatively informed, as when we understand the context of why someone is behaving that way, or 3) scientifically grounded, as when we have professional training in human psychology.

[30] As mentioned previously, it is possible to make concepts a part of the world, but then we are using particular concepts to think about those particular concepts. We can never truly escape the bonds of perspective.

Universal concepts on the other hand, represent higher-level collective abstractions, like cat or personality in general, such that Rosie is one instance of cat and Rosie's personality is one instance of the generic concept of personality. The concept cat is a universal concept because it represents a *kind* of particular, not a *particular* individual like Rosie. Concepts are organized into particulars and universals according to what they represent, not by their intrinsic nature as concepts. Confusing the concept itself with what the concept represents leads to a very confusing ambiguity. That said, we are perfectly capable of thinking about any concept itself by casting it as a particular (eg, *the concept of Rosie* instead of *Rosie*, or *the concept of cat* instead of *cat*), perhaps for the purpose of characterizing it and asking whether such mental particulars actually exist. Such representations are often of theoretical or philosophical interest, because they explore the limits of representation. Somewhat surprisingly, the capacity for at least one level of abstraction[31] seems to be a necessary prerequisite for thinking conceptually about anything.

The categories of universal and particular are viewed here as mutually exclusive and exhaustive, so any given concept whatsoever necessarily is an instance of one or the other. In fact, the concepts of particular and universal themselves are both instances of universal, as discussed in Chapter 4: *The Kernel*.

Learning Concepts

In the broadest terms, *concept learning* encompasses any relatively permanent addition to or reorganization of background conceptual knowledge.[32] The present focus, however, is on the creation of new concepts, leaving for later chapters, particularly Chapter 6, topics related to changes in the organization of conceptual knowledge.

The previous sections offered a proposal for how concepts, as part of the conceptual system, might emerge from percepts and episodes fashioned in the perceptual system. In concise terms, a binding of perceptual features representing a perceptual object generates, during extended tracking, a cluster of conceptual exemplars representing the observed modes of the conceptual object. Concept learning at the level of whole (complex) concrete object concepts (eg, apple), therefore, depends on prior learning at the level of the perceptual features (eg, color) which form the basis for characterizing those objects. Once conception has available a pool of wide-ranging characteristics (based initially on perceptual features), concept learning is free to combine them in myriad ways to form new object and kind concepts. Concept characteristics, therefore, are more fundamental elements of conceptualization than

[31] Abstraction is different from generalization. A generalization of the concept *cheetah* is *animal*, but an abstraction of *cheetah* is *species*. Colloquially one could say cheetah is a kind of animal, but an instance of a species. Stimulus generalization is common in perception, but abstraction seems exclusively conceptual.

[32] It may be possible to contrast concept learning with concept forgetting, where individual concepts become effectively detached from other background concepts. However, background knowledge reorganization following forgetting would still count as concept learning.

whole object concepts, and this section discusses how conceptual characteristics may be formed and learned.

One useful way to look at the perceptual system is as a large, but essentially fixed, collection of state spaces, where each state space is made up of one or more quality dimensions. For example, the perception of color involves such a space containing dimensions for hue, saturation, and lightness, as described in Appendix A: *Sensation*. Similarly, the perception of sound involves frequency and loudness dimensions (among others). Each dimension is more-or-less linear, and reflects an ordered continuum of real values expressed in terms of neural firing rate. One neuron, exhibiting a range of firing rates, is capable of encoding one dimension, in a minimal configuration with no redundancy. A low firing rate corresponds to one extreme of the dimension, a high rate to the other extreme, while intermediate rates correspond to graded rate values between the two poles.

Each dimension is assumed to share roughly the same neurological structure and operation, as just described. But the meaningful interpretation of the activation level on any perceptual dimension will be different, depending on what the dimension (ie, minimally, a single neuron) connects to in the underlying sensory system and in the overlying conceptual system. The *perceptual interpretation* of a particular dimensional activation depends on which sensory organs the dimension is connected to (eg, the dimension for sound intensity is connected to the auditory organs). And likewise, each activation level along the dimension may potentially receive a *conceptual interpretation* as a discrete assessment of the value or quality of sensation. Awareness that the color pink is present involves both a perceptual and conceptual interpretation of the activation of a focal point in the color state space, corresponding to the dimensional values of a reddish hue, above average lightness, and medium saturation.

The conceptual system is tightly coupled to the various state spaces in the perceptual system, it is proposed, via a set of *primitive concepts*. In the previous example, the primitive concept pink (as one node in a vast network of primitive and nonprimitive concept nodes) may activate when the individual firing rates of neurons representing the three color dimensions are all within the frequency bands jointly specifying the pink region of the color state space. One function of the concept pink, therefore, is to act as a pink detector. Such a concept is called primitive if it links to at least one state space. Nonprimitive concepts connect only to other concepts. However, primitive does not necessarily entail a link to a *perceptual* state space. The conceptual system is apparently able to construct state spaces independently of perception using certain concept learning processes. Instead of being fixed and given state spaces as in perception, these nonperceptual (but nevertheless preconceptual or core domain) state spaces are constructed on the fly as needed. Any two currently existing concepts can theoretically be juxtaposed to form the poles of a new preconceptual dimension, with intermediate values corresponding to graded blendings of the polar concepts. This general idea is behind some computer graphics programs which are able to take two pictures of very different kinds of things, like a dog and a cat, and gradually "morph" from one to the other. In fact, human color vision is based on the operation of so-called *opponent cells* which fire differentially for red/green and for yellow/blue light frequencies. The poles establish the essence of the interpretation of the dimension. This design is readily extended for nonperceptual arbitrary opponent concepts as well. More commonly however, conceptual opponent poles seem selected so as to represent simple, univariate conceptual dimensions like short/long, dirty/clean, slow/fast, or even complements like dead/alive (exhibiting minimal interpolar variation). Such primitive featural concepts are eminently useful for describing all manner of nonprimitive concepts that can be characterized in terms of discretely graded qualities or quantities. While the number and kind of perceptual state spaces is constrained by evolution, the number of possible nonperceptual state spaces is effectively unlimited.

In the above proposal the initial work of concept learning is to recruit a new preconceptual neuron to serve as a dimension for a pair of primitive polar concepts.[33] Then, once the dimensional-polar connection is in place, the dimension may further become conceptually linked as an instance of a higher-level spanning concept denoting an attribute. For example, the short/long dimension may become linked as an instance of the newly generated concept length, potentially even before its link to a linguistic label.[34] Ultimately, attributes are used to characterize and organize every concept in a concept network. Although there are many technical nuances to the proper representation of these kinds of attributive concepts (discussed later in Chapter 3: *Concept Characteristics, The Structure of Characteristics*; Chapter 4: *Properties & Attributes*; Chapter 5: *Attribute Values*), they seem to provide the foundation for representing conceptual knowledge about anything.

To summarize the core processes involved in learning new concepts, it seems useful to identify two reciprocal processes—one generating new primitive concepts from perceptual features, the other generating new primitive concepts from other primitive concepts. Assuming the various perceptual state spaces are given by evolution, concept learning initially occurs when a perceptual dimension (eg, brightness) is analyzed into (usually) two polar extremes (corresponding to high & low firing rates). Each extreme may then generate a new primitive concept (eg, dim and bright, but without explicit linguistic labels), which may subsequently become attached to various complex concepts (of objects, substances, situations, etc) as part of their complex descriptions (eg, the Sun is bright). Further concept learning may occur if the primitive pair becomes unified under an abstract attribute concept (eg, the concept of brightness). Newly generated complex concepts, can then recruit more potent collections of attribute and attribute values in their descriptions, depending on the kind of complex thing being characterized. Note in passing that a full characterization in terms of an attribute plus attribute value is a more powerful representational strategy than simple value assignment, although such nondimensional associations are often useful as a precursor to deeper analysis (see also, Chapter 4: *Properties & Attributes*)

In the inverse direction, concept learning conjoins existing primitive concepts into preconceptual dimensions. Let's assume we already have concepts for two distinct perceptual dimensions: namely size with values small and large, and weight with values light and heavy. And underlying each pair of values is a perceptual dimension corresponding to size and weight. Concept learning processes then

---

[33] A more likely scenario would involve recruiting a small population of neurons to markedly increase the reliability and accuracy of the dimensional representation (using the mean and standard deviation of the population values).

[34] However, linguistic labels are likely involved in most of the learning of conceptual attributes. Repeated labeling (eg, "what *color* is that?") can direct attention to the relevant values (eg, "red", "blue", "green", "yellow", etc), eventually leading to the mental juxtaposition under the label "color"—ref (Smith, Gasser, & Sandhofer, 1997).

might create new primitive concepts corresponding to values which are small&light, small&heavy, large&light, and large&heavy. By arranging the new values such that large&light and small&heavy form two poles of a new preconceptual dimension, concept learning can be thought of as implicitly suggesting the concept of density as a potentially interesting attribute, unifying the new values. In fact, the remaining values (small&light and large&heavy) could be found by thought experiment to be consistent with the new dimension, since they logically fall between the two extremes of low and high density. The significance of any new learning, of course, depends on how useful the new concepts are. In this case, since the concept of density is applicable to such a wide variety of objects and substances, the allocation of the additional conceptual resources is fully warranted (and probably reinforced).

Concepts, Beliefs, & Thoughts

Intuitively, concepts are the elements of thought. Individual concepts combine in meaningful ways to produce whole thoughts. It is the content of thoughts that typically express our ongoing interpretations of the world, and explain our various behaviors. Before directly examining concepts in more depth, it may be useful, therefore, to consider the context in which concepts function—namely thoughts. But what are thoughts? They seem almost as opaque as concepts, even though we seem more occupied with whole thoughts than individual concepts in nonreflective everyday experience. Thoughts are purposeful and directly useful, in a way that isolated concepts out of context are not. It takes some effort to deconstruct the content of a spontaneously occurring thought into its constituents, something like focusing on the meaning associated with individual words rather than the meaning of a whole sentence. So in an intuitive sense, thoughts, as units of purposeful behavior, may provide a useful starting point for better understanding concepts.

On occasion you may notice one of your shoes is untied. Switching from a first-person to a third-person perspective, you may also notice it is you having the thought about your shoe being untied. Whereas simply *having* the thought allows you to employ the thought without explicitly thinking about it, switching perspective allows you to focus on the thought itself for analysis.

A thought is generally expressible via language, yet does not seem to be an explicit part of language itself, as say a sentence is. Linguistic expressions are an aspect of a closely connected yet different representational system. We can mentally verbalize the sentence "my shoe is untied", but the thought, as a causative factor, seems to precede and trigger the verbalization. Mental verbalization is more like the unconscious activation of a premotor program by a thought, where the program is inhibited from actually issuing the speech commands. A thought also seems to possess a subject matter or content which can vary from one thought to the next. It points to something meaningful beyond itself, in the same way tracks in the woods indicate the presence of certain animals. The content of a thought is important for predicting its effect on subsequent thoughts, such as wanting to avoid tripping or remembering to buy new shoelaces. The thought itself, as distinguished from the content of the thought, thus is a kind of representational pattern (of neural activation) serving as host for the content. Such patterns associated with meaningful content are often called *intentional*, to distinguish them from meaningless (but sometimes purposeful) patterns, like those found in a brick wall. Thought contents can also precipitate action (in this case perhaps tying your shoe), but do not seem, at least at first glance, to be the same kind of physical phenomena that participate in normal causal happenings, as say when lightning causes thunder. They are interpretations and exhibit an optional quality that seems to require a willful translation from potentiality to actuality to produce a concrete behavioral result.[35] Summarizing, then, a *thought* is a kind of intentional conceptual mental pattern of activation, or in brief, a temporary *conceptual pattern*,[36] realized in some portion of the brain's neural circuitry. Characterized in this way, it is analogous to a temporary active *impression* produced by the perceptual system.

Brain scans indicate different areas of the brain become active when subjects think different thoughts. The active areas consistently correlate with specific functional domains such as paying attention, planning, remembering, willing, visualizing, etc. Furthermore, the areas associated with a thought are often distributed widely in the brain. A thought then is associated with the temporary activation pattern of selected, distributed neural circuits in the cortex and limbic system, which varies depending on content. The content is derived from the activation pattern by interpretive processes operating in various functional mental domains (eg, for processing numbers, visual objects, words, emotions, etc). That interpretation can then trigger further thoughts (as well as perceptual or motor activation patterns) through inference or association.[37] The content we ascribe to those thoughts is the interpretation we give to those particular activation patterns at that particular moment.

Two distinct kinds of thoughts become apparent if we give attention to the memories from which thoughts can arise.[38] Using the untied shoelace example again, one kind of thought (eg, that shoelace came untied yesterday too) arises from observing the particular situation and matching it with the memory of a prior situation. Such memories are often called declarative (or explicit), since they are

---

[35] The problem of how the intangible content of a thought can translate into concrete action is one aspect of the infamous mind-body problem. In this text mental phenomena like concepts, beliefs, and thoughts (viewed as mental states and events) are classified as fully physical phenomena, along with the usual physical phenomena of everyday experience like concrete objects and their physical properties. See Chapter 3: *Concepts & Existence* for discussion of the physical reality of mental/psychological phenomena.

[36] Technically, a conceptual pattern is not the same as a conceptual state, as the terms are used here. Most philosophical talk is about mental states, however there is an insidious ambiguity in the word "state" which can lead to confusion. A state is sometimes informally viewed as a *continuant*, and other times as an *occurrent*, two fundamentally different kinds of things. The difference is subtle, but a continuant fully exists at a time, while an occurrent fully exists in (or over) time. A pattern is a continuant and a state is an occurrent in this book—see Chapter 6: *Explanations for the Top-Level Epistemological Concepts* and *The Organization of Situation (ie, Event) Concepts* for further discussion.

[37] While most thoughts are unconscious, the presence of certain temporary activation patterns possibly becomes the neural correlate of the *experience* of having certain thoughts, namely when the patterns are evaluated by the experiential (or feeling) processes.

[38] Note that the colloquial idea of memory, as a place where individual pieces of information are stored (as in a dictionary or catalog), is not supported by basic research (Nadal, 2007). In this text the term "memory" signifies the integrated knowledge structures which hold the information, without implying that the information is individuated within those structures before being accessed. When new information is added to a knowledge structure, the entire structure changes to accommodate the new information.

readily reportable and expressible in language. We can also think about activities like bending over and retying the shoe. The source of these other kinds of thought are implicit or procedural memories which are skill-oriented and difficult to express precisely in language. There are probably other useful categorizations of thought, perhaps corresponding to the various kinds of memories they are derived from (see figure below), but the focus of the current study of conceptualization is limited to the former kind of thinking and representation. The idea of *concept* we are ultimately aiming at is related to declarative thoughts.

Squire's Taxonomy of Memory

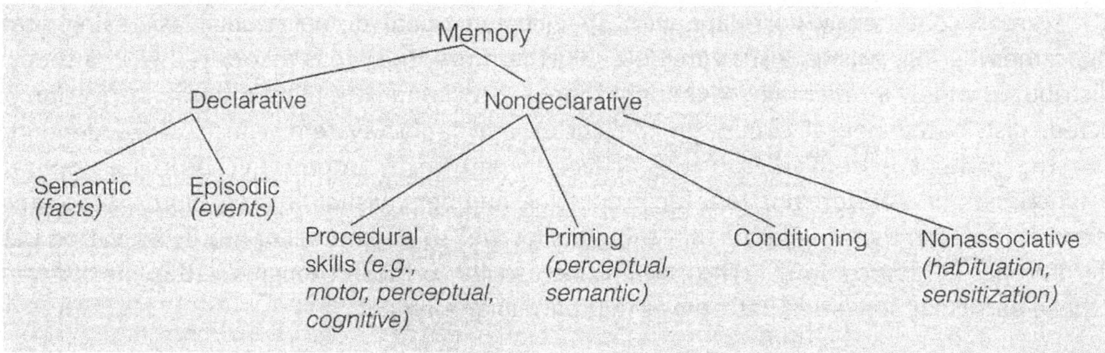

Based on extensive neuropsychological research, Larry Squire has posited that memory comprises two fundamental types: declarative (explicit) memory and various forms of nondeclarative (implicit) memory, each of which may be associated with discrete cerebral structures and processes.

Figure 5: Taxonomy of Memory (Squire, 1993) in (Sternberg, 2003)

Those prefrontal and allied cortical circuits involved with explicit declarative thinking, as opposed to perceiving or acting, seem to form part of a distinct conceptual system for manipulating the content of thoughts. The conceptual system, like the perceptual and motor systems in the cortex, contains various components.[39] There is a *working memory* component for holding current thoughts, and there is a *long-term memory* for holding background knowledge. There are also operational components that control the transfer of information between working and long-term memory (usually referred to as *short-term memory*),[40] the processes of inference and knowledge organization, the integration of goals, and the

[39] All cortical systems are apparently constructed from the same ubiquitous micro-columnar cortical circuits. Each micro-column contains roughly 100 interconnected neurons, and is about the width and length of a one-eighth inch long human hair.

[40] Short-term memory is viewed as something like a waiting room for consolidating information, before protein synthesis establishes a more permanent trace in long-term memory. Rehearsing a new phone number involves working memory between rehearsals (lasting several seconds), remembering that you had eggs for breakfast yesterday summons a short-term memory (lasting hours to days), while remembering the name of your childhood sweetheart calls up a long-term memory (lasting years).

interfaces with perception and action. The conceptual system sits (metaphorically) on top of the lower-level perceptual and motor systems and selectively mediates between them via conceptualization.

As previously mentioned, the content of some thoughts may become conscious. A thought possibly becomes conscious through proxy-like activation of the thought in working memory. That is, a conscious thought may occur when a dynamic pattern becomes active in working memory by means of cognitive copying operations from the active knowledge in long/short-term and perceptual memory. Focused expression of the thought in working memory could be a prerequisite for becoming aware of it. Our conscious experience of the serial, rather than parallel, nature of thought probably arises from limits on working memory capacity. Most of our thinking, however, is unconscious.[41] Unconscious thoughts are likely produced and manipulated continuously by those same cognitive operations, but without involving working memory.

Thinking is a highly dynamic process, with one thought (more accurately, its interpretation) typically triggering and reacting to many other thoughts in different cortical regions. On the other hand, we associate knowledge with the more-or-less static brain structures that support thinking. *Knowledge* is what we call the default potential content of those representational structures, if they were to become activated with little or no context. Knowledge, therefore, is mainly encoded in the connections among neural circuits, as opposed to the fully context sensitive activation of those circuits. In other words, your current background knowledge about horses is encoded in the configuration of the conceptual neural substrate, and when the substrate is activated and interpreted in certain ways, you have various thoughts about horses. The realtime pattern of the thought depends on which parts of the substrate are connected to each other and are simultaneously active, as well as on the strength of those connections and activations. The underlying static representational structures contain all the knowledge which may become selectively accessible through activation.

Long-term memory apparently encodes the spatially distributed static configurations of potential activation. That the configurations are static and enduring, rather than dynamic and transient as in thought, suggests they are realized in neural structures. Certain connections between neurons, among all possible connections, become primed for activation, such that they respond in approximately the same way to similar stimuli.[42] These preferential static configurations are normally interpreted by the executive processes as background *beliefs* (and wants, hopes, attitudes, etc). Beliefs (and knowledge more generally) have static content in the same way thoughts have dynamic content, producing meaning when each is interpreted. Meaning then arises from the interpretation of active or passive patterns in neural structures. Meaning is not an inherent property of a structure or representation; it is a byproduct of an interpretive process, which may produce different meanings both from the same underlying structure, and even the same activation pattern, depending on interpretive point of view and context. If our background beliefs and knowledge are based on the way long-term memory is

[41] A simple example of an unconscious thought might be the inference that allows us to "get" the punch line of a joke.
[42] However, other neuron circuits in the prefrontal cortex are apparently multifunctional, since they can rapidly switch functions depending on the task at hand (Duncan, 2010).

structured—ie, which neural circuits are causally affected by (or linked to) which other circuits—their meaning in context develops from how the structures are interpreted and ultimately used.

In the previous discussion, the word "content" functions liberally and vaguely as a term of art. However, the distinction between a particular concept, belief, or thought, and the *content* related to that concept, belief, or thought will become important in what follows. One point of view might claim that content is simply whatever information is or can be derived from a process of interpreting a representation, usually taking context into account. That is, the content of a concept couples with the potential for someone or some subpersonal process to interpret the concept in the current situation in service of a current purpose. "Content" is then a nominal kind term, rather than a natural kind term, having little or no intrinsic meaning above and beyond its use in association with interpretation. It is like the word "gift", which broadly means whatever is given (ie, fills a role in a giving event), rather than a certain kind of object possessing certain intrinsic properties.

But content seems to have somewhat deeper, if ill-defined, connotations as well. It seems to express the potential significance of the pattern in a representation to the interpreter. And that significance depends on the competence and performance of the interpreter (eg, on the interpreter's background knowledge and ability to use that knowledge), as well as on the quality and structure of the pattern in the representation. The proposal developed later, beginning in Chapter 2: *Signs*, is that 1) content only applies to the interpretation of representations, 2) different kinds of interpreters at different levels of perception and conception operate on different kinds of mental representations, 3) meaning is a side-effect of certain kinds of interpretive processes which can analyze and effectively translate between different mental representations, and 4) content and meaning, although ineffable, may turn out to be measurable in terms of the *information* gleaned from interpreting a representation either under ideal circumstances or in practice.[43] Conception and perception, therefore, are chiefly viewed as systems of multiple meaningful representations, each suitable to a useful way of modeling the world, and which can often be intertranslated among themselves in limited, but useful, ways.

This tack obviously creates a problem with grounding our use of conceptual representations in the world, if meaning depends only on concepts in our heads.[44] However, the grounding problem is mitigated by realizing that our basic-level concepts (as part of our background knowledge) were created in the first place, and subsequently adapted, via evolution and causal interactions with the world. We are all endowed with crucial conceptual operations and concept placeholders at birth that embeds conceptualization in the context of our ecosystem. We can then create and combine new concepts in many creative ways, and use them to generate new beliefs and thoughts as needed. Although our use of mental representations necessarily involves abstract models, their basis is nevertheless grounded in

[43] In the proposed analysis it is important to keep separate the distinct ideas of the *neural substrate* (as matter or substance), *concepts* (as patterns or representations in, or configurations of, the neural substrate), and *content* (as information which can be gleaned from concept representations). All three are assumed to carry some kind of theoretical and functional justification.

[44] A similar complaint is often leveled at dictionaries, which explain the meaning of one word only in terms of other words. Putnam (Putnam, 1975) and Burge (Burge, 1979) have famously shown that the "meaning" of linguistic terms, and probably thoughts, depends, at least partly, on our external, nonmental environment.

the real world. This viewpoint probably falls under the rubric of *conceptual realism*, since it takes believing and thinking as real, physical, brain states and events, respectively. But the reality of what is believed or thought (unfortunately including the content of that very belief!) is left open for debate.

The following figure highlights the major ideas in this section. Concepts are static constituents of beliefs, both of which, as permanent representational patterns, may acquire meaning via interpretation by high-order executive processes. By analogy with computer information processing, the conceptual representations constitute data, content is the total information extracted from the data, the interpretive processes are programs, and meaning is the result of interpretation (perhaps effecting conceptual changes) produced by the programs operating on the data. Moreover, background concepts and beliefs, when activated, also exhibit temporary representational patterns which may be interpreted. These dynamic patterns we call thoughts, and their interpretation is thought content. The foregoing outline for conceptual representations generally parallels that for perceptual representations. In perception, percepts are the static constituents of episodes, while impressions are the dynamic patterns of percept and episode activation. From the point-of-view of conception, perceptual episodes and percepts represent the multiple encounters with an object (usually called exemplars), which subsequently become encapsulated in the composite concept representing an enduring individual.

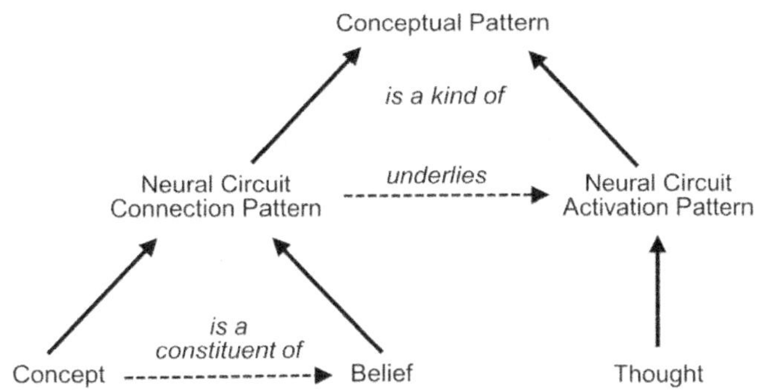

Figure 6: Key Concepts in Conceptual Information Processing

The Dynamic vs Static View of Concepts

We started this exploration of concepts by proposing that concepts were elements of thoughts. But if thoughts are transient activation patterns, then concepts become identified with parts of the activation pattern. This dynamic way of looking at concepts is generally affiliated with the tenets of Cognitive Linguistics (Croft & Cruse, 2004). In essence, concepts then are generated on the fly as needed in different situations, and not explicitly stored in conceptual structures. Concepts are produced from historically stored and current perceptual events, but there is no separate system of abstracted knowledge. Meaning then is only construed in use, as originally suggested by Wittgenstein. This idea of construed meaning is consistent with the theme of interpreted content developed previously, but it twists the normal meaning of "concept" from something we seem to possess before thinking about it, to something we generate when we need it. While the proposal for analyzing a whole thought pattern into dynamic conceptual subpatterns seems reasonable (akin to the analysis of continuous speech into phonemes or a scene into objects), there are complex subtleties in how the part/whole relationship might play out.

First, is a thought simply composed of an organized collection of individual concepts, like a sentence is composed of a sequence of words? The meaning of a whole thought then would be a straightforward function of the meaning of its constituent concepts and their interrelationships. This kind of composition exhibits a relatively loose form of integration, where the elements are clearly individuated as parts of the collection, and the whole is easily analyzed into its parts and their relations.[45] The analogy is to a chessboard, such that each piece retains a persistent identity, and functionally contributes to the meaning of the net board configuration during a game.

Or perhaps a thought reflects a somewhat tighter integration of individual concepts. In looking at a bird's wing, all of the feathers work together to enable lift, but they densely overlap. The feathers are still individual entities and can be separated out, but their individuality is not immediately apparent when they are incorporated into a wing. If the concepts in a thought are like this, then the challenge is to find a procedure which can reliably extract them.

At an even higher level of integration, the original concepts which go to make up a thought could become altered in the process, such that their before and after descriptions are significantly different. Like a molecule formed in a chemical reaction, the atoms that combine to produce the molecule are no longer complete atoms after the reaction. They have shed electrons in the process of binding together, and as a result the properties of the molecule are typically very different from the properties of the original atoms. In this case some method for predicting how concepts change when they are combined is needed.

The next level of integration might be likened to a watercolor painting. When applied to the canvas, the original color bleeds out into neighboring color regions in an indeterminate way with unpredictable borders. The core region retains the approximate identity of the original brush stroke (eg, shape & color), but details are lost because the context is so variable. If variability of context is a major factor affecting the combination of concepts into thoughts, then perhaps a statistical description of the combination process is the best that can be hoped for.

At the highest level of integration, even the identity of the source concepts is lost. This situation is like a homogeneous mixture (or perhaps superposition) of ingredients. When water, lemon juice and sugar are mixed to produce lemonade, the original ingredients become one with the final product. They have been transformed (something akin to metamorphosis) and spread throughout. The number of original ingredients cannot even be counted following combination. Applied to a thought, the result might be characterized as a Gestalt, where the combination of the original concepts has produced a new

---

[45] An even looser integration is technically possible, where there is no organization among elements—ie, a set; but the meaning of a thought clearly depends at least on the ordering of its constituent concepts.

emergent concept. Such combinations are amenable to analysis, but any simplifying background assumptions of independence are unfounded. Such tight coupling between concepts would completely overshadow the underlying properties of the constituent concepts themselves.

Appealing though they may seem, considerations such as these will not be explored further here. It seems closer to the everyday meaning of "concept" if we identify concepts with an underlying source of the activation patterns of thought. That is, not as elements of thoughts, but as elements of background beliefs.[46] Beliefs and their constituent concepts then would represent the *potential* for activation of thoughts and elements of thoughts. This approach highlights the fact that a given concept (eg, dog) is associated with numerous background beliefs (eg, that they normally bark, have fur, have four legs, etc), only some of which may become active in a given context. A concept seems to encompass more than the occasion of its use. This more static declarative-oriented perspective on concepts will be adopted from this point on.

To continue the study of concepts as elements of beliefs, we can ask what is it that all beliefs about some concept (eg, dog) share. From the perspective of the neural substrate, it must mean that all beliefs about dogs (eg, dogs are animals, dogs bark, dogs chase cats, people own dogs, etc) converge at a nexus. The nexus resembles the hub of a wagon wheel, where the spokes are the individual beliefs, and all of the beliefs have one conceptual node in common, called *dog,* at the hub. This does not mean, however, that the nexus is the concept. The nexus simply serves as an access point for the concept. The full concept consists of the beliefs, which we usually interpret as the concept's characteristics. In fact, a concept typically includes other related beliefs, further joining the directly attached beliefs together into a whole, as the rim on the wagon wheel joins the spokes together. As an example, consider the concept of an arch. The directly attached beliefs encode propositions like: an arch has a left side, an arch has a right side, an arch has a lintel. The remaining beliefs about arches would join with the other beliefs to form the whole concept: the left side is vertical, the right side is vertical, the lintel is horizontal, the left side supports the lintel, the right side supports the lintel, the distance between the left side and the right side is (approximately) the length of the lintel, etc. These beliefs circumscribe the concept.

The above model of a concept implies that every node in the set of beliefs comprising a concept is itself a nexus for a different embedded concept (ie, serves as a hub within a hub for other beliefs, some of which function at a primitive structural level). Intuitively this seems approximately right, as we are normally able to focus attention on one particular aspect of a concept at will, and intuitively sense that there are other, mostly unconscious, details supporting any interpretation. However, apart from active focusing, beliefs about a concept likely become both integrated among themselves and segregated from other unrelated beliefs, based on the strengths of their connections. The connections within a concept may come to possess a relatively low threshold for activation, so that stimulation anywhere in the concept easily spreads to its other parts (and to other associated concepts), while resisting spread to unrelated, but connected, neighbor areas. Concepts probably do not overlap significantly (ie, share beliefs), since this would encourage ambiguity, but they are certainly strongly interconnected. The

---

[46] or more broadly as elements of *conceptual attitudes*. Conceptual attitudes are the brain-structural analog to the so-called propositional attitudes like beliefs, hopes, wants, etc.

spread of activation is probably controlled by wholesale biasing or by priming specific areas for easier activation, as in attentional focusing. The degree of focusing we are able to bring to bear during activation of the intricate network of embedded concepts likely correlates with the relative fuzziness of our thinking on different occasions. A couple of drinks plays havoc with our rather fragile ability to individuate thoughts.

Language

So far, the conceptual system has been described in terms of three fundamental representations central to its functioning—namely, concepts, beliefs, and thoughts. Many mammals apparently employ analogous perceptual versions of such representations in their unself-conscious interactions with common environmental objects, including generalizations and classifications of similar objects into perceptual and functional affordance classes. However, the ability to use linguistic devices (eg, words, syntax, semantics, and logic) to combine underlying concepts, beliefs, and thoughts in fanciful ways, independent of their basic use in directly interacting with the world, opens up a new realm of cognition based on a new array of representations that are fully developed only in humans. Although linguistic representations are intimately linked with their underlying conceptual counterparts, language is here viewed as a separate cognitive system, owing to its distinctive representations and operations. This section briefly outlines some of these neocognitive representational systems operating above the level of concepts. The following narrative visualizes a hierarchical framework, consistent with the other ascending and descending cognitive pathways, in which conceptual representations fit comfortably between lower-level perceptual and higher-level linguistic representations.

Our language faculty is clearly a highly integrated system, but linguistics research traditionally recognizes two fundamentally different subsystems—namely, syntax and semantics—built on different kinds of linguistic representations.[47] Yet on the most conspicuous "surface level" of language comprehension and production, above the murky depths of syntax and semantics, we readily learn to read and write *sentences*, composed of *words* combined in a simple linear sequence. Although it is common to think about words and sentences as being overt external representations (eg, as the phonological pattern in an utterance, or the graphological pattern in an inscription), they are also necessarily represented mentally, since they are the beginning and end of the language comprehension and production processes. A linear sequence of words, whether forming full sentences or sentence fragments, constitutes the first level of linguistic mental representation.

The next time-honored level of linguistic representation revolves around the grammatical rules and constraints of our native natural language. These kinds of grammatical representations are part of a combinatorial facility for parsing and generating clause and phrase structure for word sequences,

[47] There are other important subsystems related to language, as well, such as the mental lexicon (storing knowledge about individual word forms), the phonological/graphological subsystems (storing knowledge about speech sounds and writing strokes), and the pragmatic subsystem (storing knowledge about the current context of speech acts). All such subsystems seem to maintain their own representational schemas.

usually called *syntactic analysis*. Syntactic representational units consist of whole *clause forms* (corresponding to surface sentences) as well as combinations of embedded *phrase forms* (corresponding to word groupings based on parts-of-speech and grammatical phrases—noun, verb, determiner, noun phrase, verb phrase, etc). These syntactic constituents are structured representations, conventionally organized into tree-like hierarchical configurations specifying the nested relationships between words and groups of words in a sentence. When part or all of the clause or phrase structure representation becomes active during a speech act, along with contributing contextual influences from pragmatic effects,[48] the pattern of activation becomes available for interpretation (or for production, depending on whether the sentence is being received or produced), just like any other representation. The dynamic activation of a clause form during interpretation will simply be called a *parse* to distinguish it from its underlying static clause structure representation. The interpretation of a parse (unlike many thoughts) is wholly unconscious, but it may suddenly become conscious if it is deemed ungrammatical.

A clause form, as viewed here however, is more than an inert structure corresponding to a parse tree representation of a sentence. The constituents of a clause are analogous to snippets of "code"—more precisely as functions—which are executable, and organized hierarchically like a computer program built out of embedded functions.[49] The structure of the clause specifies how its constituent functions are to be applied when it executes. For example, the function CLAUSE might construct a whole clause form given a noun phrase and a verb phrase, while the function VERB-PHRASE-1 might construct a verb phrase given a verb and a noun phrase. It is the evaluation of the clause's embedded syntactic functions that ultimately is thought to correlate with the experience of a grammatical or ungrammatical clause. Put differently, the grammaticality of the clause manifests when its program is run and its functions applied to the sentential "data" (ie, the word sequence in the sentence corresponding to the clause) making up the clause form. In this execution model of sentence interpretation, the activation of the clause structure produces side effects that go beyond simple activation, as say in a Christmas tree lights model which depends entirely on analysis of the activation pattern by independent functions. By merging internal structure, function, and control in individual building blocks of phrases, evolution has created sophisticated linguistic modules that can be more readily combined in useful ways, simplifying the coordination among modules. It is a successful evolutionary theme followed for many other more observable body parts, such as hands for manipulating objects, eyes for seeing, and for functional organs, in general.

The second major language subsystem, performing *semantic analysis*, is viewed as interacting with the syntactic component to represent the logical form of a clause structure.[50] The logical form of a sentence is a sort of gateway to the meaning of the sentence, because correct reasoning and inference seem to involve some type of naturalized mental logic. Logical forms also interact with the underlying

---

[48] Pragmatic effects depend on who the speaker is, the time and place of utterance, the physical and social situation, background knowledge, etc.

[49] The artificial intelligence programming language LISP similarly evaluates data structures that consist of nested functions.

[50] Although discussed separately, syntax and semantics are best viewed as interleaved, rather than staged, linguistic processes.

conceptual system to give the forms a conceptual underpinning. The central logical form represented and analyzed by the semantic component is the *proposition*, which is a translation of a clause structure syntactic representation into a logical semantic structure containing pointers to constituent conceptual representations.[51] It is propositional representations which allow us to engage in inferential reasoning, by chaining propositions together in more-or-less (often much less!) logically and pragmatically valid ways. A similar kind of thought chaining probably occurs wholly within the confines of conception proper via simple association between concepts—for example, when the activation of one concept spills over into others loosely connected to it—but this is association, not inference.

The representational constituents of a proposition are its terms, predicates, and operators—which combine in various logical operations such as conjunction, implication, negation, predication, quantification, etc—to produce the whole proposition. All of these subpropositional constituents will be called *logical expressions*. Thus propositions, made up of logical expressions, are structured representations. They are thought to exhibit an internal hierarchical organization similar to a syntactic clause structure, since the appropriate logical structure necessarily corresponds to a correct parse in the current context; and many constituents of logical structure correspond to the constituents of syntactic structure (Caan, 1993; Soames, 2010). The proposition ultimately is the result of the translation of a syntactic representation into a semantic representation (or alternatively, of a conceptual representation into a semantic representation during speech production).

A proposition represents the logic of a sentence, like a clause represents the syntax of a sentence, but in terms of logical, rather than syntactic, operators. Examples of logical operators would include AND, NOT, SOME, etc, which specify the combination of propositional constituents in logically consistent ways. The constituents of a proposition are again viewed as executable operations, organized hierarchically, specifying the order of application of those operations when executed. However now, the execution of the proposition's logical operations correlate with the experience of a meaningful phrase (and ultimately a clause since all propositions are by stipulation meaningful). As before, the meaning of the proposition manifests when its program is run during interpretation and its functions applied to the lexical, syntactic, conceptual, and contextual "data" to which the propositional structure and substructures are linked. Executing the propositional structure also establishes any conceptual references of logical expressions like indexicals, which depend on interpretation in a given context.[52] Evaluating a proposition, thought of as a sophisticated form of interpreting the proposition's activation pattern, produces meaningful information as a side effect for the evaluator.

Propositions are also viewed as potentially having a truth value—ie, can be thought of as being true or false depending on whether the proposition corresponds to the way things are in the world. But

---

[51] Propositions are traditionally thought of as mind-independent (ie, real) entities existing in the world, but here they simply serve as mental representations, useful for explanation, whether those mental representations turn out to be real or not.

[52] For example, if someone utters the sentence "I like Jane", the corresponding proposition will contain a function corresponding to the word "I", that determines the reference concept, say Dick, given knowledge of who the speaker is. The execution of this function corresponds to the content of "I": roughly, the speaker of this utterance. When executed in context, the function will return a pointer to the concept *Dick*.

propositions are distinguished from *beliefs* in the underlying concept network. Beliefs are always subjectively assumed to be true (but to varying degrees), because explicitly representing a belief involves a commitment to the associated proposition. A proposition is subjectively true (or false) for someone if they believe it (or its negation), but the subjective truth value of an unknown proposition (like Goldbach's conjecture)[53] is not linked to any explicit underlying belief, at least for most mathematicians. In this latter case the terms, predicates, and operators making up the proposition are simply mapped onto their separate underlying concepts, but the whole proposition remains unmapped to any conceptual belief. A credulous fan may sincerely believe that Elvis lives, and rationally accept the associated proposition as unequivocally true, while a more circumspect fan, fully understanding the meaning of the claim and its consequences, may not associate the proposition with an explicit belief. Belief is ultimately relative to one's personal knowledge of the world, but at the same time, rational personal belief is ultimately grounded in common, shared (and therefore more-or-less objective) means of extracting information from the world. Some people's beliefs are simply better organized, more objective, and more consistent than other's.

This account of what propositions are has certain similarities with the cognitive-realist position advanced by Scott Soames (Soames, 2010). Propositions are real structural components of real mental events, such as when we predicate something of something else—like predicating whiteness of snow to construct the proposition 'snow is white'. The proposition then serves as a mental representation encoding its content, as it was explicitly created by interpretive processes to represent that content. The act of predicating is merely one of a number of other real mental events that includes entertaining propositions, evaluating propositional functions, and creating beliefs from propositions. Such mental events are thus particular real states-of-affairs, where propositions are real components of those states-of-affairs. But those propositions represent other states-of-affairs which may or may not be real. Given that only real states-of-affairs are true, since only they correspond with real events, represented states-of-affairs that do not correspond with real events (ie, that correspond with nothing) are judged false. This account may go some distance toward explaining how propositions acquire their truth values.

Finally, the neural activation (ie, execution) of a propositional representation in context, for want of a better word, will be called a *proposal* (or conjecture). Conjecturing a proposition is like entertaining its conceptual components and structure without committing to a belief in it. Thinking about a belief involves a commitment to what is thought, but conjecturing about a proposition does not, allowing a rich fantasy life and hypothetical reasoning. One can then process and interpret the constituent structure of the proposition without asserting or denying it conceptually.

To summarize the stages involved in natural language processing, consider a sentence from (Jackendoff, 2002) illustrating some of the myriad levels of linguistic representation:

(1) The little star's beside a big star.

[53] Every even integer greater than 2 is the sum of two primes.

At some basic level, (1) is simply a configuration of physical ink marks on a page. However, if we choose to apply our interpretive faculties to (1), we are then treating it as an inscription (ie, a representation). Since we harbor certain linguistic expectations about inscriptions, the ink marks become reclassified as letters, according to their perceptual shapes. A similar process of reclassification occurs when we hear sounds as speech units. We represent an utterance as a sequence of phonemes in roughly the same way we represent an inscription as a sequence of letters. A morphological language component is then thought to translate the letters or phonemes into words. The flavor of such phonemic representations for the given utterance is presented in the figure below, which Jackendoff labels "Phonological structure".

The active mental representation of an inscription or utterance then initiates a cascade of contextual mappings extending from the active representation of the sentence to a parse (of its word sequence), and finally to a proposal (for a proposition), whose constituents connect with concepts. The *parse* of the sentence is roughly the activation of what is labeled "Syntactic structure" in the diagram; the *proposal*, something like the activation of "Semantic/conceptual structure". (Jackendoff's model, however, does not incorporate executable units directly into these structures.) Links to underlying concepts are shown in capital letters (STAR, BESIDE, etc). Finally, a core domain representation labeled "Spatial structure" provides a grounding link between the represented concepts (eg, little star, big star, beside) and the representation of their spatial arrangement, where objects and their shapes are thought to be registered in terms of a coordinate-like system.

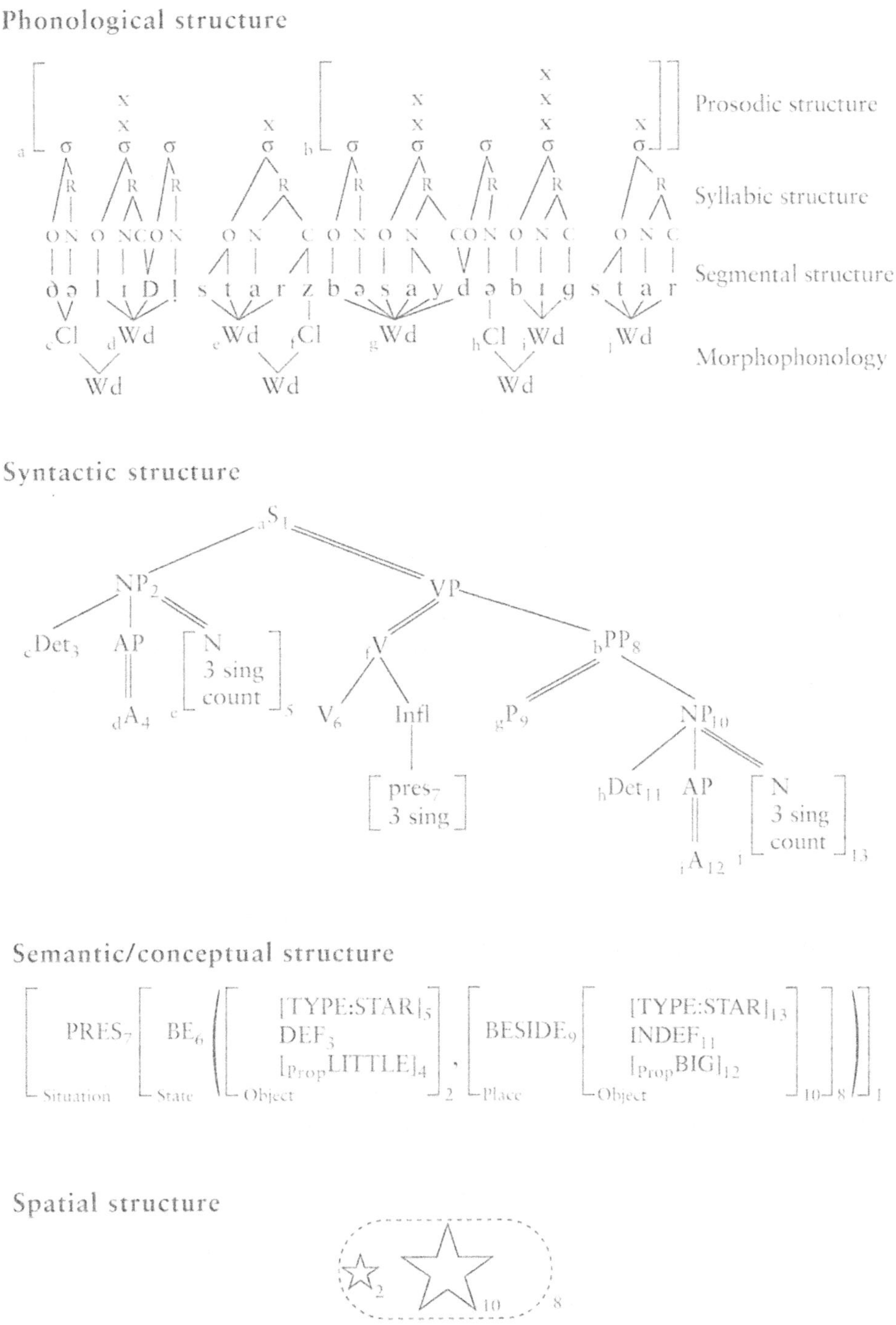

Figure 7: Linguistic Representations of the sentence "The little star's beside a big star." (Jackendoff, 2002)

Symbols & Computation

Our capacity for thinking and communicating in natural language is arguably *the* striking aspect of cognitive life that most clearly distinguishes us from other animals. However, language is only one form of our general aptitude for symbolic processing which also extends to working with diagrams, models, mathematics, games, music, and many other problem solving situations involving the manipulation of symbols. We employ individual symbols (namely, words, notes, geometric objects, tokens, etc) in a variety of ways to create complex assemblies of symbols (like sentences, musical phrases, structures, and patterns). *Symbols* constitute elements of *symbolic processing systems* that specify how they can be combined together in meaningful ways.[54]

However, it is controversial whether we actually possess an explicit symbolic processing faculty in the brain—eg, see (Barsalou, 2002) for a theory that all abstract processing is based on sophisticated extensions of perceptual mechanisms. Yet there is preliminary supporting evidence from neuroscience that some neurons can perform logic gating functions, similar to what are found in digital computers (London & Hausser, 2005). These kinds of functions are likely required to support digital, as opposed to perceptual-analog, computation. The computational effects of excitatory and inhibitory synapses on some dendrites are reminiscent of NAND gates, combinations of which are sufficient to compute any Boolean logic function. Other neural circuits in the auditory channels seem to detect only specific sequences of inputs, and therefore are compatible with the precise analysis of symbol sequences such as might be found in a sentence. If similar circuits are found in the prefrontal cortex where high-level control is exercised and can be tied to working memory, then an underlying symbolic architecture may become a fully plausible explanation for our ostensive symbol processing capabilities. The following discussion will assume we possess such an architecture dedicated to symbolic processing in order to explore the possible roots of its structure and functioning.

But before getting into a discussion of symbols and computation, more generally, it may be useful to briefly revisit the structure of *concepts* discussed previously, since the likely purpose of symbol use is to facilitate thinking and talking about concepts. As viewed here, a concept is a more-or-less fixed mental representation of something, usually of something in the world.[55] It is typically built up automatically out of recurring experiences of that something via concept learning mechanisms, and consists of a large number of *characteristics* bound together into a conceptual unit. The concept's discrete characteristics are causally derived from the lower-level graded *perceptual features* that gave rise to it. Language also contributes to concept learning by updating perceptually-based concepts (through communication), and by creating new abstract concepts, divorced from perception altogether. Speaking loosely, concepts are said to have *semantic content* (ie, meaning) when interpreted by high-level *executive computational*

[54] The intended sense of the term "symbol" above deals with the functioning of symbols in an integrated symbolic processing system (eg, words in a language). A second sense of the term relates only to solitary items (eg, the US flag as a symbol of freedom, or white as a symbol of purity). The discussion here is about the former systematic use of symbols for computational purposes.

[55] This naïve referential notion of a concept will be modified in the next chapter, but an intuitive notion is all that is required for now. The technical term "reference", somewhat unconventionally, applies in this text only between concepts, not between concepts and the world.

processes.[56] The semantic contents we create during interpretation (ie, during executive computations) are about the objects, kinds, and events of everyday experience. Accordingly, a concept constitutes a mental model of something, usually created in order to inform our interactions with that something.

One potentially useful, but admittedly coarse, metaphor for a concept in the present discussion is as a building or house with multiple rooms. Each building (ie, concept) is made up of many variously arranged and connected rooms, which stand for the variety of characteristics composing (and essentially specifying) the concept as a representation. Connections between certain rooms suggest interdependencies between certain characteristics. Furthermore, each room may have some number of lights on in it, signifying the current *level of activation* of each characteristic in the concept. Different numbers and arrangements of lights on and off in a building exemplify different concept activation patterns, and reflect different uses of the concept in different contexts, where some characteristics are more salient than others.

Activations of some concept characteristics may also trigger further activations of other characteristics, either in the same concept or other concepts. That is, the activation of certain combinations of characteristics of a concept can automatically enhance, reduce, fill in, or otherwise affect other concepts and characteristics. These kinds of triggering effects in and among concepts are context dependent, but will not be classified as full-fledged *computations* here, that term being reserved for symbolic computation alone, above the level of semantic concepts. Simple triggering is an association-driven effect such as might be exhibited in the usual interpretation of a word association task, perhaps inducing us to think about cats when we think about dogs, or about flying when we think about feathers or wings. In a more sophisticated form it may lead us to conclude that a dog's wagging tail indicates it is friendly (intraconcept association), or that a dog's looking intently at a cat indicates the cat better run (interconcept association). The basic stimulus-response nature of association (ie, simple condition -> simple action pairing) is useful for quickly evaluating simple situations, but we seem to employ more advanced symbolic processing for constructing the inference patterns involved with complex reasoning and language.

Symbols

More precisely, a symbol is a mental representation, and therefore a concept in its own right, but one possessing only syntactic characteristics, in distinction to the typical semantic concept which exhibits only semantic characteristics.[57] From a conceptual perspective, symbols are therefore classified as

[56] Currently uninterpreted, but potentially interpretable, concepts comprise the latent knowledge in our underlying *belief systems*. Currently uninterpreted concepts might loosely be said to have potential meaning, but that meaning will vary with the activation levels of the concept's characteristics and with executive biases on any particular occasion of use.

[57] This notion of an independent symbol representation is somewhat different from the venerable alternative proposal—namely, the Language Of Thought (LOT) hypothesis (Fodor, 2008). LOT seems to combine syntactic and semantic characteristics into one mental representation, which is instantiated during thought.

linguistic, as opposed to semantic, concepts. For example, words (as symbols) have syntactic characteristics such as part-of-speech (eg, adjective), number (singular vs plural), and other characteristics that specify how the word is used and combined vis-à-vis other words. Numerals and variables likewise possess syntactic characteristics that specify their symbolic use (eg, in mathematical equations).[58] Other kinds of symbols include names (like Elias), categories (like student), operators (like +), quantifiers (like some), and predicates (like happy), among others.

A symbol also possesses pointers (ie, links) to the underlying semantic concepts or concept characteristics that it may stand for. The pointers fix the symbol's range of potential application, akin to its polysemes. When the symbol is employed by executive processes, those processes are thought to follow all of the pointers in parallel to access the underlying concepts, before selecting the one best interpretation in the current context (Marslen-Wilson, 1989). Normally, the symbol itself has no meaningful content over and above its syntactic content. If the symbol itself *is* the content of interest, then the conceptual system is assumed to construct a semantic concept explicitly representing the *mentioned* (as opposed to *used*) symbol with the appropriate semantic characteristics.[59] Quotation marks are a common convention for distinguishing between a symbol and the concept it points to—eg, the symbol "cat" vs the concept cat. (See Chapter 2: *Anchoring Concepts in Personal Ontology* for a proposal of exactly what a symbol points to.)

Primitive symbols (like "troubled" and "water") are readily combined according to syntactic and semantic rules into symbolic *expressions* (like "troubled water"). And those expressions can be combined into ever larger expressions (like "bridge over troubled water"). But while each symbol in a complex expression still refers to its underlying concept, grammatically valid subexpressions (like "troubled water") will also normally point to their own underlying concepts, often created on-the-fly to express the semantics of the combination. For example, the concept underlying "troubled water" is not even about water and incorporates certain negative psychological connotations which are not derivable from either "troubled" or "water" separately. The entire phrase "bridge over troubled water" may end up, among other things, pointing to a concept representing the song written by Paul Simon.

Symbols also aid communication. Two people who use the same symbol can coordinate discussions of their semantic concepts. For example, young children often exhibit impoverished or overgeneralized concepts when compared with adults. However, if both are using the same symbol name (eg, "dog") for their respective concepts which may be markedly different, the child can update her dog concept either through overt instruction or simply by listening to adult conversation about dogs. Such coordination of concept usage, however, depends on each having previously and independently developed dog

---

[58] As used here, a *numeral* (like the symbol "3", a linguistic concept) is distinct from a *number* (like the quantity 3, a semantic concept). Both are mental representations, however.

[59] For example, in the sentences: a) π is an irrational number, and b) π is a letter of the Greek alphabet, each "π" points to a different concept—one a number, the other a letter. Using quotation marks to talk about symbols avoids this confusion in the second sentence: "π" is a letter of the Greek alphabet, where being a letter of the Greek alphabet then becomes a semantic characteristic of "π". More technically, the semantic characteristics are attached to the semantic concept |"π"| to distinguish it from the semantic concept |π| and the syntactic concept "π", as discussed later.

concepts which are more-or-less commensurate, typically consisting of stereotypical and salient characteristics. Confusion obviously arises when each is using the same word to refer to different concept representations. This kind of referential confusion is common, even in adult conversation, since a certain synchrony of concepts is required to share ideas. Normally, our built-in concept learning mechanisms provide us with roughly similar semantic concepts tied to environmental situations, if our background experiences are similar. But if those concepts are abstract, or if our background experiences are not similar, then effective communication becomes difficult.

What should be apparent from the above way of thinking about symbols, concepts, and concept interpretations is that symbols are key to the public sharing of information. Concepts are unique to each person, having developed from that person's unique experiences and interactions with the world. Concepts can be similar in two different people, but they can never be identical. Likewise, each person's interpretations of their concepts are based on internal processes which are unique to that person. Interpretations, based on nuanced concept activation levels in different contexts, and nuanced ways of processing those levels, can also be similar, but not identical. A significant problem with public information transfer by symbols is that the rich detail afforded by our semantic concepts is lost in translation to symbols. Yet the corresponding advantage is that symbols allow us to think and focus on the broad properties of and relations between concepts (ie, generalities) without taking into account all the gory details of those respective concepts.[60]

Computation

As used here, the term "computation" denotes a discrete process of deriving new information from old.[61] It embodies what is here meant by *symbolic thinking*. Given discrete pieces of information as input, a computational process produces new pieces of information as output in a systematic way. In general, computation is the basis for logical reasoning (eg, deduction), generalization (eg, induction), and explanation (eg, abduction). Previously, it was mentioned that the piecewise activation of a concept can trigger parts of other concepts and itself through a mechanism called association. Association at the conceptual level constitutes a simple form of computation, because it takes individual concept characteristics as input, and triggers other characteristics as output. But the full power of computation is only apparent at the symbolic level, where symbols (and complex symbolic expressions) are manipulated in flexible, but structured, ways.[62]

A symbolic computation is usually specified in terms of an *algorithm*, which is any specific procedure for producing outputs from inputs. Mental algorithms are presumably represented in the brain as sets of

---

[60] Of course, such abridged symbolic thinking may also lead us into oversimplification, to our own chagrin.

[61] As discussed earlier in this chapter, a basic conjecture is that perception registers continuous features (eg, degrees of brightness), while conception consolidates perceptual registrations into discrete characteristics (eg, dark, dim, bright, brilliant, etc) depending on context.

[62] The current analysis implies there are at least two distinct kinds of thinking—namely, associative and symbolic. Future research is required to decide whether there is a neurological basis for this claim.

neurally encoded instruction sequences (distinct from semantic concepts, but with flexible access to them) that analyze and evaluate the semantics encoded in those concepts.[63] The execution of an algorithm's instructions is potentially managed by native neural control circuits (eg, incorporating the NAND gates mentioned previously) which can select and run the available algorithms based on current goals and circumstances.

Executing an algorithm's instructions might involve querying concepts, making decisions based on the responses, and in due course altering conceptual semantic connections. From a global perspective, these kinds of executive processes fall under the general rubric of *interpretation*. Occurrent interpretations arise as a result of an algorithm's evaluation of concepts. An interpretation here is labeled *semantic content*, because the latent semantic information available in concepts then becomes actionable content, available to guide behavior. It is this process of interpretation which may be the source of meaning in our lives, since our mental algorithms seem designed to continuously evaluate our conceptual knowledge, status, and prospects in service of our goals.

A basic hypothesis in this text is that an algorithm's instructions are encoded as mental symbols. As a consequence, there are two main kinds of computation that need to be performed by the symbolic processing system. The first involves algorithms for combining symbols together into meaningful expressions. (The converse operation of deconstructing a complex expression into meaningful subunits is included, and both fall under the general topic of *combinatorial analysis*. We use symbol construction algorithms to generate new ideas, and symbol deconstruction algorithms to understand them.) For example, in a mathematical domain we might construct the expression "2+2" from the primitive symbols 2 and +. In pure mathematics the construction is purely syntactic, being dependent only on the syntactic characteristics of the primitive symbols and the algorithmic rules of combination. (A construction like "22+" would not satisfy these constraints.) However, while the strict combination is governed only by syntactic rules, the semantics of the underlying concepts (namely, 2 and +) will necessarily be taken into account during selection of which expression to create in the first place. Syntactic and semantic constraints are almost always in play together. As another example in the linguistic domain, consider the expression "green apple". Here, the symbolic elements are "green" and "apple", which are combined (according to English syntax) as an adjective followed by a noun. The semantic constraints on the concepts green and apple, in this context, require that green typically modify some physical object which can be colored more-or-less completely. Correspondingly, apple takes a color that is normally restricted to one of red, green, yellow, brown/black, where green may optionally signal unripe. Combinatorial processes are possibly based on constraint satisfaction algorithms, which perform unconscious searches for combinatory solutions that could satisfy multiple constraints. A person's problem solving abilities may be dependent on how many constraints can be managed at the same time (possibly in working memory), and on how many searches can be carried out simultaneously by the unconscious combinatorial processes.

---

[63] Mathematically speaking, an instruction is a function, which when computationally applied to inputs, produces outputs. In symbolic computation, not only are the inputs and outputs symbols, but the instructions themselves are composed of symbols.

The second type of computation relates to the derivation of new symbols. When the mathematical expression "2+2" is evaluated during the processing of a summing algorithm (triggered by "+"), the concepts 2 and + are accessed to conceptually compute the result 4. The concept 4 then may be translated to the symbol "4" as the symbolic result, perhaps for communication purposes. Similarly, when the linguistic expression "mother of Elias" is evaluated by an algorithm that processes possessive constructions, it should generate "Becky", based on the concepts linked to "mother", "of", and "Elias".

In the narrow context of a computation, part of the meaning of a symbol or symbolic expression (as opposed to the interpreted meaning of its underlying concept during evaluation) is the *role* it plays in the computation. The symbol performs a specific function in its relationships with the other symbols in the computation. It must satisfy the constraints imposed by the other symbols. In sentences, these roles most saliently center around the verb, which can select an agent, theme, beneficiary, location, etc, to make up a coherent sentence. But all words induce roles that other words can play in deriving the meaning of any expression. In the "mother of Elias" example, "mother of" plays the role of a characterizing permanent birthing and care-giving relation for "Elias", and "Elias" plays the role of specified offspring for "mother of".

In fact, the role of a symbol extends beyond the specific computations in which it is used (Schneider, 2011). Once an individual links a given symbol to an underlying concept, it can play the same role in any number of other computations that individual might perform. A person who knows some elementary geometry probably knows the formulas for the circumference and area of a circle: $C = 2\pi r$ and $A = \pi r^2$, where "r" is the symbol for the circle's radius. Once established, this same symbol can be used in *any* other computation involving the radius of a circle—eg, in computing the length of an arc, $s = r\theta$. For that person the symbol takes on a stable significance, since its reference is locked to its underlying concept, thereby becoming available for use in any symbolic computation involving that concept. In the mental life of any one person, thinking in symbols coordinates with thinking in concepts.

However, the situation is different when two people want to coordinate their thinking. Each person may use the same symbol in communication, but their underlying concepts will almost always be somewhat divergent, based on their past learning experiences. Since effective communication ultimately depends on sharing similar underlying concepts, some cognitive mechanism is required to determine when two concepts are sufficiently similar to be called the same reference for a given symbol. Here is where the computational roles played by the symbol in two people's thinking can provide some measure of coordination. If the symbol is (or would be) used in the same computations in the same way by both, then each can have some assurance that the underlying concepts attached to the symbol are the same for purposes of communication. Of course, it is intractable to determine the total computational role of any useful symbol; but, with patience, one can query and infer the most important computational roles of another person's symbol use in limited-scope domains. Fortunately, we seem to have powerful unconscious linguistic inference processes (glossed as intuition) to perform just these kinds of symbolic computations efficiently.

Finally, the foregoing approach to symbol representation and use suggests that symbolic thinking may have evolved for at least two distinct reasons: to support advanced forms of internal logical reasoning, and to support public communication. Although generalized conceptual reasoning sans symbols is a formidable capability, it appears limited to analyzing (via predication and basic propositional operators) the objects and events of everyday interactions with the world. If this understanding is correct, then thinking without symbols lacks the advanced syntactic features of logic (such as quantification, variable binding, and scoping) that are required for full scientific analysis and a social theory of mind. Further, it is difficult to see how abstract concepts could develop through the generalization of concept characteristics ultimately derived from perception. Abstraction is of a different order than generalization, since an abstract concept does not derive its characteristics from particular observed instances, as a generalized concept does.[64]

In support of public communication, symbols would seem to provide the most likely (or perhaps only) means of coordinating our concepts with others. A symbol exhibits objective properties which can be directly perceived, making it a sharable entity. Semantic concepts, on the other hand, have subjective properties, making them private. But if a community can collectively agree to use the same public symbols to refer to roughly similar private concepts, and have the ability to discern the conditions under which others use those symbols, then their private concepts can converge. The basic requirement is that they are active communicators willing to update their own concepts in socially adaptable ways.

The Representational Hierarchy

In the global ascending tracts for cognitive information processing in the brain, the functions of sensation, perception, conception, and language illustrate a fundamental organizing principle: hierarchical information processing is ubiquitous. In spite of evolution's opportunistic shortcuts, apparently fashioned for greater efficiency or simple satisficing (ie, good enough information processing), the overall organizational plan follows the basic evolutionary strategy of adding and integrating new faculties to the old. This hierarchical organization manifests itself in many different ways, as explored in subsequent chapters.

The following diagram attempts to locate conception in the wider cognitive representational hierarchy discussed in this chapter. It recognizes key representational elements at each level, along with a proposed vocabulary for explicitly talking about them. Each level is organized around a basic representational unit (word, concept, percept, etc—assumed to exist in long-term or short-term memory), which can combine with other units to produce a complex representation at the same level (sentence, belief, episode, etc—also potentially resident in memory). These more-or-less static representations, when interpreted by content producing processes at each level, are readily translated

[64] For example, the class *cheetah* is a generalization of individual observable cheetahs. But *species* is an abstraction of classes like *cheetah*, which are not directly observable. The distinction is subtle, because it is easy to confuse the concepts of a *group* of cheetahs (observable) with the *class* (ie, type) cheetah (not observable). A group and a class/type are different concepts. (Unfortunately, there are further complications, depending on whether words like species refer to a group, a class/type or to an individual dispersed thing, discussed in Chapter 4: *The Metaconcept Hierarchy,* and *Particular Characteristics of Universal Concepts*.)

into other representations at either lower or higher levels in the hierarchy. For example, a concept may be translated either into a percept or into a logical expression (ie, a term or predicate for use in reasoning). Translation (and attendant content generation at each level) ultimately tops out in communicable words and sentences, and bottoms out in private sensory qualities.

LEVEL		STATIC REPRESENTATIONAL UNIT	Constituent of →	STATIC REPRESENTATIONAL PATTERN	Activation →	DYNAMIC REPRESENTATIONAL PATTERN
Surface	↑	Word		Sentence		Utterance
Syntax	Flow of translations	Phrase		Clause		Parse
Semantics		Logical Expression		Proposition		Proposal
Conception		Concept		Belief		Thought
Perception		Percept		Episode		Impression
Sensation	↓	Quality		Integral Quality		Quale

Figure 8: Cognitive Representational Hierarchy

The static/dynamic distinction between mental representational patterns is framed in terms of the relative permanence of the two types of representation. A static representation sustains a dynamic representation when portions of it become activated. Static indicates the pattern is based on the connectivity among representational elements, while dynamic means the pattern is inherent in the spatiotemporal activity of those elements. The significance of the difference lies in the roles each plays in cognition. A static pattern is the basis for storing background information, which subsequently is interpreted in a more-or-less context-free manner; while a dynamic pattern gives rise to occurrent topical information, interpreted in the context of the activation (ie, including influences from neighboring activations). Dynamic representations sometimes correlate with the conscious experience of entertaining our static representations, but many dynamic representations seem to function outside of awareness, their interpreted content being employed at a subpersonal level. Static representations often correlate with what we know, whether or not we are aware of it.

The proposal that static representational units exist as distinct physical mental phenomena is controversial, but is in keeping with the meaning of "physical" as discussed in Chapter 3: *Concepts & Existence* (since these mental phenomena arise out of physical patterns in the physical neural substrate). But this does not mean they are clearly separable entities, which can be isolated from their functional environments. For example, the current analysis has, somewhat informally, talked about beliefs as if they were clearly identifiable mental representations. However, belief representations exist in an embedded neural substrate of concept nodes and links between nodes in a complex belief network. Accordingly, beliefs only become capable of individuation when a segment of a belief net is interpreted and subsequently expressed in terms of individual propositions. For example, if you witness a murder, you may unconsciously embed a belief that A murdered B into your belief network without explicitly recognizing it as one of your beliefs. Later, when asked to recall the event, you may be able to interpret your belief network and derive the relevant proposition that A murdered B, and report it with the sentence "A murdered B". In general, each person may initially create somewhat different embedded belief representations of the same event according to their unique belief net organization and dispositions. But they still may converge on the same propositional interpretation—eg, that A murdered B. In fact, many of our beliefs are not explicitly represented at all. Your belief that 1,464,378 is greater than 1,464,149 is likely derived on-the-fly from your interpretation of the two numbers, rather than from an explicit quantitative relationship between them.

To conclude this chapter: it seems our concepts serve as focal points for our mental models of the environment, other persons, and ourselves, which allows us to simulate activities and think things through before committing to action. Concepts are embodied in the current state of connectivity in long-term memory structures, which become selectively active when stimulated. These memory structures ultimately become organized into interconnected *concept networks*, which encode enduring memory traces of beliefs, wants, hopes and dreams acquired incrementally over an extended period of time. The meaning we ascribe to our concepts has many sources, but is ultimately based on how our concepts are connected together.

2. CONCEPTUAL MEANING

The Indeterminacy of Perceptual & Conceptual Content

One of our most ingrained cognitive presuppositions is that objects are "out there" in the world. This assumption is akin to the various well-known perceptual assumptions built into our perceptual faculties that enable us to construct consistent and approximate representations of our environment quickly.[65] Such assumptions give shape to the information we derive from the environment, and are part and parcel of our mental representations. The occurrent source of our object interpretations, therefore, is not only the environmental stimulus itself, but also our preconceived assumptions and other expectations, which have been triggered and shaped by the stimulus. It is our preconceptions which are subsequently amended by continued observation, and which, when instantiated by sufficient evidence, we take to be objects "out there". Raw environmental stimuli must receive some kind of an interpretation before becoming comprehensible. Occurrent perceived objects, therefore, are largely projections of our minds (mainly of our object concepts) onto the world.[66] The perception of object properties and relations are likewise projections from our property and relation concepts. Accordingly, it is technically erroneous to think about occurrent perceptual object representations as being veridical or as referring to specific *things in the environment*, as suggested by intuition, even though it is a very useful and reliable belief most of the time given environmental stabilities. Perceptual objects are tailored idealizations that simply make sense to us—customized tools that simply work for us. They are useful, but are necessarily only models of reality.

The existence of a world "out there" is not in question, but the nature of that world and of our relationship to it is. The world, in a nutshell, is ultimately ineffable due to its complexity. Although we can model it in various approximate and useful ways, its full depth, complexity, and organization necessarily outstrip the finite modeling capacities featured in perception and conception. Support for this rather unintuitive viewpoint comes from recent discoveries in physics, which demonstrate that our experiments can never hope to characterize reality as it is in itself (Baggott, 2013).[67] The best we can do is to characterize reality as it appears to us through our measurements, which is not the same thing as

[65] Some built-in perceptual assumptions (Coren, Ward, & Enns, 2004): scenes are lit from above, surfaces are usually convex (rather than concave), objects are attached to surfaces (rather than floating above), objects are usually viewed from above, objects are observed from a conventional (rather than coincidental) viewpoint.

[66] Exactly how our object concepts are *originally* created is at the heart of the innate vs learned debate. The conjecture here is that certain innate processes have been selected by evolution to operate on certain aspects of environmental stimuli to produce working concepts, much as seems to be the case with language learning and numerical competency.

[67] Experimental tests of Bell's and Leggett's inequalities unequivocally show that our descriptions of quantum particles have no meaning apart from the tools we use to detect and measure them. Since normal-sized objects are made up of quantum particles, I take it that our perceptual and conceptual faculties likewise determine how we describe normal-sized objects and events.

directly grasping reality itself. The only measurements we can make depend on our perceptual and conceptual capacities, and we have no other option than to derive objects and their properties from these limited measurements. Our situation is much like Plato's famous cave dweller, who was born in a dark cave. There is a fire deep in the cave, along with mysterious happenings, but the cave dweller is only aware of shadows projected onto the walls. The cave dweller can ascertain various properties of the shadows (eg, the area of the shadows), but these properties may have little to do with what is causing the shadows. Taking a clue from physics, it may be that the world is just a collection of overlapping physical fields (ie, specific physical attributes reminiscent of the brain's attribute maps), which our concepts carve up in various approximate ways.

Yet we apparently can interact with the perceived external world, based on our ability to notice and produce *regularities* (ie, changes and temporary states) in the environment. Perception and conception have evolved to exploit this underlying nonhomogeneity of the world, by mirroring some of the regularities internally in the form of mental representations. Conceptual and perceptual representations mediate between the flux of the world and our interpretive processes, which use the information in those representations to guide behavior.

Thus, when we talk about objects in the world, we are using mental concepts which we *assume* have specific reference in the world when activated in a particular context. Given the world is inherently unspecifiable due to its inaccessibility and infinite complexity,[68] concepts, being only finite representations and incorporating a perspective, can never exactly refer. They are necessarily generalizations. Even in the technical sciences, where precision in specifying exactly what is being talked about is paramount, the most fundamental "objects" under investigation, are always ill-defined and based on presupposition. But the referential assumption is nevertheless benign in many contexts in which we are interested, since most of our concepts are reliable and precise enough in facilitating our everyday interactions with the world. We simply have learned to make use of whatever stable correlations we can find (probably in spatial and temporal distributions of environmental energy impinging on our senses) via an evolved system of conceptualization.

Token concept activations therefore unavoidably exhibit a fundamental "indeterminacy of reference"[69] with respect to their correspondence with the world. They are only useful and approximate. No matter how much is known about any phenomenon, there is always more to be known about it; and no matter how it is characterized, it could be characterized differently, or sometimes even explained away as illusory. We apparently develop coherent conceptual descriptions of the world in increasing profundity

---

[68] Quantum mechanics is our best current description of reality, which implies an ultimately discrete structure to the universe. However, the number of possible quantum states for any interesting system is so large as to be irrelevant for describing normal concepts. Likewise, most computer theorists now believe that not all functions (eg, concepts) are computable, even given an infinite amount of time. Technically, the class of NP problems is probably beyond any means of specification.

[69] Quine originally used the phrase to talk about the relation between linguistic terms and objects in the world (Quine, 1960). There are no such things as "objects" in the world in the current account, just energy fluctuations that appear to us as objects.

(both individually and as a species), but those descriptions are, at best, only asymptotic to a presumed ideal of what "really" exists.[70]

A concept behaves like a partial specification for something and, accordingly, expresses expectations about that something. Typically, most expectations are below the level of general awareness. The expectations are embodied in a set of discrete beliefs and dispositions, which bind to a conceptual nexus (discussed in Chapter 3). Conscious and unconscious mental processes actively compare beliefs about a concept to observational regularities to determine if the two can be justifiably matched. Depending on the soundness/completeness/consistency of the constructed bridge in the given context, the match is judged good or poor or somewhere in between. Classification of an observation occurs when the best matching concept achieves sufficient activation to effectively inhibit competing interpretations, where our degree of confidence relates to how effective the collateral inhibition is.

We normally think of the so-called natural kind concepts, like lion or gold (usually opposed to more definitional nominal kind concepts, like husband or even number), as having a potentially infinite number of discrete characteristics (ie, beliefs about them), or equivalently, as potentially having characteristics of infinite complexity. The complexity of conceptual content arises from attempting to model and describe an infinitely complex stable regularity detected in the flux of the world, assumed to be caused by a natural object. Our natural kind concepts are therefore idealizations and simplifications, in spite of their encyclopedic complexity. But we nevertheless intuitively conjecture that there exists an unseen thing-in-itself, as Kant would say, causing the regularity. The conceptual nexus stands for that thing. Other concepts associated with a natural kind concept, such as its properties and relations to other concepts, achieve a certain derived existential status by their association with the thing-in-itself.

Most nominal kind concepts, however, are inherently no less complex, since they ultimately include natural kind concepts in their definitions. For example, if a husband is nominally a married man, the concept of man (and perhaps to a lesser extent the concept of being married) makes the combination complex. Virtually all useful concepts are fundamentally indeterminate.

Like words, concepts are sometimes deemed "vague", or even "fuzzy". However, it seems more accurate to say that the mental operation connecting observations to concepts is underdetermined. This means that a myriad of alternative observations can, in general, trigger or fall under a concept encoded in long-term memory depending on internal and external context. A concept simply can never be precise enough to correspond only to one particular set of observations. Concepts are tools we use for understanding, and those tools are the best we have for dealing with the complexities of the world. Moreover, we can always be fooled by appearances, showing us that our concepts are often inappropriately applied.

---

[70] This perspective on reality lies somewhere between an antirealist view like Dummett's, and a fully realist (metaphysical) view like Lowe's. Dummett likens reality to an "amorphous lump" which we slice up according to our needs (Dummett, 1993). Lowe argues reality is comprised of specific kinds of entities, discoverable independently of our needs (Lowe, 2006). The view here advocates thinking of reality as a nonhomogeneous flux of energy, only partially knowable independently.

There are other conspiring, but unconvincing, reasons for thinking concepts vague. First, the usually large number of concept associations, their intricate interconnectivity, and their inaccessibility to consciousness may give the illusion of vagueness. Second, relevant beliefs in a concept are picked out and fine-tuned when it is activated or instantiated in the current context. Each situated, instantiated concept activation then is exactly delineated by the total finite active associations it has with other concepts at a particular time. But the field of activation stretches out with unpredictable boundaries to include other related concepts thereby diffusing the focus. One final source of the impression of vagueness is probably due to the interface between our perceptual (nondiscrete) and conceptual (discrete) mental representations. Discrete concepts like red, round, and walk, are grounded in other graded representational systems like color space, shape space, and articulated body-motion space. Representations in these latter kinds of *value spaces* are part of the sensorimotor and core domain systems underlying and interacting with the conceptual system. Neither level of representation fully translates into the other, and each is relatively incomplete with respect to the whole, which includes both.

The Reference of Concepts

The previous section presents the problem of reference as it frequently appears in the philosophical literature: namely, the problem of explaining how it is that our words, concepts, or other mental states can truly pick out objective objects and features of the world, when our knowledge of those objects and features is constructed by our mental processes themselves—mental processes subject to bias, latent assumptions, illusion, and even collective hallucination. How can we trust our basic judgements involving classification and description when the only resources we have for inquiry are the perceptual and conceptual systems we are endowed with (Putnam, 1981)?[71] To help clarify what it means in the present proposal for a concept to refer to something, it may be useful to distinguish first between *representation* and *reference*. Most of the discussion of representation, however, is postponed until later in this chapter.

All concepts, in virtue of their function as tools for building thoughts and beliefs, are representations. In other words, they convey information about various (presumed) things beyond themselves. As representations, they can be used to represent all manner of objects, kinds, and situations, whether real or imaginary. What a concept represents on some occasion of use is standardly called its content. Concepts share this capacity for representing content with many other forms of representation including certain pictures, diagrams, signs, linguistic expressions, and gestures, all of which are used by agents for conveying information of various kinds.

However, there is a class of concepts which additionally can refer to things. In this context, to refer to something is to simply point to it (not to represent or describe it as being someway). The prototypical concepts which can refer are linguistic names for things, such as proper nouns (eg, Los Angeles), many

[71] If you make cookies with your grandma's cookie-cutters, you shouldn't be too surprised at the cookie shapes you end up with.

definite descriptions (eg, the author of Waverly), and some other singular terms (eg, my cat). Each of these linguistic items is a concept in its own right with descriptive linguistic content (eg, "Los Angeles" is a proper Spanish name with English translation "The Angels"). But in addition, they also point to or *refer to* other concepts which *represent* what we are talking about when we use the name. For example, the concept of the name "Los Angeles" refers to the concept of the city Los Angeles. The two concepts are distinct, but linked by reference, such that speaking or hearing the words "Los Angeles" can trigger thoughts about the city. Likewise, the concept representing the phrase "the author of Waverly" refers to the concept of a particular person. And "my cat" refers to the concept of a cat belonging to me. In this way, straightforward linguistic name concepts can be used to refer to and talk about underlying representation-only concepts (ie, those concepts whose representation does not include a name for something else). Names provide for efficient access to commonly used concepts. In lieu of a named reference to a particular underlying concept (or any concept for that matter), the only other strategy for accessing it involves instituting some sort of search for it based on a partial description.[72]

To summarize, reference, as envisioned here, is an entirely intra-mental strategy for facilitating access to mental representations. This leaves representation to mediate between the mind and the world (as discussed further in the later section entitled *The Origins and Uses of Representation*).

The "Meaning" of Concepts

The idea of meaning, in its typical application to linguistic expressions, has always been a difficult and controversial subject in philosophy and linguistics. In philosophy, one famous theory relates the meaning of a sentence to its so-called truth conditions (Davidson, 1967). One grasps the meaning of a sentence if one knows the conditions under which the sentence may be truthfully spoken. In linguistics, the meaning of a word is often related to one's ability to use it effectively in communication, so others can understand you (Wittgenstein, 1963). If you possess the appropriate skills for effectively employing a word, then you grasp its meaning. These and many similar explications of meaning, however, do not address the specifics of mental process or representation.

Within the framework of the concept model developed here, and drawing on the ideas above, concept meaning can be somewhat more precisely pinned down, but it is multifaceted and sometimes indirect. Meaning embraces 1) the mental representation of truth-like conditions as constituents of a concept (namely, integrated beliefs about it), 2) the interpretive processes used to evaluate those conditions, 3) linguistic meaning, and 4) the grounding of concepts in perception.

---

[72] The class of referring linguistic names should probably be extended to many other words besides basic singular terms. For example, common noun concepts (like "cat") would also seem to refer (to the concept characterizing a cat), if the above proposal for internal-only conceptual reference is accepted.

Integrated Beliefs

While Davidson's truth conditional interpretation of meaning is aimed at sentence meaning, it seems reasonable to extend it to the concepts giving rise to sentence meaning. But for a concept we are interested in its satisfaction conditions. If you know the conditions under which things fall within the scope of a concept, then you grasp at least part of the meaning associated with the concept. In this case, the conditions are related to your beliefs about instances of the concept, say about apples. If those beliefs are taken to include more-or-less central beliefs about them (eg, all apples normally have seeds and come from apple trees), plus other peripheral beliefs that are prototypical, statistical, or otherwise more weakly correlated with them (eg, some apples are yellow), then the sum total of all such beliefs about a concept could be said to comprise our *generic background knowledge* about it.[73] The meaning associated with the concept for each of us, then, is to a large extent wrapped up with our generic background knowledge, expressed as beliefs about concepts embedded in an interconnected web of beliefs (Quine & Ullian, 1978).

Although we seem to have an intuitive feel for our background knowledge about various generic concepts (we can often discuss or itemize their characteristics), our grasp of the extent and depth of such knowledge is tenuous. It is difficult to tell where one concept ends and another begins, because they are intertwined; and we have difficulty accessing all of a concept's branches, probably because activation spreads unevenly and depends on context. Nevertheless, it is useful to circumscribe a concept for the purpose of talking about it. Loosely speaking, a concept includes a set of connected beliefs which are local to a single *sense*, where a sense roughly corresponds to a distinct polyseme for a word in a dictionary. For example, two senses of the word "grasp" include the ideas of "physically taking hold of" and "mentally understanding". In spite of the likelihood that the latter is a metaphorical projection from the default concrete reading of the former, these are two separate concepts. The sense of a concept relates to an elemental core meaning of the concept, a core distillation which is common to a range of uses of the concept in certain contexts, and which relates to what *kind* of things the concept represents according to their intrinsic nature. Each sense serves as an invariant starting point for use in a particular context that is subsequently modulated by the details of the context. If someone says to us "My friend betrayed me", our interpretation can start with one of the invariant meanings of friend as "someone with whom we share mutual regard" and modify it as now being a one-way street, or possibly even annulled.

Given a core sense, the beliefs about a concept typically will include a large number of direct associations, glossed as *concept characterizations*: properties (ie, attributes and their values), parts, kinds (and subkinds), external relations (to other concepts), internal relations (eg, among properties and parts), contrasts (with sibling concepts), and statistical information (related to expectations), among other things. Different characterizations are relevant for different concepts, and in general they are not

---

[73]Generic background knowledge, as used here, is intended to exclude *situational background knowledge* about particular situation instances. For example, our generic background knowledge of a concept like apple is segregated from (yet still linked to) our situational background knowledge about particular apples we have seen. Likewise, it excludes *linguistic background knowledge* about words and phrases like "An apple a day keeps the doctor away".

shared among concept representations, because they are unique to a concept and subject to change as new information is acquired. The general picture of a concept that emerges from these considerations is of a *nexus*, reflecting a core sense, to which is attached a variety of background beliefs characterizing it (including beliefs about those beliefs). A concept therefore is a composite, made up of many other concepts and constraints related to its use, but individuated by its own set of characterizations.

Interpretive Processes

The second ingredient of conceptual meaning is the set of interpretive processes for analyzing and manipulating concepts. The interpretive processes are analogous to computer programs operating on concept data. Although locally algorithmic in nature, they can nevertheless be said to *interpret* the current state of activation of concepts, and to make decisions about reactivating and reconfiguring them, without positing a homunculus.[74] The process of interpreting and otherwise using concepts for various purposes gives them local meaning to each controlling process. For example, one aspect of an interpretation may lead to an evaluation of satisfiability—ie, whether or not a concept's characterization effectively matches the environmental situation, and whether or not the concept's current activation can be adjusted or otherwise stretched to match the situation better. Other interpretations may draw inferences about unseen concept characteristics, create or combine concepts, or choose between alternative concepts. Interpretation involves monitoring the activation states of concepts, activating selected portions of concepts, and noting the effects of spreading activation. The interpretation of the dynamic spreading of activation from a given concept to others shapes the contextualized meaning associated with the concept. Since the dynamic activation of a concept normally does not involve all of its characteristics, a concept's contextualized meaning in a particular use can differ significantly from its other uses, but will always include a unique, intentional invariant core meaning associated with the kind of concept it is (ie, its sense).

Although a concept's representation can be structurally thought of as the sum total of all its characteristics, meaning is not directly a property of its representation. The concept itself has no intrinsic meaning, per se. Rather, meaning is constructed from the representation by interpretive processes in a particular context of activation.[75] When the concept is triggered (by sensory stimulation, memory, linguistic expressions, etc), only certain portions of the total structure become active, depending on the learned associations within the structure, the strength of the triggering, previous activations (eg, priming), basal level tone/facilitation, related background knowledge, and the activation status of other concepts due to the current context. One could talk about a context-free, literal meaning in which the entire concept becomes activated and interpreted in isolation, but this situation is clearly an idealization, since it can never occur in practice. The closest an interpretation comes to a literal meaning is the concept's invariant core meaning. However, the core meaning is "literal" only in

[74] The homunculus we experience as a unified observing self is probably an epiphenomenon, rather than a specific interpretive process in the brain. It probably does not "interpret" things (or make choices among them) as we generally suppose, but rather attempts to "explain" things after the fact to fit in with an ongoing consistent narrative involving the self.

[75] See (Evans, 2009) for a summary of this view.

the sense that it is part of every interpreted activation of the concept in every context in which that concept is used. Similarly, the dictionary sense of a word concept does not reflect a literal meaning either. It presents a limited selection of concept characteristics useful for preliminary classification and for distinguishing the concept from near neighbors, but the context of use can modify or override any such characteristics. To summarize, neither words nor concepts have a literal meaning, ex situ; meaning is developed in interpretive processes.

Linguistic Meaning

As previously mentioned, words (and other linguistic expressions) are concepts in their own right. Therefore, they can evoke meaning, just like other concepts, but they do it indirectly. Through learning they typically become associated with other linguistic and nonlinguistic concepts. Accordingly, word concepts seem to operate at a metalevel, in their own distinct linguistic realm incorporating syntactic and morphological characteristics. Their content, however, ranges over the nonlinguistic concepts in underlying conceptual representations. While word concepts do establish connections to other word concepts (lexical relations), most of their meaning in use is based on their underlying connections to nonlinguistic concepts. For example, the word "bachelor" is nominally linked to "unmarried" and "man", which are in turn linked to the underlying concepts of being unmarried and being a man. The language faculty incorporates processes that can operate both on word concepts and on underlying concepts.

The semantic interpretation of the underlying concepts, provided by linguistic processes through words, can never be totally precise, due to the partial and ever-changing matchup between words and the underlying concepts on any particular occasion of use. Words are used over and over again for many purposes, in many ways, and their meaning depends heavily on context (ie, which mix of underlying concept beliefs are currently active). We typically use words to summon underlying beliefs from various provinces under one fleeting linguistic roof, mainly for on-the-spot communication (whether external or internal). Using Davidson's insight about truth conditions, truth then becomes a linguistic interpretation of whether the collected words that call up particular propositional contents from the underlying beliefs on a particular occasion appear to be sufficiently satisfied or integrated with all of the rest of our underlying conceptual beliefs about the world.

Perceptual Grounding

Finally, concrete concepts are directly grounded in percepts and other perceptual value spaces that give those concepts an embodied "feel". This aspect of meaning seems very different from the other aspects of conceptual meaning, because it is the only one which comes from outside of the conceptual system itself. As information flows inward and upward through the major processing stages involved with sensation, perception, and finally conception, the format of the information may change between successive representations. In particular, a radical switch from analog to digital representation is evident at the perceptual/conceptual juncture (Dretske, 1981). What were fields of attribute values in the perceptual system,[76] suddenly coalesce into discrete beliefs in the conceptual system. Experienced doctors mirror this kind of shift in information processing when they extract a set of diagnostic beliefs from an x-ray image. The mental processes of analog to digital conversion are not well understood, but they appear crucial for generalization and abstraction. Most of our beliefs may ultimately trace back to some form of analog representation of information. However, even within the conceptual system, elaborations of concrete concepts (eg, through metaphor and analogy) provide a way to extend a measure of grounding to more abstract concepts, as when we construe (mathematical) sets as containers or importance in terms of large size (Lakoff & Johnson, 1999). On the other hand, the most intangible concepts (eg, mercy) seem to have only a minimal degree of grounding, if any.

Anchoring Concepts in Personal Ontology

The core meaning associated with most of the concepts we use in everyday thinking seems to depend in an important way on certain other intuitive concepts, which structure our background knowledge of the world. These intuitive concepts (such as agent, object, substance, event, etc) are often innate or begin development very early in life, but eventually come to represent our adult ontology of the world. Seeing someone turn on the TV as an implicit event caused by an agent, and the TV itself as an affected material object, are examples of (usually unconscious) classification that confer meaning on the particular event and object through association with already established intuitive meaning. Such intuitive concepts frame interpretations of our experiences, even though they resist specification.

To a great extent, the coverage, consistency, and sophistication of our intuitive concepts and their abstract ontological relatives (such as attributes, kinds, relations, etc) will determine the depth of our understanding of the world. Refining our intuitive core concepts typically opens creative vistas. Although we tend to consciously gloss over such concepts because they are so fundamental and difficult to state explicitly, they are an integral part of our thinking. Consider for a moment what a puzzling kind of thing a hole is; for example a nail hole in a wall. It typically has a location, shape, and potentially at least can be located in different places just like other objects, but there is nothing there. It is a strange sort of thing, but we have no trouble thinking about it in reliable ways. After some study and introspection, I eventually came to understand a hole as a kind of *immateriality* (Casati & Varzi, 1994), which has spatiotemporal properties, but no intrinsic material properties. Distinguishing these two kinds of properties now seems straightforward, but realizing that they could be conceptually separated was not trivial, because materiality and immateriality are embedded intuitive concepts. Such immaterialities have a place in our underlying conceptual ontology, because they confer a specific unified meaning on uses of terms like "hole", "dent", "crack", and "valley". It is not necessary to understand the ontological details of a concept to use it effectively, but it is necessary if you want to understand the deeper significance and consequences of its use.

---

[76] An attribute field represents a distribution in the intensity of a relevant attribute or spatial frequency elements, as discussed in the Appendix.

Our intuitive ontological concepts seem relatively unorganized initially. Although we have an early intuitive feel for number, agency, event, object, etc, how they relate to each other requires further development. Experiments (Shipley, 2008) show that infants can detect and segment object motion events in a remarkably consistent fashion (possibly at several levels of granularity), and a grasp of agency is apparent by about five months of age (Carey, 2009). However, the broader understanding that agents can participate in events, and that events can be classified based on the participation of an agent (causal event) or inanimate object (happening event) develops somewhat more slowly.

As proposed in Chapter 6, our ontological structures seem to mature along two dimensions. The first is a hierarchical dimension, whereby our intuitive concepts become organized according to various *kinds* of things. The example of events becoming differentiated into causal vs happening events, or later into states vs processes vs acts, illustrates a hierarchical organization. Similar hierarchies develop for different kinds of objects, substances, properties, spatial regions, etc. Over time, the general tendency is to integrate local hierarchies into more global ones, maintaining consistency among concepts, such that lower level concepts can inherit characteristics from those higher up. Core understanding of a particular new experience or object is achieved when it can be coherently classified in terms of one (or more) of the lowest level concepts in the ontological hierarchy.[77] The more coherent the global ontological organization is, the deeper the understanding of the new experience.

The second organizing dimension reflects the ubiquitous constituency structure of the world. We tend to analyze complex things by finding ways to conceptually simplify, idealize, or subdivide them, thereby reconceptualizing them in simpler terms. Construed broadly, a *conceptual constituent* is a simple part of a complex concept: a reduced part which somehow figures in understanding the whole.[78] As such, a constituent is a generalization of the parthood relationship.[79] For example, we can simplify the concept of a causal event by decomposing it into an agent, an action, and perhaps other ontological objects and states, as in "George drove Shirley crazy". For events (and situations in general) each constituent is standardly called a *participant* in linguistics. For objects, the constituents might be the substances of which the object is composed, object characteristics, or properties. While hierarchical decomposition breaks down concepts by kind, constituency decomposition breaks them down by structure. Together they produce a useful and unified ontology as a basis for comprehending the world.

While each of us ultimately develops our own underlying ontology, shaped by evolution but reflecting our own unique experiences and perspective on the world, the tendency to organize our ontological concepts into a consistent rational structure may be universal. We are generally more successful in dealing with the world when we can effectively classify the oftentimes complex objects and events we

---

[77] The ontological hierarchy seems to bleed into lower level domain hierarchies, so the lowest ontological levels are context dependent.

[78] This idea is not the same as reductionism, where the whole is reduced to the sum of its parts, or where the whole is seen as the result of the interaction among its parts. A constituent of a complex concept is a simplification that can be effectively split off from the whole, partially reducing the complex description of the whole. The remainder is almost always another complex concept, which may or may not support further simplification.

[79] Two distinct meanings of parthood are discussed in Chapter 5: *The Primary Relations of the Material Object*. Constituency relates only to the second meaning.

encounter, thereby giving them an appropriate core meaning. There are better and worse ways of interpreting practical situations (reinforced by rewards and punishments), and there are better and worse ways of characterizing and organizing our ontologies which frame those interpretations. We can usually classify something in several different ways, but which way we choose will affect our viewpoint. When we are able to assign an experience of something (including hearing a linguistic expression) to a latent ontological concept (ie, some type of object, event, property, relation, number, agent, etc), we then say we "grasp" it. Such understanding is rudimentary, and depends on the current context, but it represents the seed from which the fuller interpretation of a meaning develops, a meaning which may grow more and more complex with each subsequent increment of meaning. Part II of this book is an attempt to uncover some basic principles for understanding and refining our innate ontological concepts, which can serve as the basis for ascribing rational core meanings to all our experiences.

Interpreting Experience

The foregoing discussion argues that meaning results when an agent is able to interpret its experiences. Such interpretations may ultimately bring about a knowledge change in the agent, as when the agent creates internal representations of experiences (eg, from active percepts and concepts) in the activation context in which they occur. But an implicit core interpretation more simply results when an experience is unconsciously cataloged as falling under at least one ontological heading in our background knowledge. The most basic form of interpretation, therefore, is probably classification.

Yet interpretation typically extends far beyond ontological considerations. In addition to the contributions of linguistic and pragmatic contexts, the impact of other background knowledge on interpretation is also unmistakable. For example, when we possess organized knowledge of specific domains (medicine, literature, politics, cooking, kinship, etc), we become able to situate and integrate the experience with knowledge in those domains, just like we do with our basic ontological knowledge. Such enhanced interpretation confers additional meaning on the experience which we can use. If I talk about the mother of my uncle's son, using your knowledge of kinship relations (and other situational knowledge), you can likely identify the specific person I'm talking about and that she's my aunt. Such specific domain knowledge is at least as important as the general ontological knowledge underpinning it. The overall picture of interpretation that emerges, then, is one which draws on a variety of types of background knowledge including at least ontological, domain, theoretical, situational (ie, particular circumstances), perceptual, goal-relevant, and linguistic knowledge. The interaction among some of these other sources of meaning from knowledge is discussed further in Chapter 6.

Although the detailed mental processes associated with interpretation remain murky, there is a long tradition of split-brain research which localizes these interpretive processes to a module in the left hemisphere—the so-called left-brain interpreter (Gazzaniga, 2002). In this model our interpreter is continually trying to make sense of the ongoing events of our experience, most of which are initially diffusely represented in our right hemisphere. But the interpreter, it is proposed, does not directly process and evaluate these diffuse representations. It rather attempts to fit those representations into various stockpiled left-brain event schemata that might make sense in the current context. The schemata assignments thus provide one way of quickly organizing the diffuse representations into potentially useful interpretations at several levels based on past experience.[80] Once potential event interpretations are available, the interpreter then appears to stitch the most likely interpretations together into event sequences, which become the episodic memory traces we all recognize as chronicling the narrative of our daily lives.

The meaningful interpretation of the representation of an experience must be distinguished from the rote *decoding* of a representation. Both interpretation and decoding involve the classification and extraction of information from a representational pattern, but in decoding the pattern is analyzed and an action selected without producing meaning. Decoding is comparable to the initial stages of a straightforward stimulus-response behavioral loop (even for complex decoding systems), where global background knowledge plays no significant role in the analysis of the stimulus pattern.

The decoding of DNA is a good example of information processing, sans meaning, at the cellular level. During gene expression, a segment of DNA is first "copied" to produce a corresponding strand of mRNA.[81] The resulting mRNA chemically expresses a pattern made up of a sequence of codons (ie, 3-letter chemical "words" like CUA and GUG). This pattern is then "translated" during protein synthesis, ultimately producing another chemical pattern in a polypeptide chain made up of amino acids like leucine and valine. The process of protein synthesis therefore does produce information content, which is "transcribed" into the configuration of the polypeptide chain for use by subsequent cellular processes. But while the overall process is exceedingly complex (much more so than the simple summary here), there is no meaningful side effect or benefit for the process of protein synthesis. The "translation" process does not interpret or otherwise use the decoded information for its own purposes. A computer, similarly, is able to decode its program instructions and data formats during normal operation. It effectively transforms input data into output data without ascribing internal meaning to the data or program, even though the data and program usually do have meaning for human programmers. Again, the primary information processing function of transforming input representations into output representations is present, but without a self-regulating interpretive process.

Although the distinction between interpretation and decoding seems relatively clear-cut in these cases, it does not take much imagination to envision a more advanced computer system which is able to learn from its input data and evaluate its actions based on stored conceptual knowledge about the world. And many animals, after all, interact very successfully with their environment by acquiring and using perceptual knowledge. The difference between primitive decoding and sophisticated interpretation is therefore probably a matter of degree, depending more on the development of useful background knowledge of various kinds, than on the sheer magnitude or complexity of information processing capacity. If the above interpretive premise of meaning is accepted—that is, the meaning of an

---

[80] However, our evolutionary imperative to assign interpretations (ie, meanings) to events is sometimes so strong as to distort primary perception and even override common sense, leading to mistakes in judgment.

[81] More accurately, an mRNA molecule is synthesized from the DNA molecule by transcription and splicing out of noncoding DNA segments.

experience arises from coherently situating it in the webs of background knowledge—it suggests further consequences, and perhaps clarifies some mysteries about the nature of meaning:

- "Meaning" is essentially a relative term, relative to the background knowledge possessed by someone. Without some background knowledge things have no meaning, since then they cannot receive an interpretation.
- The relativity of meaning does not imply it cannot be characterized, but it has no independent claim to existence as Quine long ago concluded (Quine, 1969). Meaning does not denote an objective independently existing phenomenon in the world. It is like the term "color" in that the color of an object is constructed by the mind, even though the object itself is assumed to exist independently.
- Although the substrate of meaning is a person's current entire holistic web of background beliefs (Quine & Ullian, 1978), the experience of meaning results from the interpretation of the local activation pattern of contextually selected background beliefs. Only those background beliefs which directly interface with a new experience contribute to its meaning.
- Meaning is not part of the conceptual system, per se (ie, the system for representing and maintaining concepts). Rather, it is related to a separate interpretive system, which examines conceptual representations.
- Meaning is generated when represented experiences are classified, characterized, and deemed relevant by our higher-order executive processes. Successful classification and characterization confers cognitive significance on the new experience, based on past experiences. Interpretive relevance is likely based on global criteria like consistency, semantic coherence, and compatibility with goals.
- Successful classification and characterization may ultimately result in a change of state or change in representation (eg, a new thought or belief) instigated by the interpretive processes. This change of state embodies the information content derived from the initial interpretation of the underlying conceptual representations (see Chapter 5: *Immaterial Objects & Information Content* for additional discussion) and other interpretive and contextual contributions. We normally express the interpretation in terms of conceptually discrete *propositions*, which can then interact with interpretations of background beliefs via inference; but many other perceptual forms expressing nondiscrete information content, such as schematic imagery or quantity/quality judgments, are also available to us.
- Partly contrary to Putnam's view (Putnam, 1975), meaning is clearly in the head, but the background knowledge from which meaning arises derives from the world (based both on environmental and social learning). In other words, the content and structuring of our background knowledge develops from learning about causal interactions with the world and from learning about conventional social norms (eg, word senses, names, and normal behavior). But using that knowledge during the interpretation of new experiences gives rise to the added experience of meaning. So technically the meaning of the new experience is wholly "in the head" when it is interpreted in context, even though that meaning is based on preexisting representations derived from the world.

- Because conceptual meaning is a dynamic phenomenon related to the activation and interpretation of more-or-less static belief networks, the study of meaning is to some extent separable from the study of knowledge. Knowledge relates to optimizing the correlation and organization of concepts with respect to the environmental and social world (ie, effective learning), while meaning relates to the use of that knowledge in those situations in which we find ourselves (ie, effective living). We generally refine our knowledge by adding conceptual distinctions and more reliable relations between concepts. We enhance our meaning experiences by increasing the scope of our interpretive and inference processes.
- Our implicit meanings, construed on-the-fly, can be turned into explicit conceptual representations for future access, but always with some information loss. This is the process of conscious learning, requiring the sometimes difficult encoding of meaningful experiences into explicit knowledge. However, meaning seems to have evolved initially as part of our flexible dynamic capacity for processing information online. At the conceptual level it is a high-level capacity that allows us to react effectively to abstractly meaningful events above the level of the reflex or habitual automatism; but it also often reflects a low-level capacity we share with animals, conferring a more dispositional significance to perceptual events.
- Meaning, at least at the basic ontological level, is not truth conditional—contrast with (Davidson, 1967). You don't need to know what is potentially true about an experience to ascribe meaning to it. It is enough to be able to fit the experience into your world view, no matter how fractured that world view is. Even a schizophrenic organization of information admits of meaningful, if delusional, interpretations. Truth simply confers a high reliability on the interpretation.

Word Meaning & Polysemy

Grasping the core meaning of a word in context depends crucially on assigning it to the proper ontological conceptual category, since different semantic characteristics attach to different concepts. In the case of ambiguous words, the semantic resolution is relatively straightforward. In "Carol made a deposit to her bank", the reference is obviously to the concept of a financial establishment, and not a river bank. However, virtually all words also exhibit *polysemy*, such that slightly different, but connected, senses of the same word may complicate selection of the appropriate sense. If Carol is an adult, she likely made the deposit to a financial establishment, but if a young child, probably to her piggy bank. Our subsequent inferences are quite different, depending on which sense of "bank" we select.

At an even more subtle level, we also seem to distinguish between what Cruse calls *facets* of meaning for a word (Cruse, 2000), where a facet is a somewhat finer distinction than a polysemous sense. While the use of two distinct polysemous senses in the same sentence is typically anomalous, two distinct facet interpretations can coexist peacefully. Compare "Carol withdrew all the money from her bank, and then smashed it" (conflict between financial establishment and piggy sense interpretations), but "Carol withdrew all the money from her bank, and then walked outside" (financial organization and physical building interpretations acceptable). Note that there are necessarily at least two facets in the latter sentence, since you cannot coherently withdraw money from a physical building, or physically walk out of an organization.[82] Distinct facets of meaning are commonplace, but we don't normally notice them because they are not linguistically anomalous together. Nevertheless, the differences are relevant semantically, and the different meanings are reflected in our ontology of intuitive concepts. In the double faceted sentence above, the word "bank" will link to two different ontological concepts: a material building-object and an organization, interpreted as an immateriality (discussed further in Chapter 5) with potential characteristics like net worth, employees, policies, etc. The final interpretation of bank in this sentence will include potential characteristics from both.

Individuating Word Meanings

The overriding problem with studying the meaning of words, and the underlying conceptual structures which they invoke, is their extreme variability in context. It seems impossible to precisely pin down all of the potential nuanced meanings associated with any given word. Dictionaries attempt to distinguish the primary polysemes for a word, while facets and other identified modulations—eg, see (Croft & Cruse, 2004)—can provide additional refinements of meaning, but so far there is no generally accepted theory of linguistic polysemy to say exactly what a word meaning is. Many philosophers go so far as to say there are no such things as meanings, preferring to apply Quine's dictum of "no entity without identity", claiming it is impossible to determine whether two particular expressions mean exactly the same thing (Quine, 1969). Even synonyms and paraphrases seem to contain slight differences in meaning.

The problem with saying exactly what a meaning is, however, need not defeat the useful specification of meanings. In accepting that meanings can be arbitrarily nuanced or become arbitrarily complex, we can still distinguish broader or narrower classes of linguistic meaning to a degree appropriate to the conceptual content we are intending to express. The problem seems roughly analogous, in mathematics, to trying to pin down the precise meaning of a real number. A real number, in general, will contain an arbitrary (perhaps infinite) number of decimal digits, but this doesn't keep us from saying that real numbers exist.[83] We can talk about various real numbers and organize them systematically (into classes of integers, rationals, irrationals, transcendentals, intervals, etc) even though we cannot exactly describe them. Similarly, the "exact" meaning associated with a word in context is probably ineffable. But anchoring the particular use of a word initially to a particular ontological class expressing its core component of meaning, and progressively adding other components in a systematic way determined by its context, can round out its meaning with sufficient precision for a particular occasion of use.

---

[82] More precisely, the denoted facet is a "place" interpretation associated with the space the building is occupying, since you cannot technically walk outside of a physical solid object. The bank organization (facet #1) owns the physical mortar and steel bank building-object (facet #2) which occupies a spatial region (facet #3) out of which you can walk. Yet another facet denotes the bank's staff (facet #4), as when we talk about a friendly bank.

[83] But see (Chaitin, 2005) for an alternative viewpoint.

Note that anchoring a word to a particular ontological class is somewhat different from assigning the word to a particular polyseme, as framed in a dictionary of (normally numbered) word senses. The various polysemes of a word are all connected in various ways, thereby forming a related cluster of standard usages in various contexts. But these connections among the senses can be highly diverse, ranging from relatively straightforward specializations, generalizations, and restrictions to metaphoric, metonymous, and other miscellaneous extensions (Cruse, 2000). In contrast, when a word is used in context, its core meaning can be understood as an instance of some conceptual "kind"—ie, as an instance of an ontological class (in addition to its polysemous sense). You grasp a word's core meaning in context when you (rightly or wrongly) identify an ontological class which is consistent with the core meanings of the other words in the sentence.

Relying on the assumption that the most precise meaning of a word is ultimately determined by its use in a particular context, specifying its meaning then depends on specifying its context. This basic idea views linguistic context as a series of ever widening circles around a word, where each contextual increment may add information relevant to its meaning. For words in a sentence, this amounts to analyzing the nested phrase structure, in a bottom-up fashion, similar to parsing (but separately and completely for each individual word). Each outwardly expanding phrase in which a word participates then represents a wider context of interpretation, which may contribute additional information relevant to the meaning associated with the word.

Consider the meaning of the word "ball". Out of context, the meaning of "ball" is at best ambiguous between a spherical object and a formal social occasion for dancing. However by adding more linguistic context, as in "the rubber ball", we readily select the core meaning of a spherical object, assign it a uniqueness in the current context, and additionally specify the substance of which it is made. For a noun like "ball", the lower limit of context is the noun phrase in which it occurs. The first level of contextual meaning is derived from the structure of the noun phrase, typically a determiner plus a modifying phrase.

Embedding this phrase in a sentence like "Deven threw the rubber ball" adds further information about the probable size and weight of the ball (relative to something that can be thrown). The noun phrase is now embedded in a larger verb phrase, giving rise to additional meaning derived from the verb class (in this case the generic act of "propelling" which is a kind of *transitional change* in the underlying ontology);[84] and from the argument role the noun phrase plays in the sentence (in this case the *affected* role, sometimes also called the *theme* or *patient*). The size and weight of the ball must be consistent with beliefs associated with the act of being thrown. If the original linguistic context had been even wider as in "Deven threw the rubber ball into the sewer", you would additionally know where it is located, based on information in the adverbial clause. The meaning of "ball", although it never achieves ultimate precision, nevertheless elaborates as the consideration of context widens. While all of the previous context modulations are locally linguistic in nature, there are still wider contexts involving

---

[84] If the verb had been "attended" (as in "attended a ball"), the meaning of ball (as a social dance) would instead reflect the verb class of "being present" which is a kind of *being someway* in the ontology—see Chapter 6: *The Organization of Situation (ie, Event) Concepts*.

pragmatic conditions such as speech acts reflecting the speaker's intentions, background knowledge, time and place of utterance, etc. For example, if you know that Deven is only one year old, it further limits the size and kind of the ball.

To summarize: The information embedded in nested linguistic expressions shape the meaning provided by the core ontological interpretation of a word. The final interpretation for a word in a linguistic context is a loose combinatorial function of the information contained in the hierarchical phrases in which it is embedded. Beyond the bare sentence, ever wider pragmatic contexts may further adjust a word's meaning until it reaches sufficient content to achieve its intended or practical relevance in the situation. Each content word (as well as each embedded phrase) thereby accumulates a contextualized meaning in use.

The Extension, Intension, & Sense Relations of Concepts

The discussion of concept and word meaning in this chapter, although focused, is somewhat informal in comparison with contemporary accounts of meaning. Over the last century philosophers, linguists, and cognitive scientists have developed a more formal vocabulary and theoretical basis for discussing the potential components of meaning. There seems to be a general consensus that all theories of meaning at least need to incorporate specific explanations of the *extension*, *intension*, and *sense relations* of concepts (or words, depending on emphasis). Given the previous proposal that the proper interpretation of a concept in use depends crucially on our background knowledge about it and related concepts (in addition to other background factors like context and mental state), then questions about how that knowledge is structured to support meaning naturally arise. Questions about the extension, intension, and sense relations of concepts, as the terms are used here, relate principally to knowledge structure and organization.

The extension of a concept signifies those things in the world (or even in counterfactual possible worlds) that the concept applies to. The extension of the concept cat, accordingly, typically signifies all the particular cats—past, present, and future, including cats that could have actually existed, but didn't. Likewise, the extension of the concept of my particular cat Rosie is just the flesh-and-blood Rosie herself.[85] That the extension of a concept is an important part of its meaning is especially evident for proper name concepts like Rosie, because there is usually not a lot of additional meaning inherent in a name alone.[86] But even the generic concept cat depends to a significant extent for its meaning on the

---

[85] Philosophers who believe there are other worlds besides our own, will talk about my Rosie as being only one part (or counterpart) of the whole Rosie extending over all worlds. This whole Rosie simply exhibits different characteristics in different worlds while retaining her identity. The multiple worlds view is similar to the common view in our (single) world that a person remains the same person from day to day, in spite of changes. This common view seems to imply that we think of persons (and other objects) as the sum total of their temporal parts over their lifetimes.

[86] However, there is a small amount of additional meaning regarding gender in this case, since you can probably infer that she is female from the name "Rosie".

particular cats we have observed. To represent the extension of a kind concept (like cat) is to explicitly specify its instances (like Rosie). To represent the extension of a particular concept (like Rosie) is to simply represent the concept Rosie as a kind of unique individual thing, such that each particular concept represents its own extension. Note also that since (domestic) cats comprise a separate species, the worldly population of cats (as a single cohesive group) is in the extension of the concept species. The concept cat (formally *Felis catus*) is accordingly represented as an instance of the more abstract concept species. Still other concepts like unicorn, having no extension in the actual world, require representations specifying no instances. Of course, the complete representation of all instances for a typical kind concept is impossible, but a partial representation of a concept's extension is nevertheless useful. The interpretive processes can factor in the number and scope of instances provided by background knowledge, as a significant part of developing the meaning of the concept in use.

The intension of a concept signifies the characteristics which describe it, rather than the things which satisfy those characteristics (or which are simply specified as being instances of the concept). That an adult female cat normally has four legs, meows, weighs 10-15 lbs, and that my cat Rosie is fickle and weighs 11 lb are part of the intension of my concepts cat and Rosie, respectively. In general any concept, since it is a representation, necessarily has an intensional part to distinguish it from other concepts. Even concepts which have no extension (eg, unicorn and fairy) or the same extension (eg, fool's gold and pyrite) require different intensional characteristics to tell them apart. Intensions, therefore, provide a crucial component of meaning, especially when extensional meaning is absent.

The sense relations of a concept signify those associations the concept has with certain other related concepts. These associations effectively constrain, or otherwise influence, how the concept is interpreted in use. For example, knowing that a cat is a kind of animal makes available to the interpreter all of the intensional knowledge about animals when cat becomes activated. This specific type of association (ie, the *kind of* association), is a sense relation called *hyponymy*, which allows concepts to be organized into kind hierarchies, depending on their content. Such hierarchies form the skeletal concepts in a knowledge base, where intensional characteristics and extensional instances become attached to the skeletal concepts. Another sense relation imposing a different hierarchical ordering on concepts is *meronymy*, or the part-whole relationship. Fingers and thumbs are part of hands, hands and feet are part of bodies. Knowledge about parts, for example, allows us to infer consequences for the whole based on conditions of the parts. A third important sense relation is *opposition*. Concepts which are associated in opposition are typically interpreted as a contrast field of concepts, only one of which can usually apply in any given circumstance. There are many different types of opposition, including complementaries (eg, alive and dead), antonyms (eg, long and short), disjoint siblings (eg, fish, bird, mammal, reptile, amphibian, as kinds of vertebrate), among many others. The scope of our knowledge about various topics depends to a large extent on the number of contrasting concepts we have available to us for drawing fine distinctions among possible interpretations of new experiences. Generally speaking then, sense relations provide a basis for organizing our knowledge and for situating new experiences within that organized framework.

Signs

Words exemplify the idea of *indirect meaning* in that we normally use them to trigger underlying nonlinguistic concepts from which meaning then arises. Any direct meaning associated with the word itself, as a linguistic entity, is not normally relevant unless it is the topic of discussion or being used poetically. Similarly, concepts convey indirect meaning by bearing information about subjects other than themselves—eg, information about worldly objects, object types, or various abstractions. Other such entities which we use to indirectly "stand in for" something else include images (eg, of a person), models, icons, conventional sounds (eg, a siren), gestures, certain arrangements of objects (eg, the words in a sentence), and many other forms. We call such entities *representations*, because we interpret them as representing (or indicating) something beyond themselves. For example, the siren conventionally indicates an emergency of some sort. But representational entities are peculiar kinds of things, because we do not treat them like normal everyday objects. While it is true that the representational object itself is a physical phenomenon with physical properties, such as location, intensity, and pitch (for a siren), what the siren represents (ie, an emergency situation) has different properties altogether relating to urgency, some precipitating event, and the need to get out of the way. How any physical entity can beget semantic properties unrelated to its physical properties remains a controversial philosophical problem (Fodor, 1994).

One possible approach to investigating representations is to start with a simple kind of representation, namely a *sign*. A sign is a simple representation because it signifies something else beyond itself, and that something else is typically a univocal state-of-affairs linked by simple association with the sign. The sign signifies something, but in the minimal case conveys no additional information characterizing that something, as is common with a more complex representation.

In general, when an agent comes to reliably associate one kind of situation (ie, state-of-affairs) with another, the first becomes a sign of the second (Millikan, 2004).[87] The first situation comes to signify the second situation. We say that the appearance of smoke signifies (or is a sign of) fire, that ducks flying south signifies the approach of winter, and that the experience of the color red signifies maximal activation of red/green opponent cells in the visual system (see Appendix A: *Sensation*). The link between the two situations is not fail-safe, but it carries sufficient reliability or relevance to be meaningful to the agent. Agents can glean useful information from signs.

All of the previous examples exhibit a *causal* relationship between the two situations: an effect (eg, smoke) is produced by a cause (eg, fire). In general, when one situation is known to cause another situation (whether deterministically or probabilistically), the overt observation of the effect is a sign that the often covert cause may have occurred. However, the reverse is even more informative, although normally less available since causes are usually hidden. If you happen to observe the cause (eg, fire) you can predict the effect (eg, smoke) with high reliability. Information therefore can flow in both

[87] The term "agent" is being used in an unusually broad sense to include any biological consumer of information, ranging from individual persons, animals, or organisms to internal information processing modules such as might be found in a nervous system.

directions, and either situation can be a sign of the other. Deriving information from a sign simply depends on there being a *correlation* between the two situations, rather than a more potent causal connection. If either situation generally occurs along with the second, then knowing about one can provide information about the other. Even if the correlation is small (whether positive or negative), some information is available.

All of the previous examples also involve *natural signs*, being derived from situations produced by natural physical processes in the environment. There are also *intentional signs* produced by agents for consumption by other agents.[88] Rituals and language are produced by humans for humans, bee dances indicating the location of nectar are for other bees in the hive, and my cat's meow to be let outside is an interspecies intentional sign that we both have come to understand. In a sense all intentional signs are natural signs too, since from one agent's perspective, other agents are merely part of its environment. However, an intentional sign also suggests being "selected for a purpose" (Millikan, 2004) or being produced specifically to carry natural information (Dretske, 1995). Intentional signs are selected (via evolution, culture, or individuals), and used by agents to serve agents' information processing needs. But whether the source of a sign is natural (eg, tree growth rings indicating the age of a tree) or intentional (eg, a no smoking sign indicating an area where smoking is prohibited), we must first perceive and classify the physical manifestation of the sign as an information carrying representation (as opposed to a straightforward physical phenomenon). Once we adopt this "intentional stance" (Dennett, 1987) toward the state-of-affairs, we can then interpret it as a sign, and attempt to retrieve the target state-of-affairs with which it is associated.

Intentional signs sometimes become organized into a *representational system*, where not only the signs themselves, but their interrelationships, may have meaning. For example, the signs and their relationships in a map signify certain states-of-affairs in the environment that are used by agents to get around in unfamiliar places. A representational system, as a configuration of individual signs, is structured like a collection of containers (eg, locations or variables) that can be filled in different ways or to different levels at different times, signifying different things. The relationships among the containers and their elements constitute a pattern or state of the system. The current state of the system is interpreted as being about something (its interpreted content), is configured in a certain way (its format), and has a physical medium (its substrate). For a standard map the interpreted content is information about the spatial location of environmental objects, the format is combined imagery and iconic, and the substrate is paper and ink. A basic thesis in this study is that conception is based on a representational system whose interpreted content is based on conceptual patterns, whose format is a connected network of concept nodes, and whose substrate is associative neural tissue.

[88] Again, "intentional" is used in a wide sense to cover any intended activity, whether sentience is involved or not.

Representations in General

In the wider context of the conceptual ontology developed in this book, a representation is a kind of *arrangement* of matter or energy. In a typical case of representation, a given material substrate composed of substances or parts will have a particular arrangement (or organization) of those substances or parts into a pattern. The same substrate may express different patterns at different times, and these multiple patterns are viewed as conceptually distinct from the underlying substrate. Accordingly, the pattern in the substrate is not itself a material thing, but merely the arrangement or organization of the parts of the material substrate. The previously mentioned concept of a "hole in a wall" is such a pattern, since it describes a particular arrangement of matter constituting the wall, giving rise to the concept of the hole in terms of the way the wall's substance is configured. A natural sign is a pattern (eg, a situation) in the world. An inscription chiseled in marble is a pattern in the marble. All three kinds of patterns are immaterialities (being only features of their host objects), but only the third is associated with the intentional representation of meaning. Note that the underlying substrate is always material and has physical properties. A representational pattern in a substrate is always immaterial and parasitic on the substrate. Furthermore, a complex representational pattern may have a format (eg, analog or digital),[89] and be meaningful (or not), depending on who is looking at the representation (Collins, 2010).

In the literature on semantic information theory, information is often classified as a kind of "data": namely, information just *is* well-formed meaningful truthful data (Floridi, 2005). However, it seems more in line with common usage to associate "data" with the *representation* of information, such that the data (eg, a binary sequence) can convey information. Information can be extracted from the data by an information consumer given the proper background knowledge. Data then is not information itself, but is a kind of representation (eg, possessing a digital or analog format like other kinds of representations). Information is better characterized as the meaningful semantic content of data (typically expressed in propositions or dispositions) when the data is interpreted by an information consumer in the appropriate context. In fact, (raw) data might be more usefully characterized as an uninterpreted representation.

Within this alternative framework, those patterns which are potentially meaningful to someone, are said to have an *information content*. Information content is parasitic on an underlying pattern, in the same way a pattern is parasitic on its underlying substrate. And likewise, the information content is also immaterial.[90] Particular information can be extracted from particular patterns, as when a forensic expert sifts through the ashes of a fire for evidence. Patterns in matter subsist independently of humankind, but information is only produced when a pattern is interpreted (by someone or some subpersonal process). Information is therefore always meaningful (whether true or false), whereas a

[89] The format of an old-fashioned LP vinyl record (ie, its wavy grooves) is an example of an analog representation, while a standard CD possesses a digital representation.

[90] Norbert Wiener, the founder of cybernetics, long ago famously concluded: "Information is information, not matter or energy." (Wiener, 1965)

data representation is simply well-formed (whether meaningful or not).[91] More specifically for concepts, information content is produced when conceptual patterns are interpreted, often conveniently in terms of elementary propositions.[92] On the other hand, information content is realized when a proposition (encoding a piece of information), subsequently becomes classified as a belief in the neural substrate (ie, as a conceptual representation) via long-term potentiation[93] of an active thought pattern.

To summarize, a representational system may usefully be thought of in terms of three basic elements: a material substrate, a pattern in that substrate, and the information content conveyed by the pattern. However, each element exhibits markedly different kinds of properties from the others. The substrate, as a manifestation of matter or energy, can be described by physical properties. But a pattern in the substrate is here classified as nonmaterial, being merely the particular arrangement of the matter or energy constituting the substrate. The pattern then is characterized in terms of configurational properties such as shape, value, and relative position. Similarly, the information content conveyed by the pattern is also nonmaterial. Information content seems most usefully described by semantic properties like singular/plural, containment, animacy, function, etc. There are undoubtedly other kinds of properties relevant to characterizing representational systems as well (eg, syntactic properties), but these others appear to provide support for the three main representational elements already mentioned.

Based on the prelude above, a *generic representation*, as specifically defined here, is a configuration of multi-state elements organized and selected (via evolution, culture, or an individual) to perform a given information processing function for an information consumer.[94] Before deconstructing this rather complex idea, notice that it is stated in quite general terms. This is because representations form such an eclectic group. The following table shows the breadth of some examples of common external (nonmental) representations to illustrate the idea. However, the generic characteristics of the representational systems listed below would seem to apply both to external and internal (mental) representations.

---

[91] Some alternative accounts regard information as necessarily true, in order to resolve certain logical paradoxes, but the issue is not resolved yet (Floridi, 2003; Dretske, 1981).

[92] Communications information theory (Shannon, 1948) defines information content in terms of bits (not propositions), but this is for the purpose of analyzing communications channels. Communications theory does not deal with the meaning of a message, although it does deal with its efficient encoding.

[93] Long-term potentiation/depression is the strengthening/weakening of connections between neurons, thought to be one of several bases of memory and forgetting.

[94] The idea of performing a function comes from (Dretske, 1995), of that function being selected from (Millikan, 2004), and of it being for information processing from (Palmer, 1999).

Table 1: Examples of External (nonmental) Representation

REPRESEN-TATION	REPRESENTATIONAL ELEMENT	ELEMENT VALUE	ELEMENT ORGANIZATION	PROTOTYPICAL CONTENT
TV screen	RGB triad	luminance	2D array	scene
PC monitor	pixel	luminance	2D array	scene
sentence	word (or lexeme)	particular word	sequence	extension, intension
word	letter, morpheme	particular letter	sequence	concept extension
utterance	phoneme	particular phoneme	sequence	extension, intension
sign language	sign	particular sign	sequence	extension, intension
bee dance	action	wiggle direction, frequency	feature set	direction, distance to nectar
gas gauge	pointer position	angle of deflection	linear scale	amount of gas
clock	hand positions	angle	circular scales (hr, min, sec)	current time
map	icon (road, town, etc)	particular icon	2D plane	territory, structures
photograph	point color	hue, saturation, lightness	2D plane	scene
film	frame	2D planar luminance	2D plane + time (spatiotemporal)	story
tea leaves	tea leaf	orientation, shape, color	2D overlays	future events
orrery	planet, sun, moon models	position, orientation	concentric orbits	future/past configuration of planets
tic-tac-toe board	board square	X, O, Blank	3x3 array	game progress

The following is a list of some generic traits of representational systems. This summary characterization serves as a backdrop for a better understanding of conceptual representations, and potentially for mental representation, in general:

- Representations are used by information consumers (ie, agents, sub-personal, or lesser biological processes) for information processing purposes; they are not a naturally occurring phenomenon in the nonbiological physical world. DNA and words therefore qualify as representations, but animal tracks do not, unless they are interpreted as such.
- A representation, in cognition, functions like a variable, in mathematics. Whereas a variable can take on different values, a representation is said to take on different states. Accordingly, the representation itself should be thought of as distinct from the states it accommodates (the variable/value distinction).
- A representation consists of a collection of one or more representational elements. Usually the elements are spatially or temporally organized with respect to each other — eg, into a sequence, array, network, or other configuration. A simple representation, however, could consist of a single element, or a collection of unordered elements. The organization of elements is typically expressed in terms of their proximity (as in an array), or their connectivity (as in a graph). A neural network is such a representation.
- Each representational element can be in one of a number of states at any time. An element's state corresponds to a discrete or continuous value of a variable, expressing a representational type of digital or analog, respectively. A state's type can be nominal, ordinal, interval, or ratio valued depending on the kind of information being represented (note 1). A neural network's elements are neurons, whose state is normally characterized in terms of firing rate.
- When needed for precision and to avoid ambiguity, we can also distinguish between a *representation* and a *representation's organization*. A representation's organization consists of the representational element variables and their arrangement (ie, the relations among the elements), along with a specification of the kinds of values elements can take. Any representation will exhibit a particular organization.
- The combined states of the elements constitute the summary state of the whole representation. In other words, the summary state is the *pattern* of element states currently present. All of the potential patterns expressible in a representation can be conventionally described in terms of its state-space, and the current pattern is a point in its state-space.
- Usually only some patterns, out of all possible patterns, are involved in supporting the information processing function. These well-formed patterns typically exhibit a syntactic formatting of their element states. The *syntax* acts to create higher-order syntactic features (and relations between

features) out of certain groupings of element states. For example, on a tic-tac-toe board representation, the syntax for a board square builds characters out of pen stroke elements. However, only certain stroke combinations qualify as legitimate characters — namely, X, O, or no strokes (blank).

- Each representation is paired with at least one information consumer, which analyzes the patterns appearing in the representation. While the *pattern* is technically a continuant, the consumer is a *process,* technically an occurrent. The difference between a pattern and a process is analogous to the difference between the data and the currently running program for analyzing the data on a computer, although the brain does not operate like a computer.
- A representation and its consumer operate together as a *representational system*. They arise in tandem to perform an information processing function for an organism.
- To the consumer, the pattern in a representational system is "about something", where that something is usually unrelated to the representational system itself. The aboutness of a representational system is the basis for ascribing content to its patterns (note 2).
- The term *content* implies a potential interpretation of the pattern in a representation by its information consumer. Content is the information able to be extracted by the consumer from the pattern, and depends on the successful recognition of the pattern. Strictly speaking, the representation itself does not possess content, except in the context of its information consumers, which may have different uses for and assign different meaningful content to the same data.
- The content of an interpreted representation, generally speaking, consists of two distinct kinds of information. The first relates to the *extension* of the interpreted representation, and indicates what, if anything, the content is about in the world. The second relates to the *intension* of the interpreted representation, and provides descriptive information about the content. For example, the representation "the evening star", when properly interpreted, indicates both Venus (or that particular point of light in the sky), and that it appears in the evening looking like a star.
- The pattern in a representational system may occasionally be misunderstood as being about something when it is, in fact, not about that something. In the case of faulty representation, one could colloquially say the representation is then "misrepresenting" what it evolved to represent (Dretske, 1981). But it seems better to say that the processes interpreting the representation are providing misleading information because they do not have enough sophistication to counter the temporary abnormal behavior of the system (either in the original pattern producer, in the representational substrate, or in the information consumer). The representation itself does not detect that it is misrepresenting; error detection involves evaluation of the pattern in a wider context by the information consumer.

- Typically, only some of a representation's well-formed patterns are meaningful to its consumer. Meaningfulness is a constraint over and above well-formedness. For example, of all the possible states of a tic-tac-toe board (filled with Xs, Os, and blanks), only those with the same number of Xs and Os, or one additional X (assuming X always moves first) will be meaningful for playing the game.
- The information consumer for a representation is technically a transformation process. It transforms the meaningful patterns in one representation into syntactically well-formed patterns in another representation. Simple transformations analyze the absolute state values of potentially meaningful patterns. Complex transformations often analyze selected relations among the pattern element values, ignoring their absolute state values (Palmer, 1999). Detecting relations among meaningful pattern element values is probably the basis for higher-order (linguistic, map reading, etc) information processing.
- Representational systems are causally efficacious because they involve transforming one pattern into another pattern. The input & output patterns manifest as particular physical activations of the underlying neural networks with physical consequences.
- The transformation process decomposes into at least two subfunctions: detection and registration. During detection, the consumer recognizes patterns relevant to its information processing function in the input representation based on local syntactic properties. Registration is the semantic process of analyzing and recomposing the detected information into well-formed element state values for the successor representations.
- A generic transformation can be theoretically modeled by a rule (if <conditions> then <actions>), where a rule might be realized in a variety of media including language, diagrams, neural circuits, or electronics. If the conditions of the rule match a pattern in the input representation resulting in detection, then the action of the rule registers the output state values of elements for the successor representations. Each transformation process is equivalent to numerous rules that process the input pattern and represent the relevant state values in the output pattern. In rules, the conditions and actions are normally expressed in the same syntax, but this is not a requirement for generalized transformations.
- Usually, but not necessarily, the transformation process results in a compression of the information content (Chaitin, 2005). In general, mental processing progresses from raw sensed data representations to more compact representations, corresponding to more focused use of the data. At the level of the individual agent, compact representations of data correspond to greater scope of understanding and greater ability to apply knowledge in broader contexts. Some of the highest level agentive processes consolidate information explicitly into symbolic concepts and linguistic expressions.

Mental Representation

Representations of information are common throughout the nervous system. They are perhaps most clearly exemplified in the sensory organs (eg, retina), primary sensory projection areas (eg, area V1 of the visual cortex), and the numerous attribute maps (color, contour, auditory pitch, body sensations, etc) in the sensorimotor systems. The organization of information processing into explicit representations probably simplifies complex transformations by subdividing them into stages and modules that are more easily manipulated by evolution.

The simplest information processing module (ie, a minimal representational system) consists of one or more sensory receptors connected to a transformation process, as shown in the figure below. The causal interaction between the receptors and the world produces activation of the receptors, such that the level of receptor activation corresponds to the type and intensity of the stimulation. The transformation process integrates and detects certain states of receptor activation potentially important to the organism. By performing pattern recognition or inference-like (ie, rule-like) operations on the detected activation patterns, the transformation process can register conclusions by setting the states of elements in other representations or by directly activating effectors. In general, the complexity of a transformation process varies with the kinds of patterns it needs to analyze. Detecting absolute activation templates is easier than recognizing relationships between active elements, which is still easier than analyzing well-formed syntactic groups of elements.

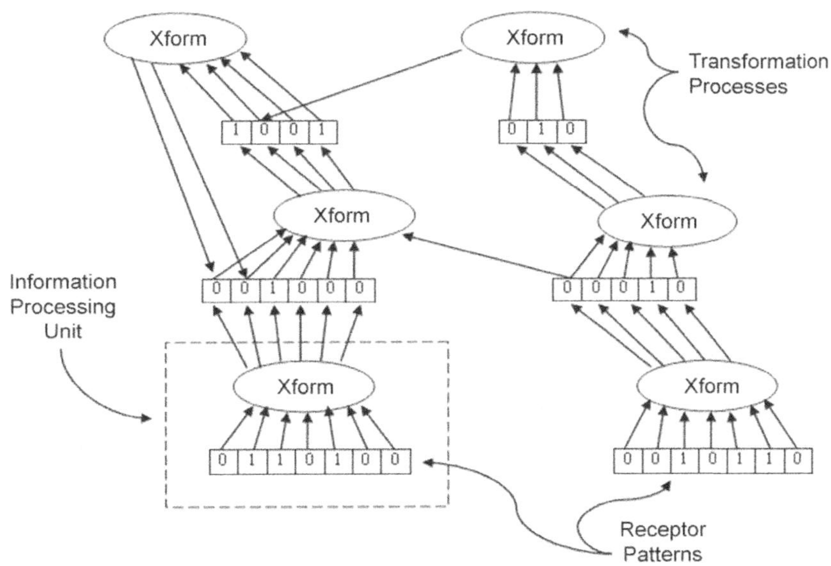

Figure 9: Information Transformation in Stages

Note however that the content of (ie, available information encoded by) a representational system, as the term is used here, is not necessarily represented explicitly inside the transformation process. For example, a simple 3x5 binary value matrix may be used to represent the digits (0-9), as on the face of a digital clock. An automatic process might then respond differently depending on which bits are currently displayed in the matrix—eg, to ring the alarm or not. However, the process itself need not explicitly represent the digits. The differential responses to digits may be implicit in the analysis performed by the process. In fact, if the process does incorporate explicit representations of its own, then it is assumed that the process can theoretically be broken down into subprocesses operating on those representations. If a process incorporates explicit rules as part of the process, then analyzing that rule content is metalevel analysis, performed by another transformational process looking at those rules as separate representations.

Information processing modules are apparently arranged into information processing channels and reentrant circuits in the nervous system. The responses from lower-level transformation processes typically feed into representational elements at a higher level, and there is also feedback from higher-level processes to lower-level representations. The responses of the processes at one level thus can adjust the state values of representational elements at other levels, leading to rather sophisticated and coordinated representations at many levels. Conceptual and linguistic representations probably reside at the top of the hierarchy, and the highest level transformation processes that analyze our highest conceptual representations may be those processes we consciously think of as aspects of the self.[95] The following diagram summarizes the perceptual, core domain, and conceptual functions as consisting of interleaved representations and interpretive processes.

The highest level mental representations, it is proposed, are organized into concept networks, where a network comprises a collection of nodes and associative links between nodes in a graphical layout. The most tightly interconnected nodes in a network encode individual beliefs and concepts, with less tightly interconnected nodes encoding coordinated knowledge over a wider domain. Interconnected nodes are not necessarily proximally located, since a network is essentially defined by its connections, proximal or remote. This means individual beliefs, concepts and domains can be distributed across the cortex. As a single representational element in a representational system, a node is therefore specified by its associations with other nodes in the network and by its level of activation.

A subset of connected nodes representing a concept (or a belief) becomes active bottom-up when a subconceptual consumer process determines that its satisfaction conditions (eg, perceptual, core domain, or linguistic cues) are sufficiently met in the current context, and triggers the concept. Similarly, a top-down consumer may also activate the concept, perhaps as part of a filling in or hypothesis generation operation. Finally, a concept may be triggered through spreading activation within a network via association with related concepts. The active concepts in a network constitute an activation pattern (or dynamic state) of the network representation, which then is available for interpretation by the higher or lower level consumer processes.

[95] Consciousness is probably not a necessary characteristic of the highest level of interpretive process, but it may have evolved for other advantageous reasons in our species. The concept of a self possibly evolved out of interpreting psychological representations of conspecifics, reflexively applied, and may depend on consciousness.

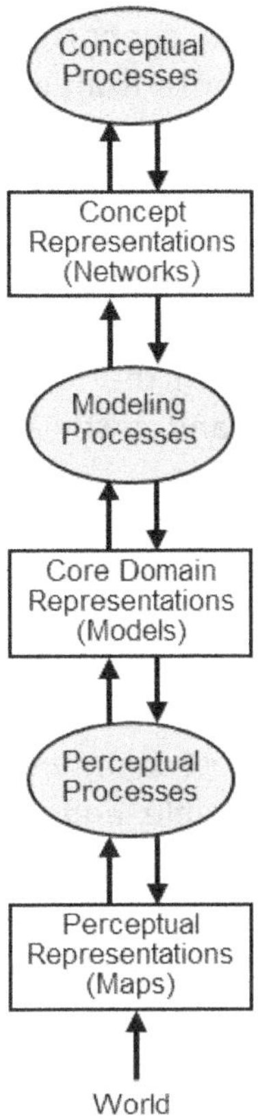

Figure 10: Interleaving of Representations and Processes

This kind of network representation is markedly different from a representation like an image, where the implicit connectivity between representational elements is fixed in a 2D array, and the state of an element is a luminance (or other attribute) value. A node element's state in a concept network representation directly incorporates relational values (rather than simple attribute values), because the node's activation state is affected by its explicit connectivity to other nodes,[96] and which of those other nodes are active. The capacity to represent relations directly is a much more powerful design than simple attribute representation.[97] It permits the representation of general relational knowledge among objects (relational: "Nynne loves Drea"; attributive: "Drea is 12 years old").

[96] A node's connectivity to other nodes is explicit only relative to the prior example of implicit proximal connectivity between elements of an array. Concept nodes in the brain may or may not be connected by neural axons. Alternately, they could be connected via synchronization or some other coordinating mechanism.
[97] In machine learning the capacity to deal with relational data is a whole field of study known as inductive logic programming. Most commercial applications of machine learning (eg, data mining) are still attribute based, although some newer parallel distributed processing connectionist architectures show promise for analyzing relations (Rogers & McClellan, 2008; Thornton, 2000).

Given that a mental representational system supports content, that content (ie, interpretation) is subject to error. Representational error can arise from errors in fixing element states, and interpretive error can arise from detection or registration errors. In this event the agent may come to harbor false information, but not know (or care) that it is false. The situation is analogous to having a map with some errors and inaccuracies, or attempting to read it in dim light, or while under the influence. If the map and lighting and processing conditions are normally sufficiently good to provide some utility, then the representation is still useful. Likewise, concepts confer an evolutionary or other practical advantage, but do not necessarily represent truthfully. In the long run true beliefs would confer the greatest advantage, but pursuing truth is a difficult and costly endeavor not usually warranted for practical everyday affairs.

Conceptual representations in humans have likely evolved for more than one reason. Although the primary reason may have been to enhance our overt perceptual interactions with our environment, another reason seems to involve communication. If the only purpose of conceptualization was for environmental interaction and problem solving, then the meaning of our concepts would likely relate only to our own subjective activities. To support communication, the content of some of our conceptual interpretations must also be publicly accessible and understandable by others. Portions of our conceptual representations thus adhere to conventions which we assume are widely shared by others in our community. Using conventions, we become able in a sense to read each other's minds with a measure of reliability, and thereby to more effectively communicate and cooperate in complex activities.

To summarize the chain of ideas associated with a mental representation: A mental representational system consists of one or more representational structures feeding into an interpretive process. The conceptual representational structures (organized into beliefs and concepts) host patterns of activation (called thoughts), which induce a meaning content assigned by an interpretive process. The interpreted content of thoughts (only sometimes codifiable in language) thereby provides information in a form useful to an agent.

The Origins and Uses of Mental Representation

How do mental representations (eg, percepts, concepts, and words) initially form and come to stand for things; and how do we use those representations to image, think, and talk about things once they are formed? Such questions concern the basic relationship between the mind and the world, and have been central to much philosophical debate in the 20th century. Traditionally, these questions have clustered under the topic of *reference* (eg, how do certain mental terms, like "Mount Everest", come to refer to particular physical features in the world), but here they fall under the heading of *representation*. In the proposals developed in this text, representation deals first-and-foremost with the relationship between the mental and the physical, while reference deals with mental-only relationships (for reasons discussed in the earlier section of this chapter entitled: *The Reference of Concepts*). Thus while concepts can refer to one another (an instance of reference), the extension of a concept under a given interpretation can signify a worldly object (an instance of representation).

Regarding the origins of representation, it seems useful to review one of the most basic learning mechanisms, namely classical conditioning, as exhibited in Pavlov's famous experiments with dogs. Pavlov conditioned dogs to salivate when they heard a buzzer, by always (and only) buzzing when they were fed (Rescorla, 1988). After some training time, the dogs began salivating whenever they heard the buzzer (whether there was food or not). The dogs had learned to associate buzzing (stimulus events) with salivating (response events). Further experiments showed that over a period of time each dog had acquired the capacity to respond to any of a variety of similar buzzing events in the same way. The dogs had generalized the individual stimulus events into a mental category which was used to trigger a response.[98] It is these kinds of categories that mentally stand for (or represent) the physical events which brought them about. The same type of stimulus generalization can also occur when a single object is repeatedly presented (ie, when the stimulus events are recurrent presentations of the same object), in which case the mental stimulus category may come to represent an object category. Note, however, that "object" here does not imply it is the same unique object persisting over time between presentations, as is commonly the case with *conceptual* objects. Any sufficiently similar different object will also participate via stimulus generalization in the formation of the object category.

Stimulus generalization is an automatic process that *causes* categories to emerge spontaneously. Repeated similar stimuli cause a change in the neural substrate thereby linking them together and enabling future classification of such stimuli. It is because the categories so formed are caused by those stimuli, that the categories are said to *represent* the stimuli. However, there is still something of a mystery in going from the causal representation of the proximal stimulus (eg, a round-shaped orange-colored region in the visual field) to the causal representation of the distal stimulus (eg, a basketball) at a later stage of perception. There must be a progression of intervening representations which are all linked in a causal chain from proximal to distal stimulus representation. But at what point do the representations switch from representing a simple circumscribed region in an attribute map to representing a perceptual object out in the environment?[99] Since each representation is paired with at least one interpreter (ie, the transformation process which analyzes the encoded information conveyed by the representation and sends the results to other representations), what the representation represents is local to its embedded interpretive functions. At some level in the representational hierarchy, the interpretation must change from proximal to distal. However, these terms are global

[98] Whether those categories were explicitly or implicitly realized does not affect the conclusion that the dogs came to classify similar buzzes together, facilitating the salivating response. See the following section on *The Three Standard Approaches to Concept Representation* for examples of explicit category representation (eg, a prototype) and implicit category representation (eg, multiple exemplars). In both cases something becomes encoded in memory representing a category.

[99] One possible explanation for the proximal/distal representational distinction involves the direction of the causal chains. Perhaps bottom-up causation creates the proximal representations, while top-down causal models of the environment create the distal representations. Where the two meet may produce the distal representation of the perceptual environmental scene we are all consciously familiar with. In this explanation, our causal models of the environment become conformed to the actual perceptual data at the juncture.

descriptions of an interpretation, rather than local descriptions. From a strictly local perspective, the progression could involve a sudden change or a gradual change. But in any case, the causal chains of lower-level representations feeding into higher-level representations, and higher-level into lower-level, would seem to remain functionally unbroken. Each intermediate representation becomes a local coordination point for integrating new information bottom-up with prior information top-down. Our highest-level conceptual representations and interpretations are thus at least partly[100] causally based on the more localized representations and interpretations below them, stretching all the way back to our sensory contact with the flux of the world. But once established, the higher-level representational models can influence information processing at progressively more detailed lower-level representations by adjusting the interpretations (eg, sensitivities and expected content) of those representations.

Turning now to the uses of representation, it is relatively easy to see how the representation of a similarity class of perceptual objects provides a useful basis for future recognition of comparable objects. The utility also extends to the potential re-identification of a single individual, given memory of multiple occurrent episodes in which the individual appears prominently. Once recognition is triggered, an appropriate response becomes possible. Moreover, representation is useful in imaginary thought, as when a representation is mentally activated and interpreted, independently of perceptual input. The representation can stand for things which are not present. And chaining representation activations together into a script may provide a useful "what if …" simulation of events as a prelude for guiding action. But the real power of representation is probably best expressed via the construction of new representations based on components of the old. As a structured pattern, a representation contains components—ie, features (for perceptions) and characteristics (for conceptions)—that convey perceptual or conceptual content when interpreted. Mixing, matching, and comparing descriptive components of different representations enables the construction of new representations that may have little resemblance to their perceptual origins (think unicorn or a speculation about a new element in the periodic table). In fact, even features and characteristics may combine according to certain constraints to produce new integral features and characteristics, as explored in Chapter 1: *Concept Learning*. At a more abstract level, collections of similar representations can themselves become collated together into a class representing their instances. The concept of a species is one possible example of a noticeably abstract representation whose conceptual instances are representations of distinct animal classes—assuming "species" is interpreted as a kind rather than an individual term (see Chapter 4: *Particular Characteristics of Universal Concepts* for the distinction). Given the many possibilities representation affords for adapting to a wide variety of potential stimuli, it is hard to overestimate the importance of representation in mental life, even though the precise causal connections between representation and interpretation remain somewhat mysterious. (See Chapter 3: *Conceptual Identity* for a specific proposal about how we assign meaning to our concepts.)

---

[100] Allowing for ancillary biasing inputs

A Simple Example of a Hypothetical Hierarchical Representation

This section looks at how an agent might incorporate a set of mental representations, organized as described above, to build strategies for one of the examples mentioned in the preceding table of representations: the game of tic-tac-toe (TTT). People probably do not organize their understanding of the game this way, but the example illustrates the basic ideas of hierarchical representation.

Playing a decent game of TTT entails an ability to analyze board situations. But when we look at the board, the image projected onto the retina is one composed only of small dots of luminance. The image of dots making up the pattern of the board, Xs, and Os must be interpreted in order to permit strategizing at higher levels of analysis. The interpretation of the board situation can be thought of as following several stages in succession, where each stage of analysis generates representations that consolidate information received from prior stages. The highest level of representation forms the basis for choosing the next best move. The representational stages discussed here parallel some of the stages in primate vision, but are used only for illustration without appreciation for how humans really do evaluate a board situation.

This simple example takes the input image of the TTT board as a 9x9 matrix of pixels (see top representational matrix in Figure 11 below). Each pixel has a value of 1 or 0, where a player's mark (ie, an X or O) is composed of a pattern of pixels. Each board square (where marks are entered) is a 3x3 matrix, and therefore there are the usual 9 board squares. The pixel patterns for an X, O, and b (a blank square) are respectively:

1	0	1
0	1	0
1	0	1

1	1	1
1	0	1
1	1	1

0	0	0
0	0	0
0	0	0

In the example shown in the following figure, the board situation contains an X in the center square, an O in the upper left corner, and another X in the lower right corner. The first representation is the 9x9 pattern of pixels corresponding to this board situation, as might be encoded in a simple retina-like array.

The information in the retinal array is initially interpreted by processes which detect stroke patterns of pixels in each square. Stroke patterns correspond to the individual strokes that compose a mark, for example, two diagonal strokes characterize an X, and an O consists of four strokes, two vertical (left and right) and two horizontal (upper and lower). When a stroke detector detects the presence of three collinear pixel elements at particular locations in the image by matching the conditions of its rule, it becomes active and registers the presence of the stroke as an element at the next representational level. The absence of any strokes also registers explicitly (#), since this condition is indicative of a blank square. Once all stroke detector rules have registered their values, the original image information has undergone a transformation, and comes to encode new feature information at the stroke level.

Given information about strokes, the next representational level attempts to build information about whole character marks—ie, X, O, and b (blank). A square with two diagonal strokes is most likely an X, one with appropriately placed vertical and horizontal strokes is an O, and a blank square has no strokes. This is the first symbolic level of representation, as it depicts the board situation without features that are irrelevant to strategizing about game play.

The last representation shown recasts the previous symbolic board representation in terms of another symbolic representation—namely, the contents of the eight rows (three horizontal, three vertical, and two diagonal) related to achieving the goals of winning or avoiding losing. Winning means being able to detect when there are two Xs and a blank in a row, while avoiding losing depends on being able to detect the row situation of two Os and a blank. Detecting other intermediate board situations that are relevant to the game goals (eg, empty rows or forks) is also facilitated by the row representation, and can be used to formulate a strategy at still higher levels.

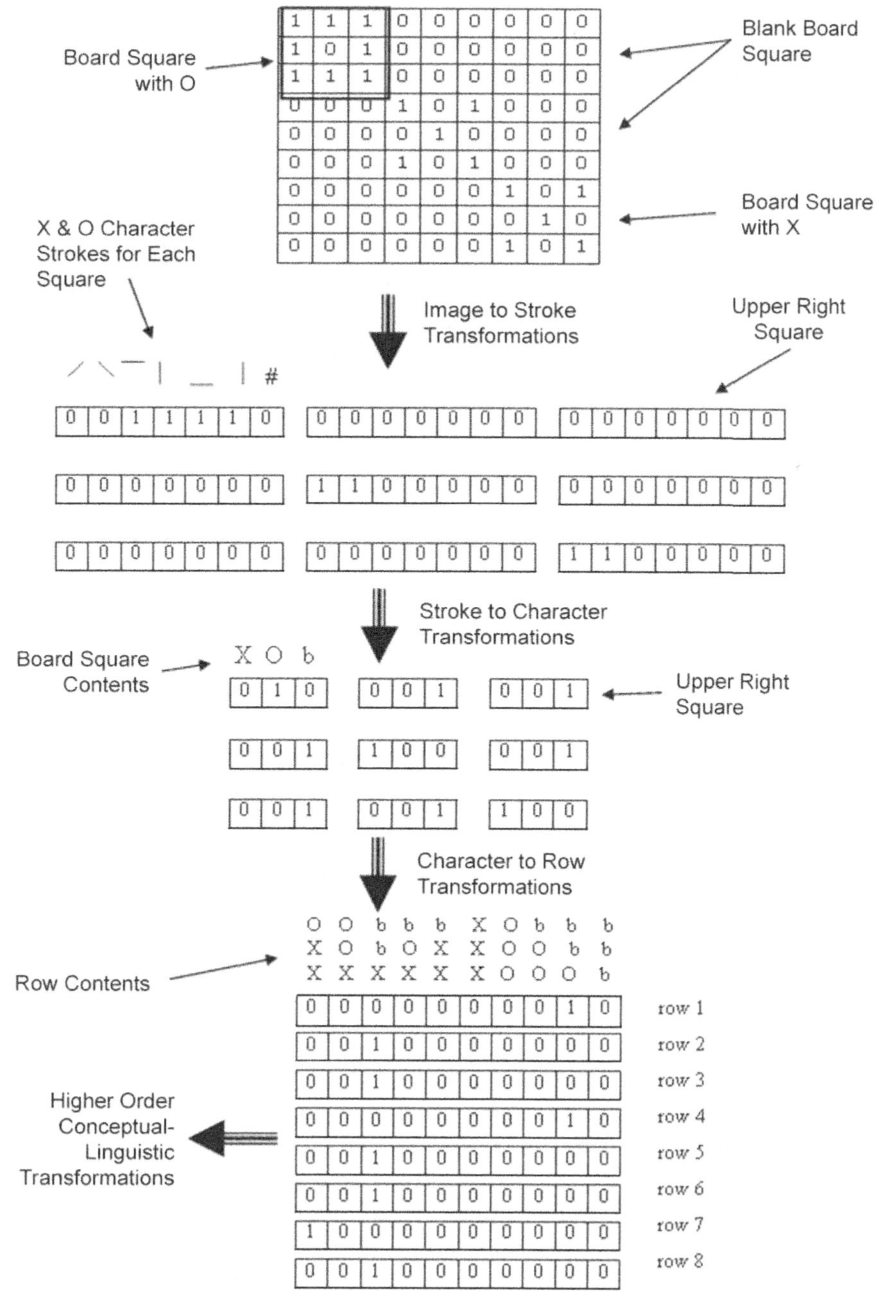

Figure 11: Hypothetical Representations for Tic-Tac-Toe

Higher levels are not shown on the diagram, but the progression of stages must eventually top off at the executive functions. The executive processes will make the final decision for the next move, although a lower-level process can also initiate moves if a straightforward board situation presents itself. Placing an X in the center square if the board is empty could be such a situation, since it requires minimal analysis to produce a good move. Lower-level processes tend to handle responses automatically, while upper-level processes are called upon mostly for planning and prediction.

The staged progression of representations from raw data to features to objects to relations among objects is generally characteristic of how information is processed in the brain.[101] Each stage takes the patterns of the previous stage and transforms them into a more useful form. The transformations essentially extract useful building blocks of information, based on the relationships among elements in the representation of prior stages. It is also common for the information at successive stages to exhibit a consolidation or compression, such that higher-level representations are more concise. In the example the potential information content can be measured at each level. At the image level the potential information content is 9x9 = 81 bits, while at the stroke level the potential content is compressed to 9x7 = 63 bits. At the character level the content is only 9x3 = 27 bits, but at the row level it goes back up to 10x8 = 80 bits. Greater compression of information generally means that a representation has a wider scope of application, and at the highest executive and symbolic levels in humans, possibly correlates with the relative depth of understanding of a situation.

Representationalism

The prior outline of mental information processing assumes that representations exist as explicit entities in the nervous system. The states of our representations correspond to our mental states. This view of mental processing is often called *representationalism*, because it posits that information-bearing mental entities are involved in interpreting the content of thoughts, beliefs, concepts, etc. The conceptual framework proposed here might legitimately be called representationalist, since it reifies concepts (including relational concepts) into an explicit network of nodes and links.

Jerry Fodor is the main proponent of latter day representationalism, as described in his relatively comprehensive theory of the Language of Thought (LOT). According to Fodor (Fodor, 2008), thought is a lot like language, with structured sentence-like mental representations conveying its content. It readily explains why it is so easy to translate thoughts into language and back again, due to the similarity of structure. A concept net likewise intertranslates with language, but its network structure is not like a collection of individuated linguistic premises from which conclusions are drawn. A concept net is more like a web of semantic connections, organized around the objects of belief (ie, concepts) rather than lists of sentences expressing beliefs. The links between the objects of belief are represented explicitly as associations, rather than implicitly via individual and predicate names and predicate argument positions as in language. A concept net therefore functions more like a graphical grammar, as opposed to a linear (linguistic) grammar as in LOT.

[101] The actual stages of processing in the nervous system are, of course, complex by comparison, and involve numerous parallel channels that interact with each other, and feedback channels to earlier stages.

In distinction to representationalism, there is a decidedly antirepresentationalist movement underway in artificial intelligence and in the philosophy of thought, probably beginning with Davidson (Lepore & Ludwig, 2005), that rejects the existence of mental representations and related explicit schemas on which mental processes operate. Instead, the antirepresentationalist proposal argues that information is embedded and holistic within the mental processes that use it. Mental processes are still envisioned as modules chained together in myriad ways, but there are no intermediate representations holding information for other processes. The processes simply communicate directly.

The basis of the antirepresentationalist movement has its roots in the generally recognized inability of symbolic and connectionist representations to confer meaning on content. Computers are powerful symbol manipulators, loaded with internal representations of all sorts, but they cannot understand what the symbols mean. Likewise, connectionist learning networks have performed some impressive classification feats, but few claim they "understand" anything in the usual sense.

Meaning in the widest sense seems somehow related to being an evolved and situated agent in an interactive environment. The representationalist generally assumes meaning arises out of physical interactions with the world—interactions usually characterized in purely causal naturalistic terms (like energy transfers and chemical reactions). Representations thus acquire their interpreted content from causal physical sources. But there is a problem in assigning meaning to content based on causal interactions with the world. No one doubts that environmental energy impinging on receptors trigger chemical cell reactions in neurons that chain together to produce complex causal patterns of activation we call thought. But there is, as yet, no fully developed theory about how those causally generated activations in a representation can come to have subjective meaning and subsequently generate causal effects based on content rather than the representation itself, independent of a homunculus.

Meaning also has to do with regularities or norms of use (eg, for assertions, their correctness or incorrectness; for promises, their sincerity, etc), not just the activation pattern of the thought itself. The same thought (eg, "my shoe is untied") can have different meanings to the same agent in different contexts. Causal processes may trigger a thought, but the thought's meaning seems based on a prior understanding of its relationship to other thoughts,[102] and of what correct thinking is all about. Davidson likens meaning to knowledge of a theory of truth (ie, of correct thinking), which allows us to reliably judge the truth or falsity of individual thoughts. That is, we understand the meaning of a thought when we understand the conditions that would make it true. One benefit of having this implicit operational "theoretical" knowledge without explicit representation is the power it gives us to directly apprehend the world. But it leaves unanswered the question of how the process of apprehending is accomplished.

[102] For example, to understand the meaning of 'blood is red', it seems necessary to understand both blood and redness in a wider context than this one thought. The other potential characteristics of blood act to fix its meaning (viscous liquid, oxygenating function, etc), while the other potential things that can be red fix the meaning of redness.

The conceptual framework proposed here is, at first glance, representationalist in spirit. However, if antirepresentationalism is taken as applying only to mental processes at the highest levels of agent functioning, as Davidson seems to do, then the framework is not inconsistent with antirepresentationalism. As stated previously, concepts in a concept net then do not have meanings directly associated with them, but rather meaning is developed from the concept representations in the context of their use by those higher level processes. In this view meaning arises from a process of interpretation rather than being a property of a representation, which makes more sense anyway, if Davidson's metalevel portrayal of meaning as a theory of knowledge is accepted.

The Three Standard Approaches to Concept Representation

Research over the last fifty years has converged on three distinct forms which concepts are thought to take, depending on the kind of information stored. Information about individual objects and events are probably stored as *exemplar concepts*, common object and event categories as *prototype concepts*, and causal knowledge about how things work as *theoretical concepts* (Murphy, 2002). Generally speaking, all of the higher conceptual processes such as categorization, object identification, inference, concept learning, generalization, drawing analogies, and concept combination are thought to operate on concept representations stored in one or more of these three forms.

Exemplar Concepts

An exemplar concept is usually viewed as a concept related to an individual particular thing. It contains specific conceptual knowledge about that one distinctive thing. However, the intuitive interpretation of an exemplar concept as a representation of an individual seems to slightly miss the mark, as suggested in Chapter 1: *Perceptual & Conceptual Objects*. Experiments show that an exemplar is probably not a unifying covering concept like a summary characterization of an object. The experiments instead indicate that the various occasion observations remain separately accessible (Nosofsky, 1988; Barsalou, Huttenlocher, & Lamberts, 1998), suggesting that an exemplar is associated with each separate encounter with the individual thing. An exemplar concept, therefore, is here understood to represent one aspect or mode (ie, a way of being or presentation) of the particular individual, rather than the individual itself. Put the other way around, the typical individual is exemplified by multiple modes, which subsequently allows recognition of that same individual in different circumstances through the matching of represented modes to observations. Both individuals and their exemplars are particular concepts, with exemplars situated at the lowest level of the conceptual hierarchy. Accordingly, exemplars contain only the specific (ie, constant) characteristics acquired from the snapshot-like encounter, rather than summary generalizations with variable characteristics extending over a range of possible values, as is distinctive of prototype concepts (taken up in the next section).

Exemplar concepts are commonly formed as projections from percepts, with inference (a conceptual operation) filling in additional conceptual characteristics. However, exemplars can also represent nonperceptual modes, such as the English language characterized in terms of Old English, Middle English, and Modern English, thought of as more-or-less distinctive ways the English language has been portrayed over the years. Accordingly, an exemplar is here regarded as a *constituent* of an individual object. But a mode is a different kind of constituent from other common constituents, like the parts of an object. A mode represents the entire object on some occasion, rather than just a spatial part on that occasion. And although a mode is like a temporal part (ie, stage or phase) of an enduring object, a mode does not have a duration and is not subject to change over time (otherwise it is a different mode).[103]

Storing each and every encounter and each and every change to an object as a fully independent exemplar concept is probably not feasible, even given the seemingly measureless capacity of long-term memory. The internal structure of an individual concept possibly economizes on storing redundant characteristics by establishing dependency links between exemplars (thought of as sets of constant characteristics). In this proposal an initial so-called "object file" (Kahneman & Treisman, 1984), interpreted here as a gateway to the modes of an enduring particular individual, becomes populated with a set of characteristics that serve as a baseline for subsequent amendments. Only the changes to an object's characteristics need be noted in an amendment, if the amendment is linked back to the baseline exemplar. The baseline exemplar can thus become linked to a succession of incremental changes, forming a chain of dependencies. The chain reflects the same information content as fully described modes of the changing object without massive redundancy. A new encounter with the same object will then form a new object file with its own dependency chain, but the two can be merged, if it is determined that the two object files contain redundant exemplars for the same individual. Alternatively, the two object file representations may simply be annexed to each other, if their respective exemplars do not overlap significantly.

Having once acquired a suitably broad range of exemplar concepts, each tagged to indicate the individual which it describes (eg, this is Fido, that is my car, this is something I've never seen before, that is like some things I've seen before but don't know what they are, etc), we can then analyze a new exemplar by comparing and contrasting it with known exemplars. The comparison is usually automatic, unconscious, and fast, although deliberation is occasionally called for. Roughly speaking, whichever existing exemplar concepts the target most closely matches, determines its object assignment.[104] Under the exemplar paradigm, categories of exemplars are therefore not explicitly represented, but rather implicitly derived based on the similarities among previously identified exemplars.

---

[103] Technically, a mode is a continuant, rather than an occurrent.

[104] The best performing exemplar model developed so far is probably the Generalized Context Model (Nosofsky, 1986), which represents each exemplar concept as a point in multidimensional space, corresponding to the values of its characteristics. The similarity of a new exemplar to an existing exemplar is a function of the distance in the multidimensional space between them.

Prototype Concepts

While exemplar concepts are normally associated with individual things (eg, Fido), a prototype concept is said to represent a class of things (eg, dog).[105] The prototype concept, accordingly, is a single schematic representation, whose characteristics describe the range of potential properties and relationships that instances of the class can have. Particular sets of characteristics become correlated with each other (ie, regularly appear together) during concept learning, and those characteristics then constitute the prototype (Rosch, Mervisa, Gray, Johnson, & Boyes-Braem, 1976). Dogs generally have a tail, can bark, and usually have an owner, and these typical characteristics become attached to the prototype instead of to our concepts of individual dogs. Individual dogs may optionally inherit those characteristics by default.

Calling such schematic concepts "prototypes", however, is something of a misnomer, given that the standard dictionary meaning of prototype relates to an individual (prototypical) item. As discussed by (Murphy, 2002), the psychological meaning of such concepts is better applied to representing categories than individuals. In the most comprehensive models—eg, see (Smith, Osherson, Rips, & Keane, 1988)—the information contained in a prototype concept goes beyond the so-called *typical characteristics* which are inheritable by default instances of the concept. For example, the prototype concept of a bird certainly includes information about typical birds like robins and sparrows (fist-sized, eats bugs, can fly, etc), but also may include information about nontypical birds like penguins (can't fly) and eagles (eats meat), since they are birds too. This kind of knowledge is exceptional and statistical in nature, and provides information about how the various characteristics are distributed across the instances of a class. Considering size, we seem to intuitively appreciate that only a few birds are very small (eg, hummers, warblers), a lot of birds are medium sized, and a modest number of birds are relatively large. For apples, we know most are red, some green, and fewer yellow. Given some relevant background knowledge, we often use such estimates of typicality and possibility to infer unobserved characteristics of objects, once we know the object is an instance of a class. From this point on, the entrenched term "prototype concept" will be stretched to indicate the broader coverage of a whole category to include prototypical and nonprototypical instances based on statistical information.

Prototype concepts are also quite useful for classifying new objects. Instead of comparing a new object with a potentially large array of exemplars to decide if it is likely to be a member of a class, the new object can also be compared directly, characteristic by characteristic, with explicitly represented class prototype concepts. For the classification task, the *diagnosticity* of a prototype's characteristics becomes important, and we seem to intuitively possess such diagnostic knowledge. We know a red breast is highly diagnostic for robins (even though females do not have it), while having feathers (a sure bet) is not.

---

[105] In the current proposal however, a prototype can also represent a class of one thing (eg, the class whose one and only instance is Fido). In this case, the prototype is a summary of the individual's characteristics, rather than the collection of its exemplars (ie, modes). The representation of an individual as a prototype concept is discussed further in Chapter 5: *Particulars as Unique Universals*.

Like the exemplar models, most prototype models are based on similarity measures between two prototypes or between an instance and a prototype. The various cognitive processes organize concepts in relation to each other and make decisions based on how similar concepts are to each other. To assess similarity, the various characteristics of each prototype concept may be independently compared and weighted, taking into account (for a given task) each characteristic's importance and relevance (typicality, diagnosticity, perhaps salience, etc). The (usually linear) sum of all the weighted evidence then determines the strength of the conclusion, whether for classification, confidence assessments, subsequent inferences, or other conceptual activity.

Theoretical Concepts

Exemplar and prototype concepts are probably acquired through straightforward and direct interactions with the world. They seem to primarily encode statistical information (ie, probabilities) gleaned from numerous encounters with common objects. Much (but not all) of this information is perceptual, and therefore mostly deals with the surface appearance of things, rather than the deeper nature of things. The information about what makes something what it is (eg, its origins, internal mechanisms, functions, causal relationships, etc) is thought to be captured in a third conceptual form, a *theoretical concept*.

Theoretical concepts are often invoked to explain not only our knowledge of technical concepts like force (in physics), carbon (in chemistry), and precedent (in law); but also for concepts like personality, taxes, and adolescence embedded in our quotidian folk theories of life. Such concepts are often relatively abstract, and we use them mainly to explain why things are the way they are, and how they came to be (Murphy & Medin, 1985). If we have functional, causal, and mechanistic knowledge, we can also explain how something works (or failed to work), and perhaps how to make it work better. The importance and intuitive force of our theoretical concepts in attempting to explain everything under the sun, whether right or wrong, is hard to underestimate.[106]

Theoretical knowledge, in general, is also more central to the core meaning associated with a concept, while exemplar or prototypical knowledge is more peripheral. It aims to describe the inherent nature of something. That water is H_2O addresses its nature, while being a clear, tasteless, potable liquid is important but peripheral. We use this theoretical chemical knowledge to explain why water has the peripheral prototypical and exemplar characteristics we observe. In particular, theoretical characteristics are usually regarded as being necessary to instances of the concept (or at least as expressing a normal condition, or occasionally a sufficient membership condition), as opposed to being only likely or possible or accidental. However, note that computational simulations of theoretical concepts are often based on probabilistic models (eg, using Bayes nets), where causal probabilities are assigned to various characteristics of a concept—eg, that birds have feathers and hollow bones tends to causally support their ability to fly (Gopnik, et al., 2004). Probability estimates figure prominently in all three conceptual models. What seems to distinguish theoretical models from the other concept models

[106] Yes, I too am guilty as charged.

is the presumption of a core causal nature or otherwise formal reason that explains why instances of the theoretical concept have the properties they do have.

Concept Integration

The evidence for at least the above three distinct forms of conceptual representation is strong. Yet we do not experience any sharp conceptual discontinuities in thinking about common classes (prototypes), their instances (exemplars), or cause/effect relationships (theory). The knowledge that Fido slept in the garage yesterday (specific exemplar knowledge), that dogs like to pee on fire hydrants (stereotypical prototype knowledge), and that dogs are placental mammals (theoretical knowledge), peaceably coexist in our conceptual system, and we use all such knowledge seamlessly. All of our declarative knowledge is psychologically integrated, because we possess high-level interpretive processes that range over the various representational forms and give them cohesive meaning.

That we integrate distinct forms of conceptual representation, however, does not mean we give them equal weight in our thinking. Some representations are more useful than others in different situations, and our interpretive processes sometimes strain to appreciate the relative strengths and weaknesses of each form in our conceptual toolbox. Categorizing by exemplars is weak if there are only a few exemplars, while categorizing by prototypes is weak in the absence of diagnostic information. And the problems with stereotyping via prototypes, or completely missing the point while focusing exclusively on theoretical issues, are well known. Yet, exemplars are an excellent resource for distinguishing individuals, prototypes for organizing knowledge about various domains into useful concept hierarchies and relationships, and theories for precise and insightful predictive reasoning. Our coming to appreciate such conceptual strengths and weaknesses seems to develop in parallel with our overall approach to acquiring knowledge. In our initial encounters with any new object, we have no choice but to rely on exemplar thinking, because there is little data on which to base a generalization. As we become more familiar with similar objects, we can more efficiently generalize to a prototype summarizing their observed characteristics, and usefully position the class of similar objects as a concept in a hierarchy of other domain concepts.[107] Ultimately, we can investigate the object's theoretical nature and discover how and why it behaves the way it does. Each new level of understanding does not suppress the previous levels—they ideally work together to provide the best interpretation depending on the situation and our goals. The following chapters begin to outline a unified network structure for investigating conceptual representation, simply called a *concept net*. Like our intuitive experience of most concepts, the concept net framework is integrated. A single framework expresses the information content of all three conceptual forms, viewed as a global knowledge base. Generally speaking, exemplar models are affiliated with *particular concepts*, representing both particular objects and particular object modes. Prototype and theoretical models correspond to *universal concepts*, representing kinds and their characteristics. Prototypical knowledge then consists mostly of practically useful beliefs about

[107] For example if you know that tigers are large cats and are typically dangerous, and then observe a lion, which also looks like a large cat, you can infer that it is probably dangerous too.

kinds, which are commonly described in terms of typicality and diagnostic characteristics. Theoretical knowledge is also about kinds, but expresses the core identity beliefs associated with causally or formally determined mechanisms and origins, when such knowledge is known.[108]

In the concept net framework therefore, there are only two basic kinds of concept: universal concepts and particular concepts. Prototype and theoretical models are actually describing different kinds of reasoning (ie, statistical vs causal/formal) rather than different kinds of concept.[109] However, newly acquired theoretical beliefs about a previously developed universal concept centered on prototypical beliefs, may trigger a new concept centered only on theoretical beliefs. In these cases it may make sense to talk about a purely theoretical or purely prototypical concept (eg, to usefully distinguish between common potable water and pure H_2O), but most concepts are likely a mixture of the two forms of knowledge (eg, where the concept whale has both prototypical fish-like characteristics as well as theoretical mammalian characteristics). In general, theoretical beliefs, being more stable and reliable for classification than prototypical beliefs, will usually determine the skeleton of a well-developed concept network, with common prototypes fleshing out and supplementing the skeleton. As discussed later in Chapter 5: *Natural vs. Nominal Kinds*, the psychological division between theoretical and prototypical concepts may roughly correspond to the philosophical distinction between natural and nominal kinds.

---

[108] Theoretical knowledge can be associated with concepts both in the form of core identity conditions and in the form of causal relations between concept properties. Certain causal relations between properties are deemed *laws of nature*, if the relation among properties is shown to be true of many different kinds of objects.

[109] In contrast, Machery's influential Heterogeneity Hypothesis (Machery, 2009) proposes that we have three distinct concepts for each category, corresponding to bodies of knowledge about exemplars, prototypes, and theoretical beliefs (eg, DOG_1, DOG_2, DOG_3). While his analysis is based on the kind of knowledge stored (ie, content), the basis advocated here is the more fundamental representational distinction between type (universal) and token (particular).

3. CONCEPT STRUCTURES & FUNCTIONS

A Network of Concepts

Concepts, and the operations that manipulate them, are features of an intricate system of discrete mental representation. It is the capacity for discrete representation that gives the conceptual system its combinatorial flexibility in mixing and matching concepts in so many creative ways.[110] Most of the time the conceptual system seems to function as an integrated whole, but it partitions quite naturally into two separate faculties, namely linguistic and semantic knowledge. In general, the concepts and operations of language deal with linguistic descriptions of the underlying semantic concepts and operations. It is the semantic concepts and operations, rather than linguistic expressions that are of primary interest here, although both involve discrete conceptual representations (eg, sentences vs beliefs).

A generic *composite concept* (eg, the concept of a lemon) is viewed as a collection of nodes joined together by strong associative links into a graph-like configuration. One node is typically the nexus for the concept, and serves as an attachment point for beliefs about the concept in the form of its kind, properties, and relationships to other concepts. It is also used as a point of reference to talk about the whole composite concept, but the whole concept is different from its nexus. The nexus nodes are organized into kind hierarchies (see left side of following figure) corresponding to what the concept represents, with the attached characteristics depending on the kind. (The characteristics of the [lemon] nexus extend to the right in the figure). A hierarchical organization allows for inheritance of characteristics from a more general concept, like [fruit], to a less general concept, like [lemon], without redundancy, such that the concept of [lemon] includes the characteristics of [fruit], but only implicitly. In other words, a characteristic like having seeds would be attached to [fruit], not [lemon], since all or most kinds of fruit can inherit this default characteristic.

The interconnected nodes making up the concept of [lemon] are circumscribed in the figure. The circumscribed nodes and links constitute the characteristics that describe the kind [lemon]. The various elementary beliefs about lemons are organized around the small circular dots, which represent relations: a lemon has color, a lemon has juice, lemon juice has a taste, etc. These elementary beliefs correspond to the propositional content of the concept if activated, leading to the important consequence that all complex conceptual knowledge must eventually be broken down into elementary

[110] A discrete representation (eg, a sentence) contrasts with a nondiscrete representation (eg, a painting), where "discrete" and "nondiscrete" are generalizations of the more technical terms "digital" and "analog". A sentence contains a number of individuals (ie, words), which combine together to produce a whole sentence; a painting contains unindividuated regions of color & texture (ie, paints), which combine to produce a whole painting. Being representations, both can evoke meaning beyond themselves. We make good use of both discrete (eg, conceptual, linguistic, mathematical) and nondiscrete (eg, sensorimotor, spatial, emotional) mental representations.

propositional representations in order to support reasoning.[111] The strength of the associative links among a concept's nodes delimits it from other concepts, such that the nodes and links among its attached beliefs are relatively easily activated. To the right of the figure are the various shared attribute value spaces, organized by domain, supplying specific attribute values (eg, values or value ranges) for the various characteristics.

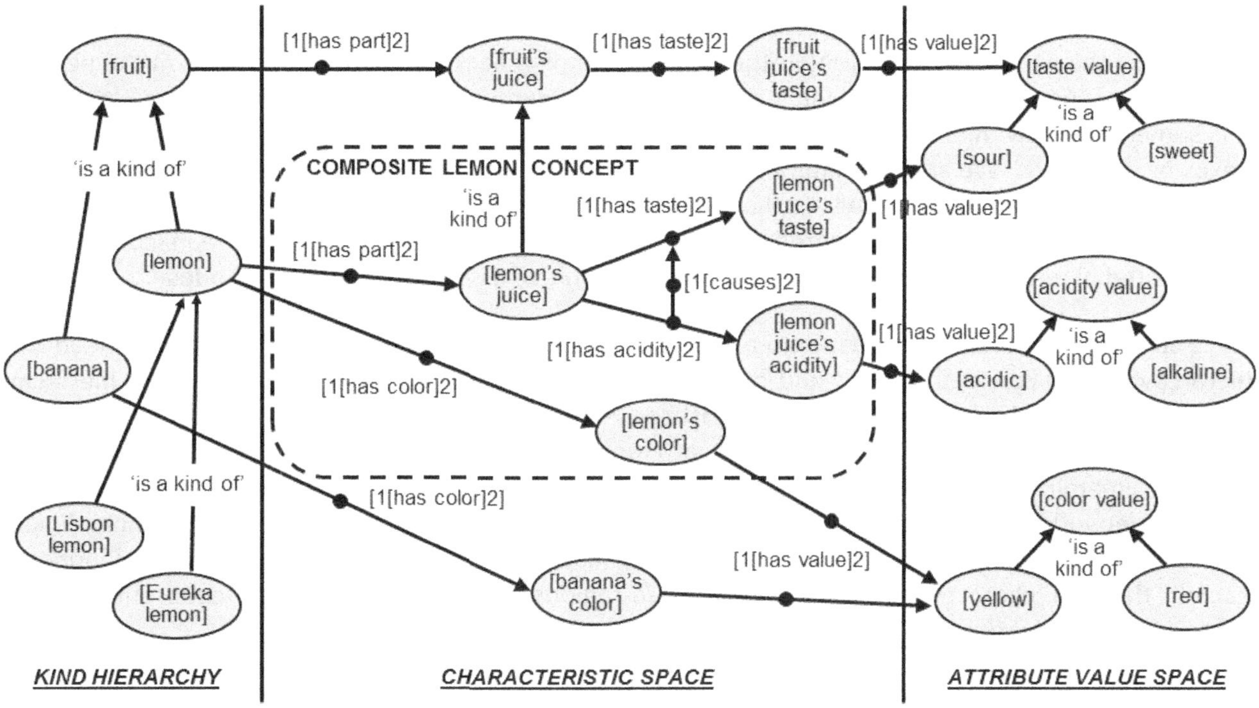

Figure 13: Simple Concept Net Example

The entire collection of associated nodes for a concept is unique for that concept. Nodes like [lemon's juice] or [lemon juice's taste] are separately represented in characteristic space, and are not shared with other concepts. The relational concepts like [[lemon][has color][lemon's color]] characterizing lemon color are likewise unique, having arguments specifically to do with lemons. This separation of concept characteristics from both the kind concepts they describe and the attribute value concepts exemplifying them allows the growth of each domain space to proceed roughly independently of the others, while remaining appropriately interconnected. For example, the concept [lemon's color] is simply characterized as a kind of [yellow], which subsequently can be further characterized if more knowledge is gained about lemon color, without disturbing the generic color attributes. The general picture of conceptual organization that emerges is of a vertical hierarchy of kind nodes, each possessing unique sideways branching characteristics that eventually either reconnect with other kind nodes in the

[111] This insight of Chomsky's became the basis for his theory of a hidden deep structure underlying the surface structure of sentences.

hierarchy or terminate in a separate attribute value space of nodes that are shared among the characteristics for all concepts. The integrated hierarchical organization however includes not only the kind hierarchy, but the characteristic and attribute value hierarchies, as well, since (as developed in Chapter 4) all concepts are primarily organized as part of one global hierarchy.

While a concept unifies a lot of information about a single kind (or particular individual) by collecting multiple nodes under one umbrella, each individual node in a concept network can also be thought of as the nexus for a different concept. For example, a lemon's juice is an individual concept in its own right, signified by the nexus [lemon's juice] in the figure, extending back into [lemon]. Its links in the network constitute its specification. Moreover, a relation and its arguments are also discrete notions. In general, it is sometimes convenient to think about any node in a concept network as representing a conceptual nexus of some sort. We are then ultimately left only with primitive links (like 'is a kind of' in the figure) joining individual concepts. All explicit concepts (whether concrete or abstract, empirical or epistemological) then are represented by the nodes, while all relationships between concepts are represented in terms of a relatively small number of primitive associations between nodes.

Since all nodes represent concepts, from a wider perspective all nodes become part of an extended kind hierarchy. Every concept is some kind of something, and accordingly takes its place in the hierarchy, subject to the usual inheritance of characteristics. For example in the figure, [sour] is a kind of [taste value], which may possess a potential range of kinds of taste values—perhaps [sweet], [salty], [sour], [bitter], [umami]. The lemon juice's taste [sour] is characterized as one of these. All knowledge in a concept network is thus integrated into a holistic structure, with primitive binary associations linking nodes vertically (according to kind) and horizontally (according to characteristic), involving all nodes in each of the kind, characteristic and attribute value regions, ultimately presenting a coherent structure that can be checked for consistency and completeness.

The bracketed labels in the figure are simply a notation to fix a convenient description for a node. Each node may also possess a separate linguistic characterization of its default interpreted content, a description made up of lexical concepts residing in the language component of the concept net. Concept network nodes maintain links with other systems as well. That concept nets are comprised of discrete nodes does not mean that they do not have associations with nondiscrete things. We also have nondiscrete impressions and evaluations of our experiences tied to concepts. But these kinds of mental activity are assumed to depend on other nondeclarative mental representations and processes—eg, perceptual and emotional experiences. The conceptual network is presumed to operate as a separate, self-contained representational system, with extensive inputs from and outputs to these other mental systems.[112] The complete, highly connected semantic network of nodes and their links embody the concepts of declarative long-term memory.

---

[112] For example, Gardenfors advocates three layers of mental representation, with connectionist network organization at the first level, geometric-like state spaces of objects & properties at the second level, and fully symbolic representations at the third (Gardenfors, 2000). Concept nets represent information at the symbolic level, with interfaces (ie, pointers) to and from the other levels. My own preference is to add a fourth linguistic level for describing information content in the symbolic level.

The Specification of Concepts

A concept encodes explicit information, information encoded in its associative links with other concepts. These other associated concepts constitute its immediate family, and determine its role or function (as opposed to its meaning, since meaning is an interpretation constructed from concept representations) in the context of the whole concept network. All of a concept's associations contribute to its current specification, but three types of association are of special importance for understanding its role.

The first is the 'is a kind of' association, specifying what *kind* of things a concept is representing. It links a source kind concept with a more general kind concept. For example, a stallion is a kind of a horse. It specifies that an instance of a stallion (eg, Trigger, Roy Rogers' palomino) is also an instance of a horse. The 'is a kind of' association situates the concept in a hierarchy of concepts, in this case a hierarchy perhaps extending upward from "[stallion] 'is a kind of' [horse] 'is a kind of' [animal] 'is a kind of' [organism], etc. The kind-of hierarchy, usually pictured as a downward branching tree of concepts from more general to more specific, establishes a skeleton of core concepts to which characteristics can be added.

The second main type of association involves the elementary beliefs used to specify descriptive characteristics of the concept. For example, "male" is a characteristic of [stallion]. A concept's characteristics distinguish it from other sibling concepts falling under the same kind concept, in this case [horse]—eg, the concept of [mare] will have a "female", as opposed to a "male", characteristic attached. Characteristics come in a wide variety of flavors including properties, parts, possessions, functions, behaviors, roles in situations, etc; but all can be represented in terms of primitive associations in a concept net. In the earlier sketch of the concept [lemon], the relationship [has part] joins the node [lemon] with [lemon's juice] to form the elementary belief "[lemon][has part][lemon's juice]". The relationship [has color] joins [lemon] with [lemon's color] to form the belief "[lemon][has color][lemon's color]". Note that a relationship like [has part] is not a direct associative link between two concepts like 'is a kind of' is. The linkage is mediated by a separate universal concept, namely the relationship concept [has part]. Thus, while the linkage between [lemon] and [lemon's juice] is indirect via [has part], the link is direct between [lemon] and [has part], and between [has part] and [lemon's juice]. These two links are primitive associative, expressing the elementary belief that in this instance of the relation, [lemon] is the first argument and [lemon's juice] is the second argument of [has part]. The two elementary beliefs are then "[has part] 'has argument 1' [lemon]" and "[has part] 'has argument 2' [lemon's juice]". It is usually simpler just to think of a concept as having characteristics, as opposed to being an argument in a relation, but the relational perspective readily generalizes to any sort of semantic connection between any two (or more) concepts. After all, the belief is actually about both a lemon and its juice together, and whether we focus on a lemon as having a part which is its juice, or on a lemon's juice as being part of a lemon, is a rhetorical issue.

The third important type of association is 'is an instance of', which specifies the epistemological nature of the concept itself. Particulars like |Fido| (where the vertical bars signify a particular concept) are always instances of universals like [dog]. However, the association can also hold between universals. For example, while [cheetah] is a kind of [animal], [cheetah] is also an instance of [[species]] (where the double brackets symbolize a higher level of universal abstraction). This means that the concept [cheetah] inherits its characteristics from [[species]], not from [animal]. For example, [cheetah] will have a characteristic population range (a typical characteristic of a species), but individual cheetahs, like Charlie, do not. The concept |Charlie| inherits characteristics from [cheetah], not [[species]].

It seems reasonable to adapt the notation of set theory to specify concepts, even though a set and a concept are not the same thing. (A set is a concept, but a concept is not necessarily a set.) Sets are often specified by simply listing their elements—eg, {|Larry|, |Moe|, |Curly|}. The corresponding concept, of which Larry, Moe and Curly are instances (not elements), is then designated [|Larry|, |Moe|, |Curly|]. Instead of listing a set's elements, it is usually more convenient to specify a criterion for set membership. For example, the set of even numbers is better specified as those numbers that are divisible by 2: {x: x is divisible by 2}, read as the set of all x (the domain of numbers is assumed) such that x is divisible by 2. The corresponding concept of an [even number] then is described by the characteristic [[even number][is divisible by][|2|]].[113] The set of even numbers, as a concept, is specified as {even number}, which is an instance of [set].

Concepts usually have conventional linguistic labels/names/descriptions—eg, [horse], [even number]—but labels play only a mnemonic role and carry no formal significance.[114] All information is carried in the associations a concept has with other concepts. Labels conveniently denote concept nodes, but it is a category error to confuse a concept (a mental entity) with its name (a linguistic or symbolic entity), or its potential core content (an propositional entity). However, the confusion becomes palpable when we represent names and propositional content as mental entities too. Then we are talking about the concept of a [name] or a [proposition].

This kind of confusion in exactly what we are talking about probably arises from the slackness of everyday language, a confusion tied to the use/mention distinction widely discussed in the philosophy of language. We *use* a label, name, or other expression in language as a representation to *mention* something else. For example, I use the name "Angus" in language to mention (ie, talk about) my pet cat in the world, named Angus. I use the label "cat" (eg, in the last sentence) to talk about one or more cats in the world, in general. Moreover, I use the labels "|Angus|" or "[cat]" to talk about the concepts in my head that represent Angus or cats in general, respectively. And I use the labels "the content of |Angus|" or "the content of [cat]" to talk about the meaning of these concept representations. These distinctions are very difficult to maintain in normal dialogue, but are important for specifying concepts. Perhaps the best way to distinguish such concepts is to ask what each concept's characteristics are. Worldly cats meow and have tails, the concept [cat] is connected to the concept |Angus| and participates in certain neural activation patterns, and the content of [cat] is an intangible informational kind of thing potentially expressible in terms of elementary propositions describing worldly cats. The rather remarkable upshot is that any thinkable concept can be characterized, not just those that deal directly with the world, if sufficient care is applied.

---

[113] assuming numbers are particulars, and not universals as is sometimes claimed.

[114] On the other hand, a linguistic description of a concept node developed by the language faculty using the lexicon, does have linguistic & semantic meaning when interpreted.

Concept Characteristics

Composite concepts encoded in long-term memory seem to have various characteristics attached to them. We know that dogs have body parts (eg, legs), behaviors (eg, barking), and attributes (eg, shape) over and above being a distinctive kind of thing (eg, animal). Such background knowledge of characteristics is tailored when a concept is used (only certain characteristics are activated and modulated in a given context), but concept characteristics are the grist for thought, supporting many kinds of inference about objects and their relations generally.

It is often said that concepts have both core and peripheral characteristics (or essential and accidental properties, or definitional and encyclopedic entries). However, most contemporary philosophers since Quine's analysis of the analytic/synthetic distinction (Quine, 1951) regard these as graded tendencies, rather than strict categories. Given that concept characteristics are primarily used to match up concepts with the world and facilitate our interactions with it, a functional analysis of characteristics seems more appropriate—ie, a classification based on how we use characteristics to reason about concepts. Four qualitatively different types of characteristics are discussed here—viz, necessary & sufficient characteristics, dispositional & potential characteristics, stereotypical & cue characteristics, and exemplar characteristics. These four types of characteristics exhibit decreasing degrees of centrality with respect to the concepts they become attached to, such that necessary and sufficient characteristics are maximally central, affording the most reliable inferences, while exemplar characteristics are the most peripheral, asserting weaker beliefs about occasional or temporary situations. Each characteristic is assumed to carry a tag, which specifies the kind of characteristic it is. The tags are then available for use by various executive processes to correctly and appropriately reason about the concepts to which the characteristics are attached.

Given such a broad range of possible characteristics for any concept, how then is it possible for two people to coordinate their use of a concept label? Obviously their concept characteristics will differ depending on the extent of their respective experiences, and as a result they will likely construe somewhat different meanings from their use. The meaning of the word "water" to a scientist (as made up mostly of H_2O molecules, based on necessary and sufficient chemical characteristics) is certainly different than the stereotypical meaning of "water" to an untutored mind (as made up of ever-divisible portions of liquid water-stuff), but they can still talk meaningfully about it together. Likewise, even though the concepts of a child have not yet matured, adults and children can nevertheless roughly agree on meaning. The requirement for coordinating meaning then does not depend on having identical characterizations for concepts. It merely depends on having sufficient overlap between characteristics, and on understanding that others may possess characteristics that we don't. Scientists can (usually) communicate with nonscientists, since scientists know about stereotypical characteristics as well as necessary and sufficient characteristics. Nonscientists also appreciate that scientists sometimes have theoretical concepts with causally necessary and sufficient characteristics to go along with common prototypical knowledge about stereotypical characteristics. And adults (usually) make allowances for children's' partial and subjective knowledge. Problems in communication can, and often do, become severe if two people assign different intuitive ontological categories to a crucial concept. However, intuitive ontological meanings (as discussed in Chapter 2) relate more to our innate, mostly shared, core domain knowledge than to acquired conceptual knowledge.

Necessary & Sufficient Characteristics

We generally think about things in the natural world as if there were "something" that makes each thing in the world what it is—some essence associated with its nature. For example, our concept of a horse likely includes many nonessential characteristics relating to its appearance, such as having four legs, having a tail, being able to run fast, capable of being ridden, etc. But it also seems to include so-called *identity conditions* dealing with what it means to "be" a horse, such as being descended from other horses and/or having a certain DNA content. Changing its nonessential characteristics does not affect what it is, but a horse is not really a horse unless it satisfies certain identity conditions, even if it looks, acts and is used like a horse. We also acknowledge concepts whose essence seems fixed by convention, rather than nature. That a stallion is an uncastrated adult male horse is determined by stipulation, rather than some natural essence of stallionhood. Even artifact concepts like chair seem to possess a certain functional essence, in this case something a person can sit on or perhaps something at least created for the purpose of sitting. We may not know what the identity conditions are for various concepts, but we nevertheless seem to assume they exist. While this knowledge is subject to change, it is usually quite stable for natural, artifactual, and conventional kind concepts, natural kinds being ultimately fixed by science, artifactual and conventional kinds by stipulation.

Our propensity for presuming the existence of essential characteristics is perhaps more of an idealized psychological/conceptual organizing principle (Carey, 2009), rather than an established scientific fact, since it hinges on an omniscient view of reality. There is scant scientific evidence for essential characteristics even for natural kind concepts (Dupré, 1993). Most everyday things are not classified according to essential characteristics because our categories are usually tailored to a particular purpose or limited by ignorance. Common prototypical characteristics are more useful than essential characteristics for everyday reasoning. Yet we also strive to understand and organize our knowledge in ever wider contexts of application and generality. Science is the formalization of this desire to discover the invariant nature of things. In accordance with the essentialist cognitive principle, concepts in a concept network generally are assumed to carry identity conditions, whether or not those conditions are known or specified. Technically speaking, a concept's identity conditions are modeled as *necessary* to it—ie, instances of the concept could not lack any of its identity characteristics, even though it could have different peripheral characteristics in various real or hypothetical circumstances. Accordingly, I say that I am the same person today as yesterday, or counterfactually that if I were rich I wouldn't be as happy, because my identity characteristics do not change in spite of other changes. The overall organization of a concept net hierarchy depends on the identity conditions of the concepts in it, since of all the various types of characteristics, identity characteristics provide the most stable basis for organizing knowledge. There is always allowance for changes to identity characteristics as new information becomes available,[115] but such changes are expected to be rarer than for other kinds of characteristics. A concept net therefore represents the best current understanding of concepts from the most useful, most global point-of-view. That we do not know identity conditions for most concepts, and that people may differ in their identity assessments, does not keep us from hypothesizing or learning better classifications.

The term "identity condition" here ideally denotes a *jointly necessary and sufficient characteristic* for classification and prediction purposes. Such a characteristic, if known, most clearly establishes the applicability of the concept. The chemical elements, such as [gold], [oxygen], [uranium], each exhibit a jointly necessary and sufficient characteristic related to the nuclear structure of their atoms, namely atomic number. If something has an atomic number of 79, for example, then it is an instance of [gold]; and if something is an instance of [gold], then it has an atomic number of 79. If something is exactly divisible by 2, then it is an [even number], and vice versa. If something is an unmarried man, then (in the default sense of the word) it is a [bachelor], and vice versa. Generally speaking, jointly necessary and sufficient characteristics are the most invariant form of description, but even these can be waived under certain conditions. When we say "Mary is a bulldozer" a key identity condition on [bulldozer] is being flouted.

But specifying fully necessary and sufficient characteristics for most concepts is an ideal, rarely achieved in practice. In practice, identity conditions will typically include characteristics currently thought to be either *individually sufficient* or, more likely, *individually necessary*. A sufficient characteristic is enough to completely determine a concept, but there may be other sufficient characteristics for the concept as well. For example, the generic concept of [jade] is technically a disjunctive concept that covers gemstones composed of either nephrite or jadeite. Knowing that a stone is composed of nephrite is sufficient to conclude it is an instance of [jade], but knowing a stone is [jade] is not enough to conclude it is composed of nephrite—it could be jadeite instead. The pattern for an individually necessary characteristic is similar. While the condition of being warm-blooded is a necessary characteristic for the concept of [bird], because all birds are warm-blooded, there are other warm-blooded animals besides birds, namely mammals. Likewise, the condition of being male is a necessary characteristic for the concept of [stallion], because all stallions are male. To summarize the levels of identity conditions: jointly necessary and sufficient characteristics constitute the most central (most invariant) form of concept specification in a concept net, with individually sufficient and individually necessary characteristics, respectively, being slightly less central. Identity conditions generate the core hierarchical skeleton architecture of a concept nets.

[115] For example, the identity conditions of whales changed when they became reclassified as mammals rather than fish.

Dispositional & Potential Characteristics

Many concepts also possess what might collectively be called *normalcy conditions*, which seem to presume some standard way of understanding the concept. Such characteristics express what is a normal interpretation of a concept under default contextual conditions. A *dispositional characteristic* specifies the tendency (but not necessarily) for an instance of the concept to express a particular disposition, while a *potential characteristic* marks a tendency (but not with sufficiency) to classify something as an instance of the concept. Dispositional and potential characteristics are therefore analogous to necessary and sufficient characteristics, but carry somewhat less weight in drawing reliable inferences, particularly in contexts deviating from their default context. Normalcy conditions participate in so-called defeasible or default reasoning about concepts, such that instances of a concept are expected to follow the norm unless there are reasons to believe differently.

For example, a dispositional characteristic for a dog is that it has four legs. This is part of the normal default meaning associated with the concept [dog], even though it is not necessarily true (as an identity condition). The source of this characteristic is still causally based on the dog's DNA, like it is for a necessary characteristic, but it expresses only a tendency toward the normal. An instance of a concept need not be normal; circumstances can cancel normalcy conditions, as when a dog loses a leg in an accident yet remains a dog. Cancellation may further postulate some reason or explanation for deviation from the norm. The explanation will show how the dispositional condition is blocked, or at minimum, that there is some unknown blocking mechanism.

As an example of a potential characteristic, consider the inference from having feathers to being a bird. This is a potential characteristic because if you find an animal with feathers, you can reliably conclude it is a bird, but only under normal conditions. In the context of a hoax, you will be led astray, since having feathers is not a sufficient condition for inferring bird under nonstandard conditions. Knowing that this is a potential characteristic may lead you to check for abnormal circumstances. In general, the potential characteristics for a concept theoretically derive from its identity conditions, just as for dispositional characteristics, and likewise assume other blocking conditions do not apply.

One particularly important class of normalcy characteristics, closely related to identity conditions, are physical laws, like Newton's law of gravity. Physical laws provide us with the best currently known descriptions of natural kind concepts. Technically they are not necessary or sufficient characteristics, because they are idealistic and approximate descriptions (ie, models) that apply only if interfering background circumstances can be practically ignored. In Newton's case, his law of gravity was superseded by Einstein's more accurate model of general relativity. In the example of Kepler's first law of planetary motion, a planet moves around its sun in an elliptical orbit. But small deviations from an ellipse are caused by other orbiting masses, the fact that the sun is not exactly at the focus of the ellipse, and uncountably many other effects. In fact, circumstances may even conspire to override or invalidate the law in a particular situation, as would happen if Jupiter were to wander into the inner solar system. Physical laws, including the more probabilistic biological principles, are here accordingly regarded as dispositions (salt normally dissolves in water, neurons normally fire action potentials, etc) and potentialities (human DNA normally produces a heart, turning on the ignition key normally starts the engine), which are valid descriptions for certain concepts in "normal" contexts, and included in the normalcy characteristics for those kinds of objects.

Stereotypical & Cue Characteristics

Stereotypical characteristics for a concept circumscribe what seems reasonable to believe about the concept. They form the basis for prototypical concepts expressing mostly statistical information gleaned from concept instances. These conditions are likely developed from frequent encounters in everyday dealings with things, and therefore are fairly reliable much of the time. Our stereotypical knowledge about apples may include the usual red, green, and yellow apples based on frequent exposure, but apples can more rarely be brown or even black (if rotten), a possibility easy to overlook. The relative frequency of occurrence of each color is probably encoded in the characteristic, with red being most typical, followed by green and yellow to a lesser extent. Stereotypical characteristics participate in default reasoning like normalcy characteristics, but they usually have a weaker attachment to their base concept than identity or normalcy characteristics, and are more easily overridden by conflicting information with less justification. Stereotypical characteristics in prototypical concepts are analogous to necessary and dispositional characteristics in theoretical concepts, in that they express the probability of a characteristic, given an instance of a known concept. Some of our entrenched conceptual stereotypes are a symptom of over reliance on stereotypical characteristics to the exclusion of identity or normalcy conditions, which are based on causal rather than statistical reasoning.

The *cue characteristics* for a concept specify useful observational characteristics for recognizing or classifying instances of a concept, where the characteristic can vary in both salience and diagnosticity. Having a mane is a strong cue characteristic for the concept of [lion], because adult male lions almost always have a distinctive mane. Seeing a mane in the appropriate context, you can reliably conclude [lion]. (It is not a normalcy characteristic because not all lions normally have manes, namely females and cubs. And it is not a stereotypical characteristic of [lion], as the term is used here, because male lions are relatively rare, compared to females and cubs.) We are particularly sensitive to many subtle cue conditions which act as activation triggers for the concepts they are part of. Cue conditions are highly correlated with observations from past successful classifications, but can easily lead to misclassification, particularly in novel circumstances. Over reliance on cue conditions makes one vulnerable to jumping to conclusions. Cue conditions give rise to so-called typicality effects, where we judge some instances of a concept to be better instances than others (Murphy, The Big Book of Concepts, 2002). For example, most people will say a robin is a better example of a bird than a penguin, while admitting that both are birds (the ability to fly being a strong cue condition). Cue conditions seem to carry a salience or weight indicating how important the characteristic is for recognition in general, in addition to how diagnostic the characteristic is for the category. Most cue conditions derive from perceptual features such as shape, size, color/texture, environmental setting, spatiotemporal features, etc.

Exemplar Characteristics

Exemplar characteristics express knowledge pertaining to particular objects and their particular properties and relations. While the other types of characteristics apply to universals (ie, categories), exemplar characteristics apply to particulars (ie, individual things) by encoding situational kinds of information. That my kitchen floor is currently dirty (but will soon be swept clean), and that Pegasus has wings are exemplar characteristics attached to my particular concepts of |my kitchen| and |Pegasus|, respectively. Note that causal knowledge about particular objects can also be included, such as that |this billiard ball| caused (or can cause) |that billiard ball| to move. Exemplar characteristics are commonly used to (re)identify individuals, and to predict what those individuals will do on various occasions, based on past experience.

Summary

The above types of characterizing conditions seem to exhibit a relative ordering with respect to their degree of centrality or utility in reasoning about the concepts they are attached to. Furthermore, they seem to support a variety of different strategies for reasoning about different kinds of concepts. In normal development, we seem to start out in infancy organizing concepts around highly tenuous and often accidental exemplar characteristics. Although we retain some of these rough-and-ready characteristics over the course of further development, maturing requires the progressive acquisition of the more useful kinds of characteristics—eg, cue for effective environmental interactions, stereotypical for beginning complex social interactions, normalcy for more reliable understanding and explanations, and finally necessary & sufficient for scientific and philosophical inquiry. The diagram below shows this relationship between these functional clusters of concept characteristics.

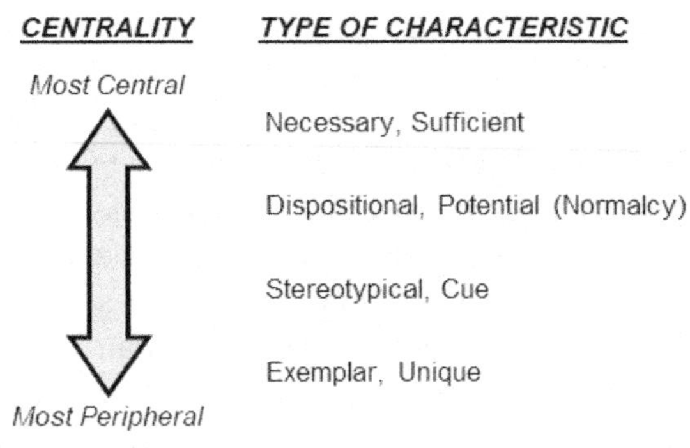

Figure 14: Relative Centrality of Concept Characteristics

The Structure of Characteristics

A characteristic, as defined here, is an integrated unit of encoded information attached to a concept and structured in a particular way. It consists of a set of associated beliefs which often chain together, end to end, such that the chain constitutes the whole characteristic describing one aspect of the concept. One end incorporates the concept which is being characterized, while (for universal concepts) the other end is usually one or more concepts specifying a distribution of potential values associated with the characterization. The picture of a spider comes to mind, its body being the main concept, its legs representing characteristics, the individual segments of its jointed legs representing associated elementary beliefs, and its feet covered with hairs, depicting values that can be individually or jointly activated by instances—apologies! The significance of the potential values for a characteristic is based on the cluster of specific values for the characteristic exhibited by instances of the concept (for prototypical characteristics—eg, most tigers have stripes, but there are albino tigers), or on modal possibilities (for theoretical characteristics—eg, many birds can fly, but some can't). An instance of the concept will generally express one of those values for that characteristic.

As a straightforward example, consider encoding some basic knowledge about what songbirds eat, in order to characterize one aspect of the concept [songbird]. In this case there is only one elementary belief, which is organized around the relation "[songbird][eats][???]"[116]. Assuming we know the kinds of things songbirds eat, a basic characterization becomes something like "[songbird] [eats] [[seed],[insect],[fruit]]". Translating to colloquial English, we are saying that songbirds eat seeds, songbirds eat insects, and/or songbirds eat fruit, where the three alternatives enumerate the possible values of this characteristic. Technically, the comma between alternatives is an *inclusive or*, since any particular songbird may eat more than one kind of food. Now when we observe a particular occasion of Tweetie eating some seed, it fits with our knowledge of songbirds, because Tweetie (as we already knew) is an instance of [songbird], the particular substance he happens to be eating is characterized as an instance of [seed], and the observed episode is an instance of an eating situation whose participants are |Tweetie| and that particular instance of seed, say |seed portion #6|. In this case, |seed portion #6| is an instance of [seed], and [seed] is a kind of [[seed],[insect],[fruit]] as well as a kind of [plant part].

The first thing to notice about the characterization "[songbird] [eat] [[seed],[insect],[fruit]]" is that it expresses selectional restrictions, which is in line with how we use concepts to talk about one thing and not another. It places broad restrictions on interpretations of the characterization, such that only some situations count as legitimate instances of songbirds eating. The selectional restrictions allow us to reliably recognize instances of songbirds eating, to predict what a songbird will eat, to hypothesize that an unknown bird may be a songbird based on what it is eating, and perhaps to question our background knowledge if we observe a songbird eating something else.

Second, the range of potential values expresses what have previously been called stereotypical conditions for the characteristic. The alternatives seed, insect, and fruit are based on past observations

---

[116] A relation is technically composed of a relationship—eg, [does eat]—plus variable arguments—eg, [songbird] and [???], as discussed subsequently in Chapter 4: *Relations*.

(or other learning) and we expect most songbirds (at least by default) to conform to our preliminary expectations. Considering other, perhaps biological, possibilities requires further effortful analysis. Moreover, we seem to develop a natural feel for the relative distribution of the potential alternatives over many observations. In my experience (whether right or wrong), the most common songbird food is seed, followed by fruit (eg, berries), and less commonly insects. Likewise, I judge red apples as more common than green, followed by yellow, and so it is with knowledge about many things. Our knowledge about the *salience* of a particular characteristic's value is related to how often we encounter a value (its frequency of occurrence or familiarity) compared to its alternatives, and how noticeable it is. To account for our knowledge of the distribution of salience, a relative weight can be assigned to each alternative, so they become comparable. My knowledge about what songbirds eat then would look something like "[songbird] [eats] [[seed](.6),[fruit](.3),[insect](.1)]". An alternative characterization with one very high salience will tend to become more readily noticed, and may take on the role of a cue condition for the characteristic. The color orange is a cue condition for a male robin's breast because of its very high salience, other colors being practically nonexistent.

Third, a diagnostic assessment is also easily added to a concept's characteristic, indicating how useful the characteristic is for distinguishing instances of the concept. In the songbird example, an index might be added after the relationship to signal that what a songbird eats is not very diagnostic for distinguishing among the different varieties of songbird: "[songbird] [eats](.3) [[seed](.6),[fruit](.3),[insect](.1)]". On the other hand, how it sings probably is highly diagnostic.

Fourth, the identity and normalcy characteristics for songbirds are obviously tied to its song (hence the name). Identity conditions applying to any songbird (eg, possessing DNA which codes for a syrinx, the song organ, whether expressed in the phenotype or not), and normalcy conditions applying to all normal songbirds (eg, sings a well-formed distinctive song) seem to function more like yes/no conditions, rather than graded tendencies like diagnosticity or salience. Since identity and normalcy conditions also have the highest diagnostic value, it may make sense to simply assign a diagnostic value of 1.0 to these kinds of characteristics as a first cut at knowledge representation.

And finally, the general format of the example characteristic above is intended to serve as a baseline model for the majority of concept characteristics. Specifically, it is a statistical representation which is thought to be gradually acquired via inductive generalization over exposure to many instances of objects of various kinds. It is probably the first kind of generalization learning that occurs in children as applied to perceptual objects, perhaps beginning as early as nine months (Baldwin, Markman, & Melartin, 1993). Such statistical representations of knowledge appear to significantly predate the learning of more sophisticated logical and causal representations (Leslie, 2008). The conjecture here is that logical (eg, quantified) and causal representations of knowledge, such as "All people die", "Some birds don't fly", and "germs cause disease", can be straightforwardly added to the baseline statistical representations (which are initially glossed as statistical generic expressions like "people die", "birds fly", and "germs correlate with disease"). The statistical baseline representations may thereby form the basis for our prototypical and exemplar concepts, while the logical/causal additions (ie, further markers for all, some, most, normally, causes, two, no, etc) may provide support for formal reasoning about theoretical concepts, as outlined previously in Chapter 2: *The Three Standard Approaches to Concept Representation.*

The example so far describes a characteristic consisting of one elementary belief organized around a single relation. A more complex chaining of beliefs can accompany attributes and their valuations. One practical birding guide to characterizing songbirds suggests the color groupings [blue], [black & white], [brown], [red & orange], [yellow]. These categories represent potential values of the color property of songbirds, often useful for species identification. Still longer chains are required for some characterizations, as attributes themselves can have attributes, in general. Such an attribute, for example, might further characterize the luminescence of songbird colors.

For some special hierarchical concept characterizations, the beliefs can also grow into trees (or even lattices, with bridging beliefs linking concepts between branches of the tree). The parts of an object usually have such a hierarchical structure. One characteristic of a songbird is its syrinx, a singing organ which has subparts serving different functions (according to one theory). A sketch of this characteristic might look like:

```
[songbird]
    [1[has part]2]   [syrinx]
                         [1[has part]2]   [syrinx's left-side muscles]
                                              [1[produces]2]  [low-frequency sounds]
                         [1[has part]2]   [syrinx's right-side muscles]
                                              [1[produces]2]  [high-frequency sounds]
```

Lastly, beliefs can form bridges between whole characteristics. In this way characteristics can become correlated, as when birds' ability to fly, having feathers, and having hollow bones are causally linked together. Such causal knowledge often qualifies as proper scientific discovery, but we seem attuned to detecting such regularities in forming most of our intuitive basic-level concepts as well.

Typicality (aka Prototype) Effects in Concepts

It is now widely recognized that the majority of conceptual category assessments are not all-or-nothing affairs. Some members of a category are judged better members of the category than others. Robins and sparrows are prototypical examples of birds, while penguins and chickens are not. Other examples, like hawks, seem to fall on a graded scale between the two extremes. Yet all are unequivocally birds. Bats, on the other hand, seem a lot like birds, but are not. Most people even judge small numbers like 2, 4 and 6 to be better examples of an even number than numbers like 4378 or 916.

Although typicality effects have proven difficult to quantify precisely, and are variously modeled in the literature, the effects are readily demonstrable. Consider just two characteristics of birds, namely size and flying ability. We can simply assign potential attribute values for bird size ranging from small, medium, large; and for flight as can fly and can't fly.[117] Then prototypical birds correspond to those that are small and can fly (eg, finch, robin, jay). If we order the attribute values such that they increasingly depart from the prototype values, the corresponding ordering of bird types agrees fairly well with experiments in which subjects grade them as increasingly poorer examples of birds:

<u>prototype</u>

small, can fly	finch, robin, jay

<u>once removed</u>

medium, can fly	owl, parrot, seagull
small, can't fly	island rail

<u>twice removed</u>

medium, can't fly	chicken, kiwi, steamer duck
large, can fly	condor, pelican, flamingo

<u>thrice removed</u>

large, can't fly	penguin, ostrich, dodo (extinct).

Moreover, it is easy to see how a small flying mammal, like a bat, could be mistaken for a bird. Additional information, such as the salience (eg, frequency/familiarity of observed attribute values) and the diagnosticity measures included in the previously discussed concept model, will further bias the proto-typicality ratings, but the simple presence/absence of just a few characteristics is enough to explain some of the graded effect.

The source of the tension between consistent conceptual categories and typicality effects may stem from the interplay between three of the primary functions served by conception—namely, recognition, prediction and classification. In recognition, we are attempting to match an observation with a learned category, for the purpose of shaping a timely and appropriate response. Noticing certain salient characteristics, like stripes or spots, triggers different concepts, like tiger or cow. Recognizing a tiger then produces a different reaction from recognizing a cow.[118] This facility for quick and suggestive recognition based on appearances is of great survival value, and our conceptual representations seem tightly integrated with perception to accomplish this function.

In prediction, we use learned knowledge about concepts to anticipate unobserved or future characteristics. Once we have concluded something is a cow, we expect it will produce milk even if we have never observed this cow before. Such expectations often guide us in fulfilling our goals. In the

---

[117] A more realistic account would also distinguish very small (eg, hummingbird) & very large (eg, ostrich), and perhaps awkwardness (eg, albatross), but this is not necessary to demonstrate prototype effects.

[118] The corollary to recognition of a category instance is identification of a particular individual. We may have learned to identify a particular tiger or cow, as well as recognize that each is a tiger or cow using roughly the same matching strategy based on similarity.

most difficult cases, we need to know what something is, rather than merely how it appears, in order to make reliable predictions based on causal principles.

Classification serves yet a different purpose. Classifying concepts helps us organize our knowledge into a useful framework supporting deliberative reasoning, as opposed to simple association used for perceptual recognition and prediction. If we are familiar only with tigers, but have been told that both tigers and lions are big cats, we tend to infer that lions share some significant characteristics with tigers. Classification depends critically on what something is, as opposed to how something appears, and the characteristics relevant to classification are usually latent, rather than salient. In fact, recognition and prediction seem almost like primitive conceptual elaborations of perceptual detection/recognition/prediction processes, and may have arisen as a first evolutionary step toward the more sophisticated functions of classification, which are often far removed from perception.

These kinds of overlapping conceptual functions may help explain why we often think ambivalently about categories, why we know chickens are birds but rather poor examples of birds. Certain characteristics (eg, bird DNA, as an identity condition, or feathered, as a normalcy condition) are key to category membership, but various other characteristics (eg, nesting in trees, as a stereotypical condition, or ability to fly, as a cue condition), are only suggestive.

We seem to maintain an operational separation between different kinds of characteristics. If so, prototypical intuitions may arise from weighing and combining evidence from the peripheral characteristics, for and against, to arrive at an overall degree of goodness of fit for a category (or an individual conceived of as a category of one). The total number of relevant peripheral characteristics that match and don't match certainly contributes, but high salience also correlates with prototypical assessments. For example, consider the characteristic of an apple's color. To a first approximation, an apple's possible color values may register as red, green, yellow, and brown (for rotten apples). Such color values reflect experiences with apples, and accordingly we have likely acquired some knowledge of the natural distribution of these colors. Red apples are the most common, making them prototypical, followed by green and yellow, which score lower. Brown apples are like chickens—members of the category, but decidedly atypical. When we see an atypical example, we can usually still classify it as an apple, but its prototypicality will be downgraded by its atypical color value. Alternately, category membership questions more reliably depend on the satisfaction of core identity and normalcy characteristics, where identity characteristics require unequivocal satisfaction and any normalcy characteristics that are not satisfied must be justifiably explained away. Thus a chicken, satisfying the core identity and normalcy characteristics for a bird, is decidedly a bird, but it scores low on the weighting of peripheral characteristics for birds, creating a certain tension for category membership.

Two other possible sources of typicality effects relevant to the previous concept model concern the applicability and the diagnosticity of a characteristic. Oftentimes it can be difficult to decide if a characteristic applies to an instance of a concept. Having feathers is normal for birds, but if an exemplar has a body covering that cannot reliably be called either feathers or nonfeathers, its membership in the bird category will be biased according to the degree of resemblance to feathers. It is questionable then whether the characteristic applies. Although such situations might loosely be included in typicality effects, they are probably better classed as issues of confidence. Likewise, characteristic values that are highly diagnostic for a category provide a stronger bias toward membership than somewhat less diagnostic values.

Finally, relative diagnosticity may have a wider scope than basic prototypicality, since proper diagnosis involves differential comparisons among all alternatives in a concept's contrast set—eg, for [bird], the contrasting vertebrate concepts of [fish], [mammal], [reptile], and [amphibian]. In other words, the characteristics of fish, mammals, reptiles, and amphibians may require consideration in ultimately deciding if something is a bird in borderline cases.

All in all, appreciating the basis of our concept categories can be an insightful, if messy, business. But it tends to undercut confidence in our intuitively clear concepts about many things.

Representing Typicality Effects in Concept Networks

As alluded to previously, there are probably many sources of typicality effects that shape our everyday conceptualizations. Some of these sources probably arise from the structure of concept representations themselves, while others likely are the result of conceptual processes operating on those representations. This section proposes four ways that typicality effects could arise from the underlying structure of concept representations (the former source). Two of those ways relate to the hierarchical organization of concepts (the generalization/abstraction or kind/instance hierarchy), and the other two to how concepts are characterized.

Given that concepts are effectively organized into kind/instance hierarchies, it follows that that concepts lower in a hierarchy are more specific than those higher in the hierarchy. For example, an alarm clock is a specialization of (ie, a kind of) a clock. Representing this knowledge makes the association between the concepts [alarm clock] and [clock] explicit in a concept network—namely, "[alarm clock] 'is a kind of' [clock]". It is then relatively straightforward to represent that an alarm clock is a typical kind of clock by attaching a number in the interval (0,1] to the association—in this case a number close to 1, such as 0.9. An atypical clock, like an atomic clock, might have a number closer to 0, like 0.1. This simple scheme is consistent with the typicality effects initially discovered in the 70s (Rosch, Cognitive Representation of Semantic Categories, 1975).

This same idea can be extended down to the particular instances of a concept. If there is a particular instance of an alarm clock, represented say as "|my alarm clock| 'is an instance of' [alarm clock]", then attaching 0.9 to this association indicates that my alarm clock is a very typical alarm clock. Alternately, a number like 0.1 would indicate it is a rather atypical alarm clock.

One significant advantage of this simple scheme is that it also allows typicality computations over hierarchically related concepts. For example, given that an alarm clock is a typical clock (0.9), and that my alarm clock is a typical alarm clock (0.9), it is easily concluded that my alarm clock is a fairly typical clock (0.81) using a multiplicative rule (0.9 x 0.9 = 0.81). Alternately, any atypical concept in a hierarchical chain of concepts will significantly lower the overall typicality. See the following diagram for examples of this and other similar computations.

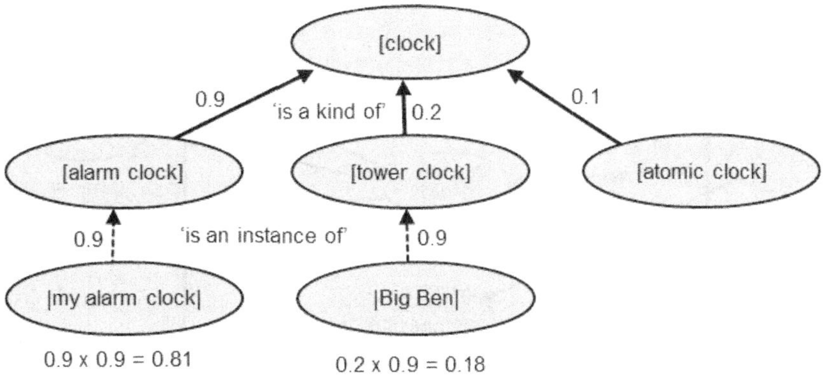

Figure 15: Concept Typicality in a Hierarchy

In addition to the above *hierarchical* sources of typicality effects, concept *characteristics* can also be more-or-less typical. For example, my impression of the color of apples suggests that red apples are more typical than either green or yellow varieties. Assuming the relation between an apple's color and a color value, say red, is represented as "[apple's color] [has color value] [red]", a number (0,1] can be attached to the relation, say 0.7, indicating red is fairly typical for apples. Green and yellow as apple colors will have lower values—eg, 0.4 and 0.2 respectively.[119]

The final source of characteristic typicality probably comes from the relative diagnosticity of each attribute associated with a concept. A concept like [apple] will be characterized in terms of various attributes like color, size, shape, sweetness, etc. Some of these attributes are more important (ie, more diagnostic) than others for identifying apples. Color seems relatively more diagnostic for apples than sweetness because many other fruits have similar sweetness values. Numerical diagnostic values can be added to the attribute representations for a concept, just as for the attribute values described previously. For example, given "[apple] [has color] [apple's color]", adding a numerical value 0.7 to the relation indicates color is a moderately diagnostic attribute for apples generally.

Combining attribute diagnostic estimates with attribute value estimates produces typicality effects for whole characteristics, just as in the hierarchical concept cases. A highly diagnostic attribute multiplied by a highly typical attribute value results in a highly typical value for the whole characteristic, as illustrated in the following diagram.

[119] These particular numbers represent subjective non-normalized independent estimates of typicality. But in some well-understood domains, normalizing the estimates to sum to 1.0 may offer a more consistent basis for reasoning.

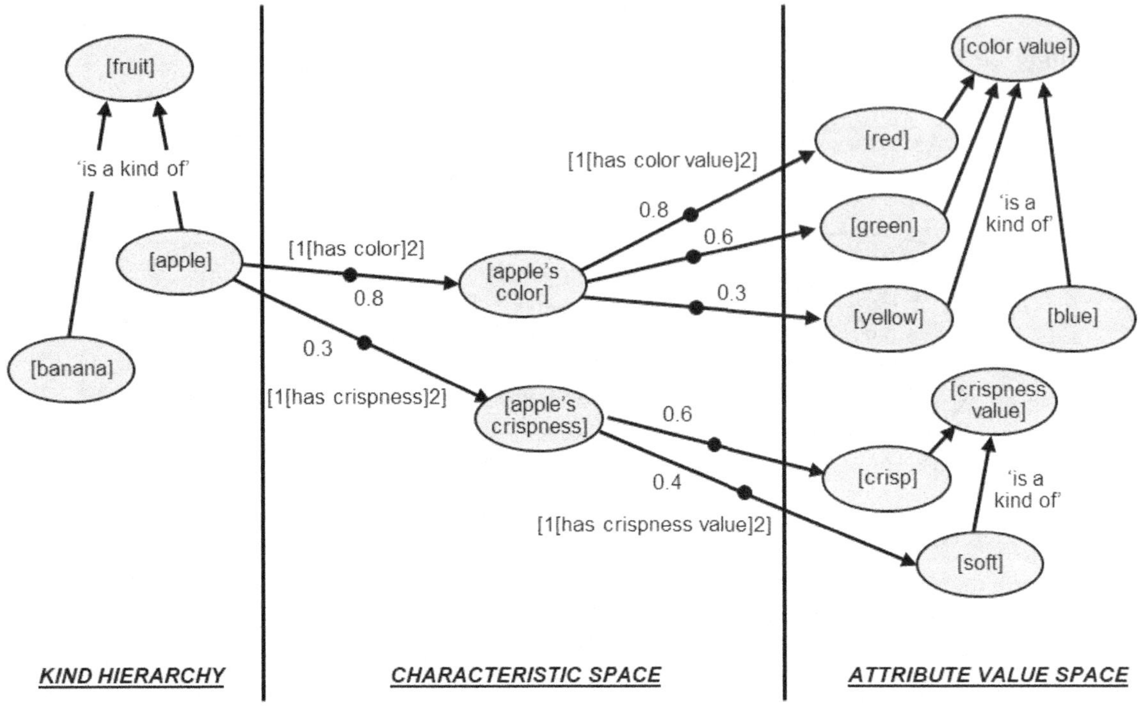

Figure 16: Characteristic Typicality

Clustering Effects in Concepts

One of the most interesting features of conceptual organization is the way our concepts seem to automatically group themselves into categories. That is, when presented with a variety of case examples from almost any domain of interest, we naturally tend to group similar examples together based on characteristics that usefully distinguish between different categories of examples. The process of acquiring categories seems to arise from our innate ability to represent individual examples in a way that supports grouping, more technically called clustering, based on past experience.[120] For example, a child's initial concept of bird, perhaps acquired at a zoo, might include observations of some songbirds, hawks, and ducks. Based on characteristics derived mostly from perceptual features, these bird instances will tend to cluster along the dimensions used to represent the observations. The following figure illustrates how a number of bird observations might come to be grouped, based on just the two characteristics of size and call (vocalization).

[120] Conceptual grouping amplifies the better studied mechanisms of perceptual grouping (Palmer, 1999). While perceptual grouping operates on more-or-less fixed features, conceptual grouping seems to operate more flexibly on any characteristics relevant to the categorization task at hand.

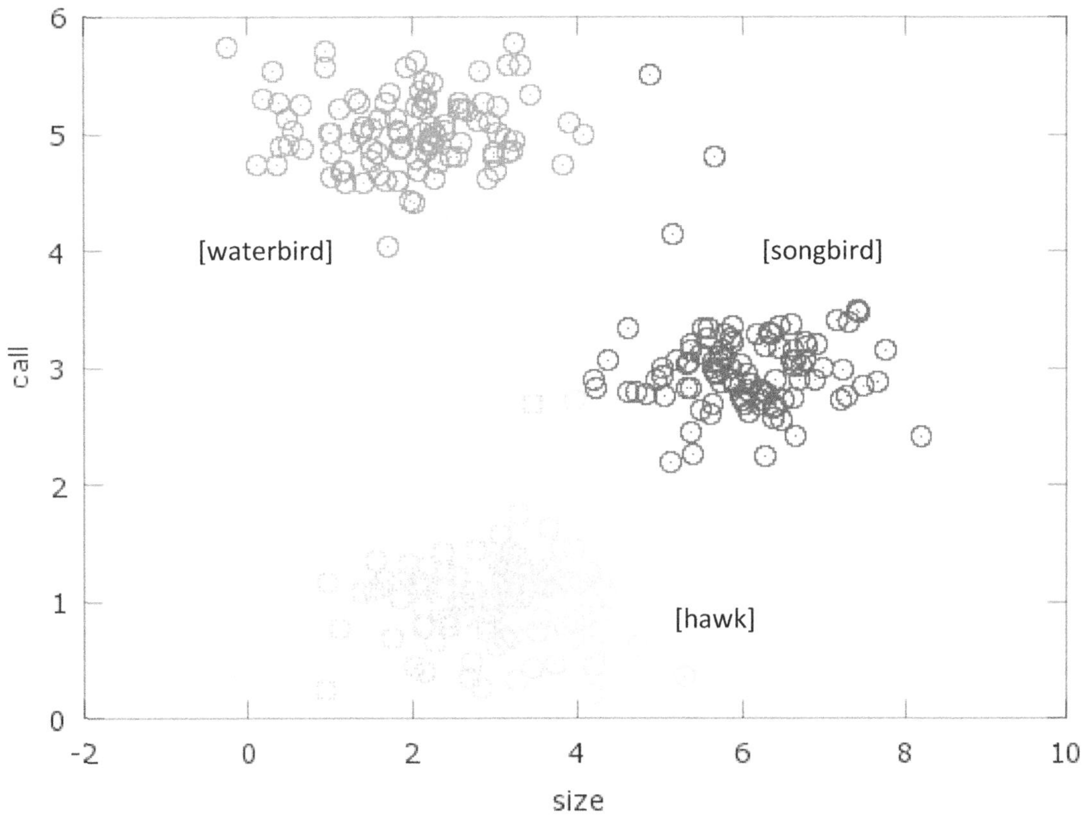

Figure 17: Hypothetical Clustering of Instances for the Concept Bird

As is evident from the diagram, effective categorization depends on being able to effectively separate the individual examples into distinct clusters. Such separation is possible only if the appropriate characteristics are selected for representation. For maximal separation of clusters, characteristics which tend to have multimodal distributions over the entire set of examples are the most useful. The characteristic of size is useful because it tends to separate the clusters for waterbirds and songbirds, even though the separation is not perfect and is relatively ineffective in separating waterbirds and hawks. At the other extreme, characteristics with unimodal or fixed (constant) distributions are practically useless, because they provide no information relevant to distinguishing one example from another. We make these kinds of selective decisions more-or-less automatically all the time based on the specific task facing us. To support such selective decisions, we necessarily retain fairly comprehensive distributional information about characteristics of all kinds in long-term memory, probably organized by domain.

The diagram above demonstrates clustering based on two characteristic dimensions.[121] Typical conceptual clustering, however, involves many dimensions, where each selected characteristic promotes some additional degree of discrimination among the examples. Each instance of a concept, therefore, can be viewed as a point in a multidimensional characteristic space, as well as a node in a concept network. The instances of a category like [waterbird], in general, will cover a specific region in the multidimensional characteristic space for a category like [bird]. A newly observed example of a bird can then be compared with previously observed examples of birds in order to properly classify the new example. New observations which are not similar to any existing cluster may provide evidence for a new category. In fact, established linguistic categories are routinely and purposefully extended to encompass new instances, as when the traditional concept of a particle was extended to include the elementary particles discovered in physics experiments. In this case a new size value of vanishingly small was added to the size characteristic for the particle concept. Such concepts often seem represented in terms of their instance clusters (as opposed to representations of explicit criteria), and therefore may best fit the *exemplar model* of classification, discussed in Chapter 2: *Three Basic Varieties of Concept Representation*.

The clustering of the instances of a concept produces a hierarchical relationship between the concept and a set of subconcepts, based on the selection of task-dependent characteristics (eg, [bird] → [waterbird], [songbird], [hawk]). For the most global and generic tasks, a taxonomic subdivision generally leads to the most useful and stable organization of categories in long-term memory, and the associated characteristics are given special recognition, especially in science. Special characteristics like mass, length, time, and number become our most fundamental, if not most straightforward, means of describing the world. Yet in most domains of knowledge, the number and kind of characteristics required to produce a useful multilevel hierarchy of concepts is large and highly domain dependent. The hierarchical structure of concepts in most domains ends up being complex, mainly because the concepts at each level of the hierarchy will typically possess very different (and usually uncorrelated) characteristics at different levels. For example, while number of legs is a useful characteristic for distinguishing birds from other animals like reptiles, mammals, and insects, it is useless for distinguishing various kinds of birds, since birds are normally bipedal. Selecting (and sometimes inventing) those characteristics that partition a concept into useful subconcepts often requires an uncommon breadth of experience and creativity.

Concepts & Existence

When we ask a question about whether or not something exists in the world, somewhat surprisingly, we also seem to be asking a question about our own internal conceptual representations. Given that we are entertaining a concept of something, we typically want to know whether that concept reliably

[121] The dimensions in this case are continuous, but discrete dimensions like number of legs are also possible. Furthermore, characteristics are not limited to ordered dimensions; they commonly include dimensions exhibiting a simple enumeration of possible values like the characteristic of locomotion with possible values walking, flying, swimming.

correlates with external reality. We say Roy Roger's horse Trigger exists, but Pegasus does not, because our concept of Trigger correlates with perceived reality, while our concept of Pegasus does not. Moreover, we are interested in the existence or nonexistence of a lot more than tangible objects. We use the same terminology to talk about the existence or nonexistence of memes (like persistent superstitions), of numbers (like smallest prime, which most mathematicians would admit does exist, whereas largest prime does not), and electrons (which are usually said to exist as quantum wave probability functions, whatever those are).

However, the previous observations about "existence" are too cursory for a couple of reasons. First they fail to clearly distinguish between the existence of a concept as an objective mental phenomenon existing in the brain, and the interpreted content of that concept as something existing in the world. In this text concepts are viewed as enduring physical patterns in the neural substrate whose existence necessarily precedes our occurrent interpretations of those patterns. As a variation on the standard philosophical theory of *conceptualism*, concept representations, in some form or other, are assumed to exist in the brain. The question of whether particular concepts correlate with reality is a different matter. Within such a "theory of mental concepts", anything that can be thought up and represented using our given conceptual faculties exists as a potential concept—even the concept of a round square.[122]

Concepts thus comprise a domain of mental theoretical objects, which we can reason about in standard ways (eg, using predicate calculus). Expressed as a formal theory, concepts (as primitive theoretical entities) become the individuals, predicates, and values of variables in quantified formulas. Within a theory of concepts, the existential quantifier, accordingly, ranges over the domain of mental concepts, and not necessarily over the content of those concepts. If a collection of formulas can then be found to (roughly) model the organization and structure of our concepts, it can serve as a basis for deriving potentially interesting new knowledge about our concepts. For example, a basic *principle of connectivity* in a concept net would seem to be that every concept is an instance of some universal concept (viz, even universal concepts themselves).[123] In other words, being an instance of some class is what connects every concept to the wider concept network, and establishes its fundamental meaning. There are many other principles that could be specified, as well, as developed in Part II: *Concept Networks*.

Turning next to the content of concepts—ie, what we use concepts to represent—existence now relates to whether we believe that content exists in whatever domain we happen to be talking about. In the standard physical domain, we want to know whether a concept's interpretation correlates with commonly experienced reality. In the physical domain for example, given the two (existing) concepts |Trigger| and |Pegasus|, we want to distinguish between the existence status of what those concepts represent, namely an existent horse vs a mythological horse. In a banking domain, |my checking account| would probably exist, but |Sherlock Holmes' checking account| probably does not. In a

[122] We can characterize a round square as being both round and square even though the characterization is inconsistent. Although the inconsistency of a round square is obvious, many of our concepts are surely inconsistent, unbeknownst to us.

[123] The connectivity principle could be formally expressed as: $\forall xy$ instance-of(x,y) $\rightarrow$ instance-of(y,[[universal]]).

mathematical domain, |smallest prime| exists, but |largest prime| does not. In order to specify basic information about conceptual content therefore, each concept will be assigned an epistemological property specifying its *actuality*, taking on one of three (or more) possible values: extensional, intensional, or spurious. [124] A concept's actuality informs the interpretation of that concept in the relevant domain by specifying whether the concept's content should be interpreted as existing or not (ie, as having an *extensional* or *intensional* denotation, respectively). The third value of actuality, labeled *spurious*, is envisioned as applying to a subclass of intensional concepts, indicating an impossible or contradictory denotation. In a physical/real domain, the concept |Trigger| would likely have an actuality value of extensional, while the concept |Pegasus| would have a value of intensional. A concept like |the horse which is a grapefruit| is spurious, even though it has a description as a concept. In a mathematical domain, |the smallest prime number| is extensional, |the largest prime number| is intensional, and |the number less than 2 and greater than 3| is spurious.

Extensional Concepts

We typically use extensional concepts to model things we think of as existing in the domain of reality.[125] But building a bridge between our concepts and reality is a tricky business, even though we often take it for granted. Reality in this context means the space-time-material-causal manifold we call the concrete world, and our extensional concepts are about the concrete objects, states, and processes we perceive in the world.[126] But different schools of thought have different explanations of how we arrive at our judgements about reality. The approach advocated here is based on accumulating evidence. If the positive evidence (or other rational confirmation) "sufficiently" outweighs the negative evidence, then the entity is deemed real and existing (ie, extensional); alternately it is deemed intensional and not existing, or the answer to the question of existence is currently unknown. Sources of such evidence typically come from direct and indirect observation, deductive/inductive/abductive inferences, trusted authorities, and compelling intuitions. For example, according to our best historical, physical, and archeological evidence, unicorns are not real (intensional), tables and quarks and photosynthesis are real (extensional), and whether Camelot is real or not is unknown. Thus, something is real for us if the preponderance of rational evidence says so, and it is mostly consistent to believe so. However, this leaves room for each person, in developing their own model of reality, to subjectively judge what is rational, what is evidence, and what is consistent. Many discussions of reality revolve around the actuality of concepts like personhood, race, aliens, and, for philosophers, universals.

[124] Other values of *actuality* might include fictional, hallucinatory, physical, or abstract depending on how much granularity and coverage is needed for adequate discussion of the domain objects.

[125] In other specialized domains, like mathematics, it is mathematical objects, like numbers, which are interpreted as extensional.

[126] However, sometimes we forget that "reality" really means our "internal model of what we take to be reality", since our experience of a comprehensive reality is constructed in our minds over a long period of time (but updated in real-time from incremental perceptual cues).

In rough agreement with Searle's characterization (Searle, 2004), physical (ie, real concrete extensional) phenomena are here distinguished as: 1) entities having a location in spacetime (whether precise or approximate), 2) entities having a material *basis*[127] (whether in mass or energy), 3) entities, at least potentially, having some kind of structure (ie, having "parts", along with their properties, configurations, and interactions) which causes the entity to have the properties it has, and 4) entities able to causally affect other physical entities.[128] If the evidence suggests an entity satisfies all of these requirements, then the entity is deemed to be a real physical entity, and the concept is labeled extensional.

In addition to the usual physical objects of everyday experience, extensional entities encompass virtually all substantiated scientific phenomena, including physical object properties, relations and states, as well as dynamic events and processes, and even some immaterial notions (like information), which still depend on a material substrate for expression. In particular the mind (in addition to the brain) is a physical entity, because it is collocated with the brain in spacetime, has a material basis in the neural networks (matter) and neural firings (energy) of the brain, depends on particular brain structures and organization which cause the appearance of phenomena like consciousness and thought (although we do not yet know how these phenomena arise), and is able, through agency (also not currently well-understood), to act on and be affected by the world.[129] According to these criteria, the mind is like many other naturally occurring, but less complex, physical phenomena—eg, solidity, liquidity, digestion, magnetism (Searle, 2004)—which result from the combined actions of many underlying particles or cells configured in certain ways. Mental psychological phenomena thus constitute a subdomain of physical phenomena, rather than a qualitatively different kind of nonphysical phenomena. The various psychological entities we talk about (concepts, beliefs, thoughts, desires, etc) are likewise mental physical phenomena subject to scientific study, since they, in turn, arise from the organization of the physical mind. Accordingly, our thoughts (as representational patterns) can ultimately react to and interact with other physical objects in the world under the same causal physical umbrella as colliding billiard balls.

---

[127] But having a material basis is not the same as being material. For example, information is not material, but nevertheless is always expressed in a material basis such as in speaking/writing or a picture, which conveys the information and so qualifies as physical.

[128] Searle's requirement in *Mind: A Brief Introduction* (Searle, 2004) is "First, real physical phenomena are located in space-time. (Thus electrons are physical and numbers are not.) Second, their features and behavior are causally explainable by microphysics. (Thus, solidity and liquidity meet this test. Ghosts, if they existed, would not.) Third, where real, physical phenomena function causally. (Thus solidity is a real physical phenomenon. Rainbows, under the description "rainbow," are not real physical arches in the sky. They do not cause anything.) And the physical universe is causally closed in the trivial sense that anything that functions causally in it must be part of it."

[129] That the physical mind "emerges" from the physical operation of the brain is not a form of dualism. Dualism does not recognize the dependence of the mind on the brain, and puts mind in a nonphysical domain.

Intensional Concepts

We do not just talk about things in the physical extensional domain, although we commonly do so by default. Talk about nonextensional things like liberty, fictional characters, numbers, fantasies, and other abstractions plays an important role in our social life and partially defines what it means to be human. These kinds of concepts are labeled *intensional*, being creations of the mind to facilitate some intentional purpose, commonly understood as having no physical basis. For example, the concept |Sherlock Holmes| is an intensional concept (in the physical domain, but not in the storied domain), with intensional characteristics such as being tall and lean (*A Study in Scarlet*, 1887).[130] Numbers and other mathematical objects are another important collection of intensional concepts (again, with respect to the physical domain) created to model extensional physical phenomena, as well as to explore the structure of abstract mathematical objects for their own sake. Of course, we may each disagree on what we call extensional and what we call intensional (eg, whether God is the creator of the cosmos, or is the creation of our own minds), but rational discussion always seems to require that we first clarify and agree on the domain we are talking about.

Our intensional concepts are typically derived from previously established extensional concepts by the mental process called *nominalization*. Nominalization turns an extensional predicate concept into an intensional concept. It involves mentally projecting the extensional concept of an entity, typically in the physical domain, into an intensional concept of an abstract pseudo entity. For example, the concept of being a (real) detective can be projected into a pseudo detective named |Sherlock Holmes|, which we then call fictional. The pseudo concept retains some salient characteristics from the extensional projection (now understood as pseudo characteristics), but it is necessarily incomplete and possibly altered, since it is only a projection created for a particular purpose. In the domain of a Conan Doyle story, however, |Sherlock Holmes| explicitly exists as a detective. In addition to such conceptual kinds (like person), other forms of nominalization involve extensional properties and relations:

is honest → honesty
is smooth → smoothness
is adjacent to → adjacency
is a dog → canine (as a noun)
has a brother → brotherhood
to fly → flying, flight

In all these cases, we conceptually turn an extensional property or relational entity into an intensional pseudo entity, which we can then treat as an object for further analysis or characterization. The abstract "objectification" of properties and relations is of particular import in scientific theorizing,

[130] Note however, that |Sherlock Holmes| does possess some extensional properties too, notably literary properties such as being a character created by Sir Arthur Conan Doyle, and being one of the most famous fictional detectives of all time. See Chapter 5: *Pseudo Concepts* for a discussion of the joint inheritance of physical and fictional properties.

because it allows the development of laws and principles summarizing the results of many concrete experiments.

Spurious Concepts

The third value of actuality relates to concepts which would otherwise be intensional, except that they are representations of impossible or null objects. This third classification is needed to distinguish between intensional concepts like |the largest prime number| and |the number less than 2 and greater than 3| within the domain of mathematics. Meinong's (in)famous "round square" is another example of a concept for which we can have a representation but which denotes no object in the geometric domain.

In the physical domain labeling a concept as spurious can be a controversial call, particularly since the choice between extensional, intensional, and spurious is based on selecting and interpreting the evidence. Evidence is notoriously malleable, leading many people to posit extensional ghosts, aliens, conspiracy theories, and even concepts![131] Collective science, likewise, has sometimes accepted phenomena initially as extensional, only later to retract the claim—examples include philogiston, the ether, and bodily humors. And in philosophy, the study of the metaphysical nature of being qua being and existence remains an active area of exploration.

Summary

To conclude this multifaceted discussion about concepts and existence, it seems appropriate to step back and reflect on the wider context in which concepts are said to exist. In general, it is assumed here that all knowledge is represented in terms of structured beliefs. And concepts are components of those beliefs, in spite of their problematic identity and individuation criteria. Further, all knowledge is organized into more-or-less separable, but sometimes overlapping, conceptual domains, where each domain is populated with presumed objects, beliefs about those objects, and ways of reasoning about them. Reasoning about biological organisms, for instance, employs different methods and criteria than reasoning about mathematical structures, as practiced by experts in each domain. Likewise, the "conceptual" domain under consideration here amounts to a partly formal, partly informal *theory of concepts*. And as in any theory, certain fundamental objects—namely certain specified concepts—are simply assumed to exist in this theoretical domain. However, we also share an intuitive understanding of a common reality domain, which serves as a backdrop for interacting with each other. Whether the proposed theory is actually a good model of conceptualization in reality depends on external evidence matching the domain model to psychological experiments.

---

[131] See (Machery, 2009) for arguments that concepts do not exist as natural kinds.

The *existence* of concepts as theoretical entities, therefore, is distinct from the proposed *property* of concepts called *actuality*. Actuality amounts to an assessment which can be assigned to each concept specifying how it is to be interpreted within a domain—ie, as extensional, intensional, or spurious. Furthermore, actuality should not be thought of as expressing a so-called *mode of being*, where different concepts represent objects possessing ontologically different kinds of being. (For example, Meinong has proposed separate realms of being for concrete entities and for abstract entities, following Plato's ancient idea of separate realms containing particular and universal objects.) Expanding the meaning of being in this way would mean adding additional tools for logical quantification over the objects in each realm. There is no need to add to the one univocal mode of being (simply called "existence") if claims to existence (or nonexistence) only apply to the objects in each domain. The existence of various objects then is dealt with separately within each respective domain, including our common domain of intuitive reality, according to the best practices in that domain. In this view there is no overarching *metaphysical theory* encompassing all other domains. There are only individual theories which we independently evaluate in terms of their correlation with perceived observation and evidence. We can expand the scope of our individual theories, and compare theories with each other, but we don't have the wherewithal to grasp or theorize about the whole of being (including ourselves) at once.

Truth

If existence is relative to a domain, as outlined in the previous section, then so is truth. In other words a well-formed proposition is true, if it correctly or adequately describes whatever exists in its target domain, otherwise false or indeterminate. And what exists differs from domain to domain, depending on the grounding assumptions for each domain. In mathematical/logical systems the truth of propositions standardly depends on axioms, and it is the axioms that provide structure and ground what exists (eg, numbers, probability) in each system. It is the same with geometric systems, typically featuring points and lines—eg, depending on which variation of the parallel postulate is in play. In physical scientific domains the grounding for truth and what exists is usually based on theoretical postulates and established scientific methodology. In fictional domains the basis for truth is the widely accepted narrative describing the goings-on in a story, plus common sense inferences. In our shared quotidian common-sense domain we usually rely on consensus within our intuitive perceptual and conceptual background assumptions about the world, including intuitive principles of causation. All truths therefore ultimately arise from sets of explicit or implicit grounding assumptions specific to our domains of interest, assumptions regarded as true by default, unless exposed as false via domain inconsistencies.

As a consequence, there is no such thing as *absolute truth* independent of the specific domain being talked about. For a similar pluralistic view see (Ball, 2017). This conclusion appears to diverge somewhat from most of the other leading theories of truth, which are not overtly domain specific. It will be convenient to refer to the above characterization of truth (by domain) as the Domain Theory for comparison with other more traditional theories of truth:

- In the Correspondence Theory, a proposition is true if it corresponds to a fact or state-of-affairs in the world. While this theory seems reasonable for physical scientific and common-sense domains, it is at least questionable for fictional and mathematical domains, where truth is derived from accepted stipulations and chosen axioms. Whether the abstract objects in mathematical domains, such as numbers and infinities, actually exist in the world, and therefore enter into facts about the world, is controversial. The Domain Theory recognizes the significance of correspondence, but it is between true propositions and postulated domain elements, rather than between true propositions and the world.
- In the Coherence Theory, a proposition is true if it is part of a whole system of interrelated true propositions. This theory comes close to the above proposed Domain Theory, if an interrelated system of propositions adequately describes the content of a domain of interest, and is based on a set of assumptions. It is the grounding assumptions that justify the truth of the whole system, instead of simple coherence. Coherence alone is insufficient to warrant truth, since a set of coherent statements can misrepresent a domain (eg, as often used to advantage in war strategy). But the Domain Theory implies coherence.
- In the Pragmatic Theory, a proposition is true if it is part of the final explanation of reality at the end of inquiry. But it may also acknowledge the practical value of truth in the here and now. This view is closely aligned with a scientifically based correspondence view, which claims to present progressively more accurate accounts of reality over time (aiming ultimately at a Theory of Everything), as well as provide practical technological advantages along the way. It is somewhat more difficult to apply it to abstract domains like mathematics, which are open-ended in the sense that, according to Godel's incompleteness theorems, some truths are not derivable from the axioms. Such truths, in principle, are beyond any inquiry.
- And finally in Tarski's Theory, a sentence is true if the referents of its terms satisfy its predicates. For example, "snow is white" is true if the term "snow" refers to snow, and snow satisfies the predicate "is white". This amounts to a domain independent characterization of truth, but many feel that it does not get at the underlying nature of truth itself, because the concepts of reference and satisfaction are not well understood.[132] However, it is clearly compatible with the Domain Theory, since Tarski's Theory could apply to any domain whatsoever.

The foregoing characterization of truth naturally leads to the idea that knowledge in a concept net should be organized according to separable domains. This does not mean that the domains are completely separable, but that there are core organizing principles that ideally specify the grounding assumptions for each domain. Truths for a domain are then based on those assumptions. Of course ferreting out those assumptions, for example in the common sense domain, is notoriously difficult. But the assumptions carve out a circumscribed set of propositions that, in effect, define the domain—a domain that may grow, and even merge with or replace other domains.

[132] The same could be said of the so-called Deflationary Theory of truth, which claims that truth has no underlying nature over and above how it is used in language. Thus, a proposition 'p' is true, iff p. For example, 'snow is white' is true iff snow is white. There is then nothing else of significance to say about truth.

As an example consider the domain of Newtonian Mechanics. Within this domain it is true that velocities add. But within the domain of Relativity, that statement is false. The conflict is resolved by recognizing that Relativity supersedes Newtonian Mechanics as a theory of motion. So the statement is simply deemed false in the wider contemporary context.

The Complexity of Concepts

Looking at concept characteristics suggests three aspects to concept complexity. One aspect simply relates to the number of characteristics connected to the concept's nexus. For example, the concept [even number] which has only one important core characteristic, namely "is exactly divisible by 2", is obviously less complex than [even number greater than 10] which has the additional characteristic "is greater than 10". Similarly for latent characteristics: [child] as a young human, is less complex than [boy] as a young male human. The relative complexity of a concept thus depends on its vertical location in the concept hierarchy. More general concepts like [dog] are less complex than more specific concepts like [collie], because specific concepts inherit the characteristics from more inclusive concepts higher in the hierarchy, and add new characteristics of their own.[133]

Another aspect relates to the number of nodes comprising an individual characteristic. Since a characteristic's nodes chain by association from one node to the next, the length of the chaining (or branching, if a tree) can vary. A complex concept typically has a large reach, a reach sometimes further complicated by bridging nodes and links between chains or branches. The simplest characteristics are probably beliefs containing only one predicate. Different contexts also seem to activate different nodes and links of a characteristic, selecting some and not others, depending on point of view. For example, a person's body temperature may be described as 99 degF or as 37.2 degC or as low-grade in different circumstances.

A third aspect of concept complexity involves the degree of interaction between a concept's characteristics. Interaction between characteristics is like the interaction between the nodes of a single characteristic, but reflects a more global correlation between whole characteristics. Such knowledge is particularly important in categorizing natural kind concepts, because certain properties tend to cluster together, often forming the basis for natural laws or principles.[134] We may not have ready access to the embedded causal or logical relations, but they are the basis for many of our intuitions about how the world works and is organized. The concept [triangle], for instance, might have three characteristics, one for each side, perhaps labeled [side1], [side2], [side3]. Each side, furthermore, could have two endpoints labeled [end11], [end12], [end21], [end22], [end31], [end32]. Then a bridge between each

---

[133] We also seem to have an intuitive feel for the *accessibility* of a concept. Highly abstract concepts like [justice], [responsibility], [irony] may be more inaccessible than complex. Their characteristics are probably far removed from straightforward concrete meanings, and seem to encompass an unusually broad range of applicability. To effectively grasp such concepts requires us to think up concrete examples covering the breadth of normal use.

[134] The inclusion of such "theoretical" knowledge, mainly expressed in subjective folk theories, is the influential proposal of the so-called "theory-theory" of concepts (Murphy, 2002).

endpoint can specify that it be coincident with exactly one other endpoint from a different side. The chaining from [triangle] to [side] to [end] to coincidence with another [end] is part of the complexity of the concept triangle. Of course, we do not normally think about triangles in this degree of detail, but the information is intuitively available to us.

Finally, a concept may be viewed as complex because it incorporates complex procedures for evaluating information. For example, recognition often involves weighing evidence and is perhaps realized through procedures or functions attached to a concept to supplement simple matching with observations. When applied to a possible instance, the functions could execute and return values relevant to deciding whether it is an instance of the concept. The function affiliated with the concept [even number] may be a procedure that can take an instance—eg, 156—and actually divide it by 2, to see if there is a remainder.[135] If there is no remainder, then the instance is an even number, otherwise not. Recognition functions for so-called natural kind concepts, like [dog], can involve weighing a lot of evidence, depending on the circumstances, although usually only a few cues are enough. Since everything in a concept net is either a concept node or a link between nodes, a whole procedure is necessarily a concept too (with appropriate characteristics), but the internals of the procedure are regarded here as nonconceptual, even though the procedures may incorporate concept references and perform analysis on concepts.

An extension of simple matching to generalized inference is needed to deal with more sophisticated reasoning. To take an example from Murphy & Medin, if you see a man jump fully clothed into a swimming pool, it requires inference to conjecture that he is drunk (Murphy & Medin, 1985). The concept [is drunk] probably does not have [jumps fully clothed into a swimming pool] as a characteristic. Rather it likely does have [strange behavior] as a cue characteristic. Once the inference from [jumps fully clothed into a swimming pool] to [strange behavior] is made, the connection to [is drunk] is straightforward. Long chains of inference, controlled by a variety of inference procedures, are probably commonplace in everyday thinking.

Conceptual Identity

When we ask whether two concepts are really the same concept in different guises, we are probably less interested in comparing two mental representations in the brain than in comparing the respective interpreted contents of those two representations—ie, their meaning or what they represent. But both kinds of conceptual identity are important for scientific purposes, since they help clarify the fundamental nature of the things we call concepts and their interpreted contents.

Focusing for the moment on conceptual structure (as a representation) rather than conceptual content (as information), it is apparent that two concepts with the same internal structure and external connections are equivalent (ie, equal representations). Their nodes and links are simply duplicates of each other. The following schematic illustrates the duplication of a concept, where the concept is

[135] Other procedures might give the same result more easily—eg, by looking only at the last digit.

anchored at both ends in the kind hierarchy and in its value spaces. Under these conditions of corresponding physiological connectivity, two distinct concept representations are deemed the same if they possess the same internal structure and the same interconnections with other concepts. This physiological state of connectivity is mirrored psychologically in the concept's characteristics (expressing beliefs about the concept) and in the concept's associations with other concepts. Note, also, that concept identity ignores the activation levels of the characteristics in the concept, since activation levels are only relevant to the use of the concept during interpretation.

This definition of technical identity between concepts, however, is of limited use, at least from the perspective of knowledge representation. Other than simple redundancy, why would the conceptual system create identical conceptual structures? Perhaps one function of duplication (copying) is to retrieve a concept from long-term memory and reproduce it in working or short-term memory, thereby making it amenable to more rapid reconfiguration. Perhaps being able to detect and remove duplicate redundant concepts left over after intermediate computations or learning is required to recycle resources. But these are operational issues. The notion of equivalence that we really care about concerns the content of concepts. Learning that my concepts of Mark Twain and Samuel Clemens are equivalent counts as genuine content knowledge, but learning that two mental representations possess the same structure is a technical achievement mainly relevant to the operational mechanisms supporting conceptual content.

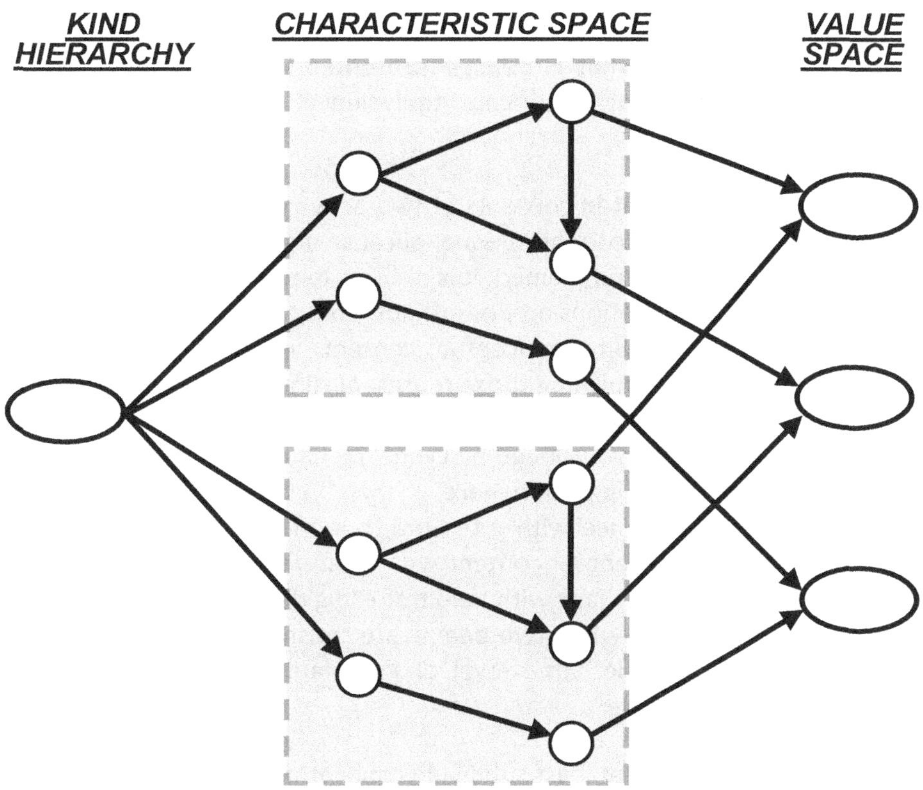

Figure 18: Concept Duplication

In practice, a concept's (usually extensive) internal associations are almost always activated with different strengths in different contexts. It is therefore implausible that different activations of the same concept could induce exactly the same interpretive content (ie, meaning), even if they were to be interpreted by the same computational algorithms. Moreover, exact synonymy is a rare phenomenon in language; and if it exists, is probably limited to narrow contexts, as in the use of "long" and "extended" in "a long time" and "an extended time". Interpretation permits extremely fine-grained nuances and elaborations when needed, but typically is tailored to the purposes at hand. Course-grained content (conveying only a small amount of information) works fine much of the time. If two senses of two concepts like [car] and [automobile] are deemed equivalent by fiat regarding their interpreted content under any and all circumstances, then this can be stated in a concept net by asserting that each is a kind of the other (ie, "[car]'is a kind of'[automobile]" and "[automobile]'is a kind of'[car]"), indicating that any instance of one is also an instance of the other. When the concepts are particulars, like |Cicero| and |Tully|, the proper relation is one of equality, rather than reciprocal kinds. Equal particulars represent the same thing, but maintain their separate identity in a concept net to distinguish each point of view.

Content equivalence is of particular importance in scientific discovery, as when an object's temperature is judged equivalent to the average kinetic energy of the molecules in the object, light is discovered to be electromagnetic radiation, the morning star and the evening star turn out to be the same celestial object, or even that my cat is Angus. The interpreted content of each such pair of concepts is the same (immaterial) informational thing, encoding what is being represented, but the characterizations are different. Characterizations are fully relevant to concept equivalence, but only partially relevant to conceptual content equivalence.

The general question of whether the interpreted contents of two active concepts (say in two different people) are the same is also an important philosophical issue, because it goes to the heart of the nature and function of concepts. As Quine long ago complained, it is difficult to talk or think about something if you can't even say when two different descriptions are, or are not, about the same thing (Quine, 1969). Can two different people really share the same conceptual content, when their respective concept representations and their procedures for interpreting those representations are personally unique? Can the conceptual content associated with synonymous expressions have precisely the same meaning (eg, is "Lynne loves Dave" really equivalent to "Dave is loved by Lynne")? Can the same conceptual content be derived from different sources? The suggestion here is that, yes, conceptual content can be individuated and shared (in general accordance with intuition), but that the theory is only partially developed in philosophy. The scope of conceptual content we encounter on a daily basis is vast, and careful study of all its subtleties in context (perhaps with help from "big data"?) will likely be required to grasp its regularities. But generally speaking, when two people are satisfied that their private concepts represent the same thing, interpreted to the same level of granularity, we can say that they are equivalent interpretations for those two people.

As a somewhat deeper foray, it seems we can at least subdivide conceptual content into two basic kinds: namely, symbolic content (associated with syntactic symbols like "dog") and semantic content (associated with semantic concepts like [dog]). (Chapter 1: *Symbols & Computation* covers the distinction in additional detail.) Considering the more straightforward content of symbols first, it appears their contents can be individuated in terms of the role they play in mental computations, plus the underlying semantic concept (ie, the conceptual structure) they point to. The contents of two symbols (in a given context) then are the same if the symbols participate in the same way in the same mental computations, *and* if they both point to semantic concept representations, whose contents are the same when interpreted in those computations. The symbols "π" and "pi" as used in mathematical computations are commonly so interpreted, assuming they point to the same number concept by convention. But of course the rub is in deciding when the interpreted contents of two different semantic concepts are the same, as discussed below. To restate the identity criterion for symbols in different words: the contents of two symbols are the same if they are fully substitutable for each other in the context of a computation, and they refer to semantic concepts which are interpreted to have the same content in that computation.

The more difficult case is in deciding when two semantic concepts are interpreted to have the same content, in spite of having somewhat different structural characteristics, activation levels, and interpretive procedures in play. This semantic case is difficult because it hinges on the relationship between the world and the information that the semantic concept carries about the world—a notoriously problematic mind-matter relationship to understand. We somehow manage to create concepts that properly *correlate* with regularities we detect in the world, so that our concepts become models of those regularities (thereby conveying information about them). Complex regularities in environmental contexts are broken down into "simpler" characterizing regularities, which are then cobbled back together into concept characteristics in mental contexts. But the details of how concept characteristics are flexibly interpreted to reliably model the regularities in different contexts remain incomplete.

But even though our concepts are created by fully causal mechanisms and convey information about the environment in their structure, they have no intrinsic content. Content is the significance or meaning of the concept to an information consumer (ie, an interpretive computational process) which can appraise and use the information for some purpose. Accordingly, the concept [dog], when interpreted in context, provides occurrent actionable information about dogs. That [dog] is about dogs, rather than something else, to the information consumer, is probably driven by the internal coordination between the causally-based acquisition of [dog], and the rewards/punishments provided by its proper/improper interpretation in learning situations.[136]

Yet improper interpretation of a properly established concept will naturally occur in some situations. In these situations even the most reliably constructed concept will lead to misclassifying objects that closely match its characteristics. At night you may reliably mistake most cats for small dogs, because the characteristics that you happen to observe match [dog] better than [cat]. In these cases you will straightforwardly interpret the triggering of [dog] as meaning a dog is present. But both dogs and cats are then reliably triggering your concept of [dog]. Why is the concept not reliably established as [dog or cat], rather than [dog], since the concept conveys reliable information about both? Even if you are reliably told that you are mistakenly classifying some cats as dogs, you wouldn't change your interpretations of [dog] triggerings to be about dogs or cats. It would seem that our interpretations are *guided* by acquired concept characteristics, but go *beyond* them by taking other factors into account as well.

One important factor could be an intuitive assumption we seem to make about interpreting semantic concepts generally. The assumption is that each concept *uniquely* expresses whatever it is about in the context of use. In other words, any given concept used in a given context is normally interpreted as being about one thing, and one thing only. This assumption possibly stems from the way we typically use concepts to express a single intention, one intention at a time. With an intention in mind, the information producer chooses concepts, both primitive (lexical)[137] and complex combinations (expressions), that aim to express that intention, where each concept contributes to the whole intention as unambiguously as possible. The goal of the information consumer is to recover that intention through conceptual decomposition, making use of the uniqueness assumption for each concept. What

---

[136] We are also able to update our concepts based on guidance from others.

[137] Each primitive lexical concept generally expresses one sense (ie, polyseme) of a word, but that sense is modulated in context in accordance with the intention of the information producer.

each concept is interpreted to be uniquely about in context is chameleon-like, but it can be thought of as being temporarily attached to the concept, expressing its inherent nature or essence (ie, its unique meaning) on the current occasion of use. More-or-less stable approximations[138] to fundamental natures are seen in natural kind concepts, like [gold], whose nature is often specified in terms of physical laws. But even a nominal kind concept, like [bachelor], seems to possess a relatively stable unique core quality centered around the meaning of *man* and *unmarried*, which distinguish it from other closely related concepts—cf (Millikan, 1999).

The uniqueness assumption is related to the idea of an innate predisposition to "psychological essentialism" in children, where they tend to assume natural kind categories have some essential property that makes things what they are. The conjecture here is that an early psychological predisposition to conceptual essences survives into adulthood, providing a basis for organizing our concepts into hierarchies by content, even though there is no scientific support for essences, per se (Dupré, 1993). Some support for this psychological conjecture comes from research on the underpinnings of psychological essentialism in children, now attributed in part to a ubiquitous human tendency to follow a fundamentally implicit *inherence heuristic*. The inherence heuristic biases us "to explain observed patterns (e.g., girls wear pink) predominantly in terms of the inherent features of their constituents (e.g., pink is a delicate color)" (Cimpian & Salomon, 2014).

This latent assumption plays out slightly differently in our conceptual interpretations, depending on whether the concept is a particular (eg, |Rosie|) or a universal (eg, [cat]). For particular (object) concepts, we seem to assume there is some unique primary essence that makes each object numerically distinct from any other particular object. Each particular object is identical to itself and nothing else. For universal (category) concepts, we assume there is some secondary essence which is shared by all instances of a category. In this case the additional essence does not confer numerical distinctness, but instead imposes collective sameness on instances, by way of duplicate essences attached to each category instance. That is, having the same secondary essence is what makes each instance a member of the category. Both primary and secondary essence assumptions, taken together, cover the interpretations of semantic concepts, whether particular or universal.

If this analysis is correct, then the content of most (or possibly all) semantic concepts is ultimately based on psychological idealization. This conclusion is strangely reminiscent of Plato's theory of ideals, in which each material object exemplifies an ideal that actually exists in another parallel realm of existence. But a contemporary explanation would have no brook with another realm, instead relying on the useful, if flawed, psychological assumption of essences. The answer to the question of what makes the content of two semantic concepts the same then, is that they are interpreted to have the same intended essence in context, even though that essence will necessarily remain conveniently ineffable.

[138] The nature of something is approximated by its conceptual characteristics; however, that nature is always subject to revision in the light of new knowledge and new contexts.

The Communication of Concepts

Language facilitates building useful conceptual representations in much the same way perception facilitates building useful perceptual representations; this is its primary function, although probably not its original function. Initially, speech sounds may have become directly associated with percepts, serving as a kind of attentional orientation to perceptual objects. One could orient a fellow hunter to a mammoth by saying its name. From this beginning it is easier to imagine other names being invented for distinct situations (eg, bellowing sounds in the distance) and then for simple associative combinations (eg, mammoths bellowing), leading to the rudiments of syntax. Once basic syntactic combination comes into play, developing an efficient way to exploit the potential for all manner of creative combination of names (via language) seems exceptionally useful. Concepts, associated on the one hand with linguistic names, and on the other with percepts, perhaps evolved as mediators, imparting a new descriptive kind of representation between the concrete and situated representations of perception and the highly abstract and arbitrary representations of language. The subsequent elaboration of conceptual operations has today developed into a practical thinking capacity for answering generalized "what if?" questions, a capacity now more-or-less independent of both language and perception.

By saying something about something in a certain way, a speaker encodes enough propositional information in an utterance to help a listener construct conceptual structures according to the speaker's intention. The speaker will assume the listener already knows and has concepts for all or almost all of the words in the utterance. The specific beliefs about those concepts need not be identical between speaker and listener, but some measure of overlap in core ontological content of concepts is required for successful communication. The speaker's words thus act like arrows aimed at the nexus for each of the listener's concepts. Stimulating the nexus for a listener's concept concurrently primes the listener's beliefs about that concept, making them accessible for inference and comparison with what else the speaker is saying.

The triggered belief activations will undershoot or overshoot the speaker's intentions in many ways, intentions which themselves may be poorly formed, so both speaker and listener typically tolerate considerable leeway in precision. But successful communication depends on recovering the speaker's propositional content. In order to facilitate interpretation, it appears that utterances (and written sentences too) organize the informational content around one kernel proposition. The other propositions embedded in the utterance generally play a supporting role, providing additional detail or qualification. The kernel proposition normally encapsulates a subject, a verb, and optionally an object, although the order of presentation may vary from language to language. If a listener can recover at least the kernel proposition, then the substance of the communication has been understood.

PART II: CONCEPT NETWORKS

Part II develops a network model for representing and analyzing concepts. It describes, first, the core structures used for representing epistemological concepts (ie, the fundamental concepts required to represent any kind of conceptual knowledge). Next, a variety of examples of domain concepts illustrate how new knowledge is added to the core. The final chapter suggests principles for organizing the knowledge in a large concept network.

4. CONCEPT NET THEORY

This chapter outlines a graphical approach to representing knowledge (ie, the content of typical declarative sentences) in a semantic network of nodes and links between nodes. The representation provides a framework into which new knowledge about some area of interest can be incrementally added, and which is amenable to being queried, reasoned about, or checked for consistency.

However, the discussion is not about an autonomous thinking agent. Such an agent would need something akin to perceptual and other support systems for successful realtime interaction with the world. The concept network described here is more like a road map, a tool for exploring the conceptual landscape. The map draws attention to landmarks and their relationships. The ultimate objective is to clarify the core semantic meaning of complex topical statements and background knowledge, by exposing the oftentimes subtle content of and relationships between concepts.[139]

The discussion pertains to knowledge entered into a "concept network" to distinguish it from other similar proposals, such as that developed by Sowa and the "conceptual graph" community (Sowa, 2000), and the "question answering" MultiNet by Helbig (Helbig, 2006). Concept networks are similar to conceptual graphs and question answering systems, but have an alternative semantic emphasis. Roughly speaking, a conceptual graph corresponds to a situation (comprising a subnet) in a concept net, with the graph expressing a predicate calculus statement. The relational structures, categories, and organization of a concept net provide the means for integrating the representations of such situations. Both MultiNet and conceptual graphs, moreover, seem to focus on representing the information contained in everyday sentences. The focus for concept nets is more on representing the ontological and background knowledge required for accurately "understanding" such sentences. The overall design probably falls within the purview of metaphysical conceptualism (Cocchiarella, 2007), which aims to represent and organize concepts as humans understand and use them (Nirenburg & Raskin, 2004), but informed by metaphysical and scientific analysis.

Basic Terminology

The broad term *semantic network* refers to any web of semantic elements and relations between elements used to represent an internally consistent body of knowledge. It consists of a directed, fully-connected graph with labeled nodes and links between nodes. The term *concept network* refers to a particular variety of semantic network that is organized according to the principles described in this and

[139] Confused nuances in conceptual meaning can easily lead to invalid conceptual representations, often detectable only with formal consistency checking. Our intuitive powers of understanding are usually quite accurate in detecting overall conceptual coherence, but the details are notoriously troublesome—we think we understand more than we do.

subsequent chapters. Each node in a concept network is interpreted as a *conceptual element* and the linking of two nodes will be called an *association*.

The generic term *concept* intuitively (and loosely) refers to any individuated topic of thought, typically composed of many conceptual elements. While this description of the gist of a concept is vague, the concepts in a concept network are not vague or fuzzy entities. Any concept is completely specified through its conceptual elements and associative links with other concepts in the network.[140] That those elements and links invariably only partially and inaccurately specify it, simply indicates a lack of complete knowledge about it. The presence of a concept in the network indicates that the concept is *recognized*, but it may be incompletely or even erroneously specified at any point in time. Examples of concepts include: my cat Rosie (a particular concept), cat (a universal, or kind, concept), Becky's books (a collective concept), is an instance of (a linking element of an association between a particular concept and a universal concept, as in "Rosie is an instance of a cat"), Houston (a place), unicorn (an imaginary kind of thing), the computation to determine if an integer is even (a function). Any distinct, explicit unit of thought (simple or complex, concrete or abstract, named or described or individuated) that can be partially specified can theoretically, if not practically or usefully, be represented as a concept. Of course, particular things which apparently exist in some sense, but cannot be designated even in principle (eg, most of the real numbers), cannot be explicitly represented in the network. In contrast, the set of all real numbers (a particular infinite set), or the general concept of a real number (a universal kind), are perfectly legitimate concepts.

An *association* is a primitive semantic joining of two elementary concept nodes, say C_1 and C_2. It corresponds to an instance of a binary relation, R, such that C_1 has association R to C_2.[141] Associations are specified via the directed labeled links between nodes. An example of an association is 'is a kind of' in "a turtle is a kind of an animal". A link between the two nodes (turtle and animal) in the network indicates that the corresponding association between the two concepts is assumed to hold. The total number of primitive distinct associations is relatively small, perhaps 40 or so, depending on how expressive the concept network needs to be. Associations are primitive because they provide the basic means for representing knowledge, and all knowledge is ultimately expressed via associations. More complex knowledge is then represented compositionally in terms of simpler knowledge. For example, a sentence such as "Drea loves Nynne", does not express an association. In this case, "loves" expresses a generalized domain situation, represented in terms of many associations, and is different from 'is a kind of' which expresses a basic epistemological relationship having to do with how knowledge is represented and organized, rather than how things are being characterized.[142]

---

[140] The parallel with the field of *structural semantics*, where the *sense* of a concept depends primarily on its relationships to other concepts, is valid; but conceptual meaning depends on more than just concept relationships.

[141] An association can be reified into an instance of a binary relation, thereby becoming a concept in its own right, but the term "association" denotes only the unreified linking of two concepts. The distinction is akin to the (philosophical) difference between *use* and *mention*.

[142] An association expresses an atomic formula in first-order predicate calculus—eg, |is-a-kind-of([turtle],[animal])| for the example above. Note that the formula itself is a concept. In general, quantified variables range only over concepts.

Each node in a concept network carries a unique *label*, so that it can be unambiguously referred to. Such labeling conventions have been introduced to make it easier to describe the contents of a concept net. In general, labels (eg, proper and common names, and predicates) are one function of language, and we make generous use of them as shortcuts for understanding our own and others' ideas. We can talk about particular concepts by using labels like |Jeff|, descriptions like |the father of Elias|, or other phrases as labels for those concepts (and their associations). Language has many drawbacks when it comes to labeling concepts, but it is better than remembering #G002974.

Useful labels attempt to convey the intuitive meaning of concepts and associations, but should not be thought of as definitive, semantically significant, or be confused with the concept itself. Labels are only rough and ready linguistic descriptions of concepts, and are optional. A labeled concept should be thought of as a shorthand representation for the unlabeled basic concept itself plus another concept representing a name or description (eg, "Venus" or "the morning star"), with which the basic concept is associated. In this way a concept can have any number of alternative names or descriptions which are (linguistic) concepts in their own right.

The operative purpose of labeling is to facilitate asserting (and sometimes retracting) knowledge to/from the concept net. Knowledge is asserted/retracted by making *statements*, which the system responds to by creating/removing the appropriate concepts and associations—eg, asserting "a whale is a kind of mammal" adds the association "[whale] 'is a kind of' [mammal]" to the concept net. But instead of using generalized English sentences or well-formed formulas from logic, we can use an informal, but structured, "subject verb object1 object2 ..." syntax as in "[whale] 'is a kind of' [mammal]". Here, [whale] is the subject, [mammal] is the first and only object, and the binary predicate 'is a kind of' functions as the verb. Statements are enclosed in quotes, just like their more complex English counterparts.

A *primitive statement* is a three-tuple which combines two concepts via association, and which can be directly asserted into or retracted from the concept network. A primitive statement expresses the most basic unit of knowledge, translating into two nodes with an associative link in a concept net. Examples of primitive statements include "an integer is a kind of a number" and "Bismark is an instance of a dog" (English versions), and "[integer] 'is a kind of' [number]" and "|Bismark| 'is an instance of' [dog]" (concept net versions). Particular concepts are delimited with vertical bars, and universal concepts with brackets, to avoid oftentimes subtle confusions. A *complex statement* is a statement that expresses a generalized situation like "Lynne loves Dave" (or any other well-formed first-order predicate calculus formula whose predicates are not primitive associations), which decomposes into primitive statements. Generalized situations are discussed in the next chapter.

The concept network, at any point in time, consists of the structure of nodes and links representing all statements that have been asserted and not subsequently retracted. After a statement is asserted into the network, it is *believed* (by the system). That is, the association (or generalized situation) between the known concepts is assumed to hold. The absence of a specified link between recognized concepts indicates that the corresponding statement is not directly believed. However, enough information may be present in the network to derive the statement indirectly. For example, even though the statement

"Socrates is mortal" may not be in the network, if the related statements "All men are mortal" and "Socrates is a man" are present, then "Socrates is mortal" can be derived.

Starting with a *kernel network* of primitive concepts and associations representing the epistemological baseline, the concept network grows by adding new links between existing nodes, or by adding new nodes and linking them appropriately with existing nodes (ie, by asserting and retracting statements). The kernel comprises the minimal, operational concept network, which embodies the essential mechanisms for basic knowledge representation, to which other statements about some area of interest can then be added.

Universals & Particulars

In order to begin organizing a potentially very large network of concepts, it is first useful to distinguish between "things" and "kinds of things". A kind of thing, hereafter called a *universal*, is a concept that represents a category of things, for example [platypus]. The bracket notation indicates it is a universal concept. Common nouns are common labels for universal concepts. The things covered by a category (called its *instances*) are concepts like |Polonius| or |Jeff's platypus|, which in this case are instances of [platypus]. The concepts |Polonius| and |Jeff's platypus| also happen to be *particulars*, in that they themselves cannot have instances (unlike universal concepts).

Singular terms, like proper names, typically express particular concepts in a standard interpretation. First-and-foremost as a model of objects in the world, a particular concept is thought of as possessing an infinite number of particular characteristics, the sum total of which ideally and precisely describe every concrete nuance of the object. Even though the mental representation of a particular object could never express such an infinite specification, the interpreted meaning of the particular concept nevertheless presumes all such characteristics could omnisciently become represented. Each worldly object, therefore (as mentally represented by a particular concept), is also presumed to be unique, being precisely and completely specified by all of its particular characteristics. Universal concepts, on the other hand, are here not interpreted as expressing objects in the world, but rather as expressing abstract categories of particular objects (eg, [cat] vs |Rosie|)—but see the subsequent section entitled *The Metaphysics of Universals*. In comparison with particulars (possessing only particular characteristics), universals (possessing one or more universal characteristics) are abstract specifications. That a universal concept is abstract derives from at least one of its characteristics being universal. This allows it to cover more than one particular characteristic (eg, [red] vs |this particular paint's shade of red| and |that particular paint's shade of red|).

The universal/particular distinction forms the basis for categorizing all concepts. Every recognized concept is either a universal or a particular. Below is a simple diagram of the top-level concepts in a concept network:

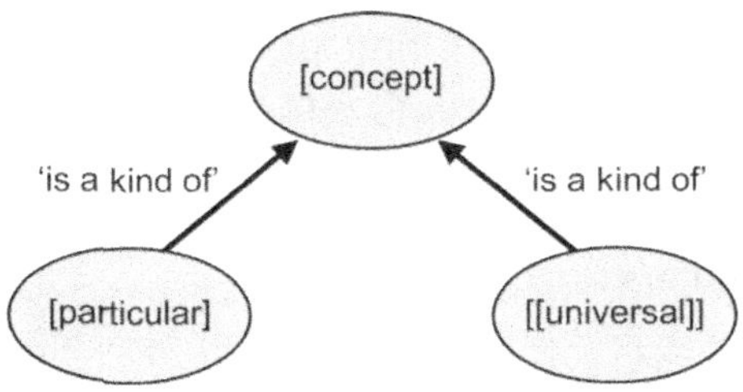

Figure 19: Top-Level Concept Net Concepts

The concept [concept] is the most general concept of all. Its meaning is similar to the general word "thing" in the sense that anything is a thing of some sort. All concepts are then of two fundamental kinds, namely [particular] or [[universal]]. Most common noun concepts like [dog], [lamp], [blood], etc, are kinds of [particular]. They have instances which typically denote nonmental concrete things in the world, such as |my cat Rosie|, |this 19th century antique lamp|, or |all the blood in Dave's body|.

On the other hand, concepts like [[attribute]], [[virtue]], and [[kind]] are kinds of [[universal]], with instances like [length], [honesty], and [dog]. That is, the various *kinds* of universal concepts can have *instances* which are other universal concepts, and the double brackets indicate the next higher level of abstraction above particular concepts. Double-bracketed concepts are more abstract than the basic universal concepts like [platypus], and are often called metaconcepts. The concept [[species]], with double brackets, is an example of such a metaconcept, since its instances (like [platypus], [cheetah], etc) are all universal concepts with their own instances. Note that [[universal]] in the above figure is such a metaconcept, where all metaconcepts are kinds of [[universal]], including [[universal]] itself as demonstrated in the next section.

As another example of a metauniversal, consider grammatical parts of speech. The concept [[part of speech]] is an abstract category covering a number of word categories, like [noun], [pronoun], [adjective], etc, where each of these is an instance of a part of speech. Furthermore, each of these categories has instances which are particular words. For example, the word "invidious" is an instance of an [adjective]—or more precisely the concept |"invidious"| is an instance of the concept [adjective]. As a test that [[part of speech]] is actually more abstract than [adjective], observe that "invidious" itself is not a part of speech. In practice what counts as a particular, as opposed to a universal, is subject to a certain amount of interpretation, often depending on utility or communicative purpose. But the choice must reflect the underlying meaning associated with the concepts as they are used—in this case whether or not the concept is thought about as having instances which themselves can have instances.

Every concept in the network, as part of its specification, must be an instance of at least one universal concept, however general or abstract that concept is. This fundamental requirement automatically joins any new concept to the existing concept net (every conceivable/representable concept is an instance of some kind), and also provides for the inheritance of characteristics from its governing universal concepts. For example, once a characteristic, such as "has a soft pliable bill", is assigned to the concept [platypus], |Polonius| as a particular instance of a platypus, will inherit that characteristic by default.

Top-level concepts like [concept], [particular], and [[universal]], as opposed to [lamp], [blood], and [flock], are epistemological (as opposed to natural or nominal domain kind) concepts in the same sense that associations like 'is a kind of' or 'is an instance of' are epistemological relationships (as opposed to natural or nominal domain relationships like "loves" or "is mortal"). Such epistemological concepts form the basis for constructing concept nets.

The Kernel

Starting with the three basic concepts and two associations in the figure above, it becomes possible to work out a minimally-sized concept network that can serve as a baseline for adding more knowledge. This *kernel*, in essence, represents knowledge about knowledge—that is, the concepts used to represent concepts in the first place, and the specifications for how concepts can potentially relate to each other.

Given that we are dividing all concepts into the two subcategories of particular and universal, it is apparent that both [particular] and [concept] are instances of [[universal]]. This is because they are both categories, and all categories are universals (as opposed to particulars). Then, what is [[universal]] an instance of? It would seem we need another metalevel concept category like [[[metauniversal]]] for [[universal]] to be an instance of. The proposed concept [[[metauniversal]]] then would also be a kind of [[universal]]—ie, falling under [[universal]] in the figure. This is because any instance of [[[metauniversal]]] must also be an instance of [[universal]]. (The only other thing it could possibly be is a [particular], which of course it is not because it can have instances.) Now, if [[universal]] is an instance of [[[metauniversal]]], and [[[metauniversal]]] is a kind of [[universal]], then [[universal]] must also, indirectly, be an instance of [[universal]] itself. For the purpose of deciding what [[universal]] is an instance of, then, we don't really need the concept of [[[metauniversal]], at least in the kernel. The updated network with instance associations added now looks like the following figure:

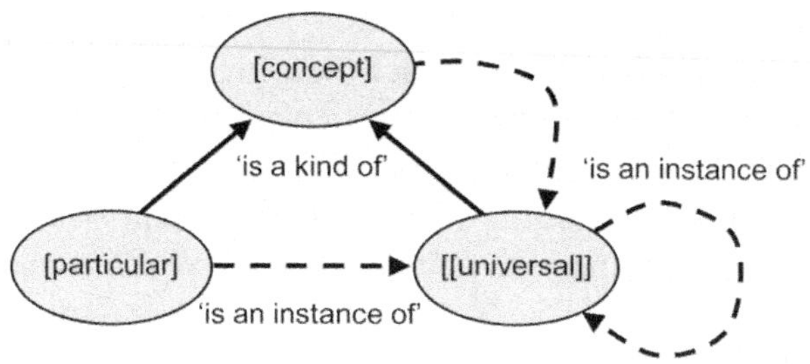

Figure 20: Associations Between Top-Level Concepts (1)

The next step is to work out the specifications for the five associations in the diagram. In other words, what sort of concepts can be joined by the 'is a kind of' and 'is an instance of' links. The meaning of these associations is different, so the sorts of concepts they can join are expected to differ.

Thinking first about the 'is a kind of' link, it should be apparent that it can only link two universals together, such that one is a subkind/superkind of the other. When we assert, for example, that a platypus is a kind of animal, we need to verify that both [platypus] and [animal] are universals.[143] Alternately, attempting to assert that Polonius is a kind of platypus should generate an error—it is an instance of a platypus, not a kind of platypus.

The specification for an association can be carried, it is proposed, by a *relation*, such that the association corresponds to an instance of that relation. For example, if we want to check whether "[platypus] 'is a kind of' [animal]" is a legitimate assertion, we can check conformance to the relation [[[universal]][is a kind of][[universal]]]. In this case both [platypus] and [animal] are instances of [[universal]], and 'is a kind of' is an instance of [is a kind of], confirming legitimacy according to the selectional restrictions. The relation [[[universal]][is a kind of][[universal]]] is not part of the above figure, but can be added, as shown in the next figure, where the abbreviation [1[is a kind of]2] label is shorthand for the above tripartite relation label. The two arguments to the relation are indicated by 1 and 2, respectively, which themselves are abbreviations for the primitive associations 'has arg1' and 'has arg2'. The relation [1[is a kind of]2] is a node just like the other concepts in the network, although it is abbreviated as a dot in the diagram. The associations labeled 1 and 2 specify the selectional restrictions on the arguments. The relation [1[is a kind of]2], itself, is a kind of [particular]. In fact, we can also add that "[concept] 'is a kind of' [concept]", since it is a universal too, and every concept (including universals) are ultimately a kind of [concept].

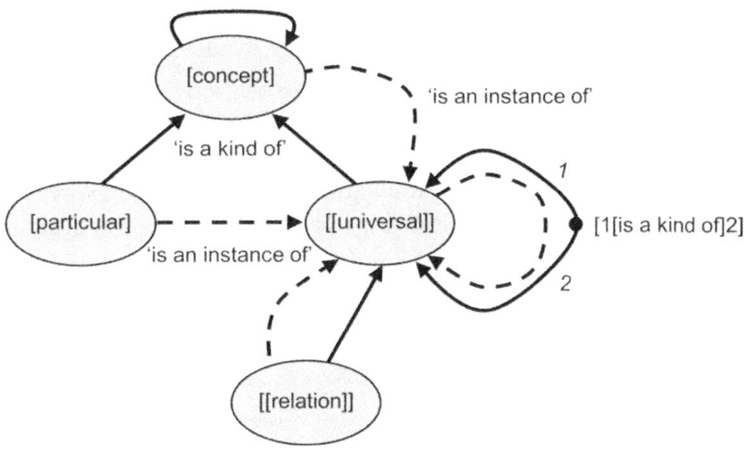

Figure 21: Associations Between Top-Level Concepts (2)

The new epistemological concept [[relation]] has also been added and tied in to the existing kernel as both an instance of and a kind of [[universal]]. We can now check the consistency of all 'is a kind of' links in the kernel by seeing whether they conform to the [[[universal]][is a kind of][[universal]]] relational specification—which they do.

143 Which is the subkind and which is the superkind is an empirical question, and is not part of the kernel specification.

In like manner, the 'is an instance of' link is seen as associating any concept with a universal (again every concept must be an instance of some universal), via the relation [[concept][is an instance of][[universal]]], depicted in the following diagram as [1[is an instance of]2]. Furthermore, the relation [1[has arg*]2] is added to authorize any instances of a relation to have any number of arguments, where [1[has arg*]2] stands for [1[has arg1]], [1[has arg2]2], [1[has arg3]2], etc. Adding these relations to the kernel finally produces:

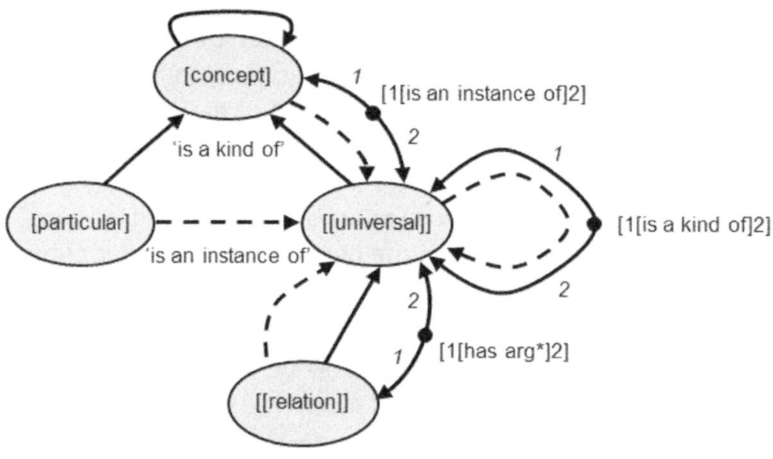

Figure 22: Associations Between Top-Level Concepts (3)

This diagram contains a complete specification of the kernel. But note again that relations like [1[is a kind of]2] and [1[is an instance of]2] are not association links. They are concept nodes which are instances of [[relation]]. Depicting them as small dots is a shorthand way to avoid cluttering the diagram. The formal concept net explicitly includes these relation concepts and all their associations, as discussed in the next section.

Constructing the kernel concept net is then a matter of asserting the following statements tabula rasa:

"[concept] 'is a kind of' [concept]",
"[particular] 'is a kind of' [concept]",
"[[universal]] 'is a kind of' [concept]",
"[[relation]] 'is a kind of' [[universal]]",
"[particular] 'is an instance of' [[universal]]",
"[concept] 'is an instance of' [[universal]]",
"[[universal]] 'is an instance of' [[universal]]",
"[[relation]] 'is an instance of' [[universal]]",
"[[universal]][is a kind of][[universal]]",
"[concept][is an instance of][[universal]]"
"[[relation]][has arg*][[universal]]".

All except the last three statements are primitive. The last three statements expand into quite a few primitive statements according to the formal definition of a relation, which is given in the subsequent section on *Relations*.

Concept Net Organization & Structure

The basic organizing principle for concepts in a concept network rests on the kind/instance associations between concepts. The figure below expands on the previous kernel with a broader range of top-level epistemological concepts, and additionally includes a few examples of lower-level domain concepts. The most general concepts are at the top with other more specific concepts branching progressively downward. Instances of any concept are also implicitly instances of all ancestral concepts above them, so |Polonius|, in addition to being an instance of a [platypus], is also indirectly an instance of an [animal], an [individual], a [continuant], a [particular], and a [concept].

Focusing on the [particular] side of the hierarchy, the concepts at the very bottom are particular instances—eg, |Polonius|. The various kinds of particular instances lie above. Focusing on the [[universal]] side, the concepts at the bottom are universal instances—eg, [platypus] & [animal]. The diverse kinds of universal instances again are arranged above all particular instances. Domain universal instances, like [platypus], are therefore also kinds of [particular], so a concept like [platypus] (contrary to |Polonius|) is an instance of a [[kind]], and ultimately an instance of [[universal]], resulting in a highly integrated and interconnected hierarchy. Focusing next on the domain characterizations, the value categories [soft] and [pliable], perhaps contrasting with other values like [hard] and [rigid], specify allowable attribute values for platypus instances like |Polonius| (or more precisely, in this case, for a platypus' bill). In general, universal attributes are attached to universal kinds, so that instances of those kinds can inherit particular instances of those attributes.

To understand the overall organization of the concept hierarchy is to appreciate the difference in meaning between the [1[is a kind of]2] and [1[is an instance of]2] relations. The English phrase "is a" typically denotes [1[is an instance of]2] as exemplified in "Polonius is a platypus" or "platypus is a species (of egg-laying mammal)". However, in colloquial usage "is a" sometimes denotes [1[is a kind of]2] as in "a whale is a mammal". [1[is a kind of]2] always denotes a relation between two universal concepts. Disambiguation usually relies on context and background knowledge.

The hierarchy ideally organizes concepts in an effective and efficient manner.[144] However, the top-level epistemological categories and relationships are subject to change to accommodate new understanding, just as lower level domain concepts are. One would expect fundamental paradigm changes in epistemological knowledge to be fewer and farther between than empirical reinterpretations of domain knowledge, but no knowledge is sacrosanct. More detailed specifications and justifications for the

[144] The MultiNet project, for example, has developed automatic adaptive procedures for doing this: (http://pi7.fernuni-hagen.de/forschung/multinet/multinet_en.html).

current top-level concepts are provided in Chapter 6: *Explanations for the Top-Level Epistemological Concepts*, and they are somewhat different from those displayed in the simplified figure below.

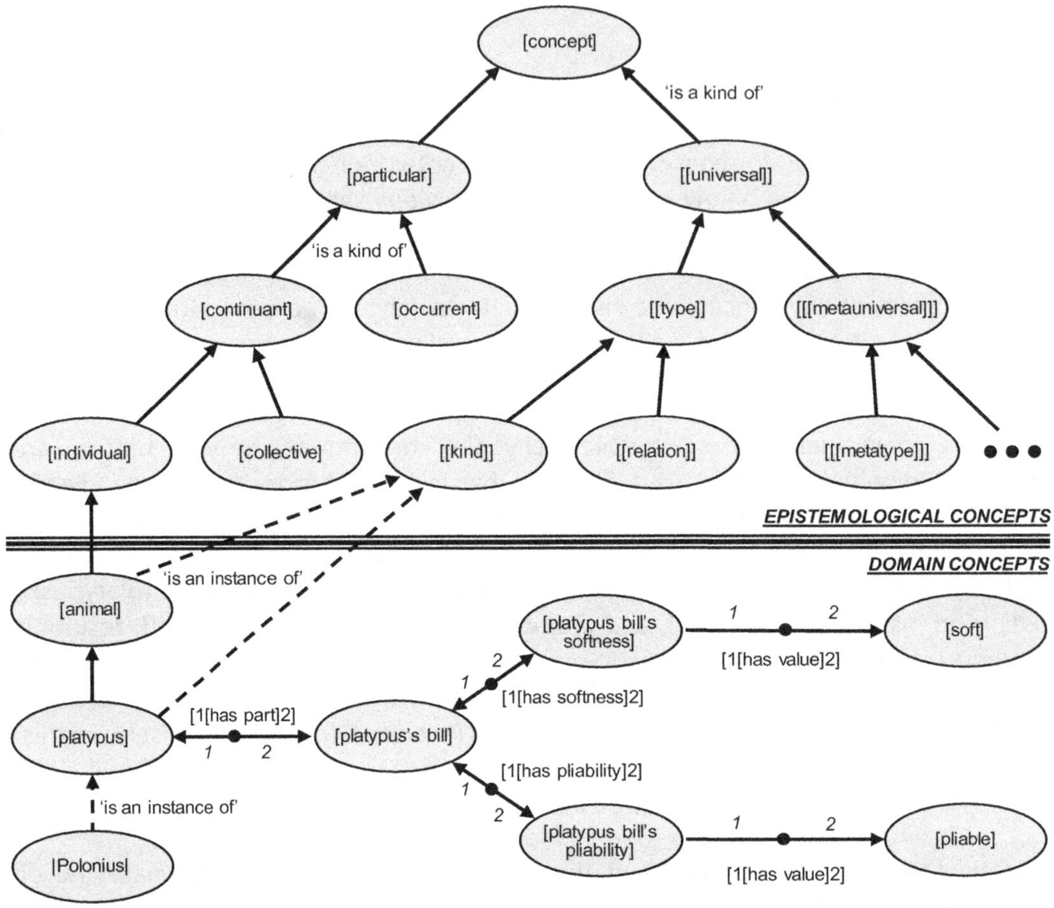

Figure 23: Expanded Concept Hierarchy

The Metaconcept Hierarchy

It is interesting semantically that the entire right side of the concept hierarchy tree (beginning with [[universal]]) forms a kind/instance hierarchy, just like the left side (beginning with [particular]). Double-bracketed abstract concepts like [[universal]], [[type]], [[kind]], and [[relation]], are all instances of an even more abstract concept, [[[metauniversal]]], in the same way that single-bracketed concepts like [particular], [continuant], [animal], are instances of [[universal]].

Reference to metauniversals in everyday speech is rarer than for universals, but still frequent enough—eg, [[virtue]], [[profession]], [[species]]. They are also commonly used for representing knowledge about theoretical topics, such as those employed in scientific classifications.

To help clarify the basic idea of *abstraction*, review again the concept [[species]], as used in biological classification systems. [[species]] has instances like [platypus], while [platypus] has instances which are particular individuals like |Polonius|. The concept [[species]] itself is an instance, directly or indirectly, of [[[metauniversal]]] from which it inherits its basic epistemological characteristics—eg, that each instance must be a universal concept. [[species]] is also a kind of [[biological kind]], along with [[genus]], [[phylum]], etc. The path from "[platypus] 'is an instance of' [[species]]", "[[species]] 'is a kind of' [[biological kind]]", "[[biological kind]] 'is a kind of' [[kind]]", "[[kind]] 'is a kind of' [[type]]", "[[type]] 'is a kind of' [[universal]]", "[[universal]] 'is a kind of' [concept]" is one example of a metaconcept kind/instance hierarchy.

The totality of concepts represented in a large concept hierarchy covering abstract metaphysical as well as everyday knowledge also can be viewed according to the level of abstraction exhibited. At the lowest level of abstraction are particular concepts, like |Polonius|, which can have no instances. Such particular concepts, in turn, are all instances of universal concepts, like [platypus], which are at the next level of abstraction. Universal concepts themselves are instances of metauniversal concepts, like [[species]], at a still higher level. Finally, concepts like [[species]] are ultimately instances of the concept [[[metauniversal]]]—and indirectly also instances of [[universal]], since a [[[metauniversal]] is a kind of [[universal]]. The number of abstraction levels actually represented in a concept network is a matter of practicality, but is theoretically unbounded—eg, extending to [[[[metametauniversal]]]] and beyond. The human capacity for grasping the semantics of abstraction seems limited to about four levels. Another way to diagram the concept hierarchy, in lieu of a single giant downward branching tree, is as a series of abstraction hierarchies, where concepts at any level are instances of concepts at the next higher level, as shown in the figure below. Each level then encompasses a subtree of the original concept hierarchy limited to concepts at that level.

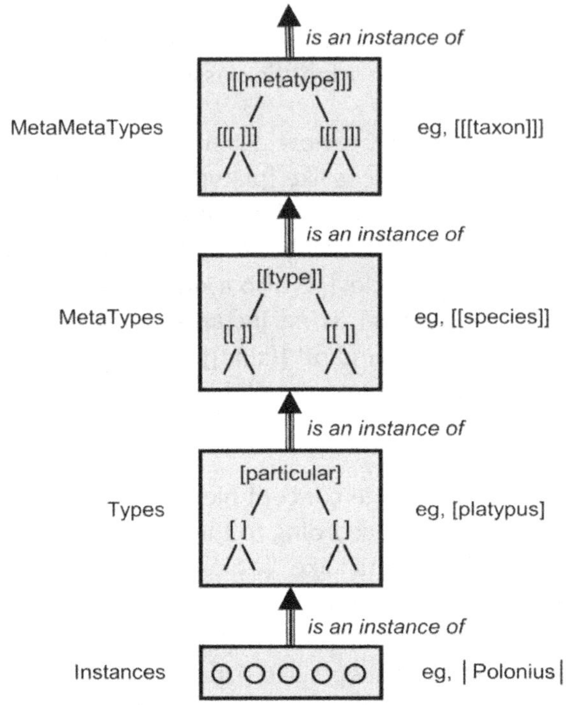

Figure 24: Hierarchy of Metaconcepts

Particular Characteristics of Universal Concepts

Most philosophers who study metaphysics seem to agree that various *particular objects* exist in the world as we experience it. And in this text, particular concepts are said to usefully correspond to those particular objects. What is more contentious, however, is whether or not there are *universal entities* in the world that usefully correspond to universal concepts. Does there exist a universal "redness" of which some particular objects (the red ones, and not others) are instances? In other words, is there, in addition to this particular apple and that particular fire engine, some naturally occurring redness entity, phenomenon, or type which manifests itself in both items, such that they appear red to us? Since we are so readily able to classify particular things according to such shared qualities, perhaps those universal entities are also real. Realism (Armstrong, 1989) claims there are both particular objects and universal entities (ie, natural classes or types or kinds) in the world, while Nominalism (Quine, 1953) claims there are only particulars. On either account, some explanation of what inheres in particular objects that allows us to group them together into a class according to a shared characteristic is needed. However, these are metaphysical issues (concerning what actually exists), rather than conceptualization issues (concerning how we represent things in our minds, whether those things exist in the world or not—the subject of this text). The distinction is important here because conclusions about potential universal objects do not necessarily apply to our concepts about those objects.

Speaking only about conceptualization then, how do we intuitively understand the meaning of particular and universal concepts? The default approach taken here is to say universal concepts like [apple] or [dragon] never correspond to worldly entities. They are simply convenient mental categories (ie, classes or kinds or types of particular things), and as such, are always abstract. Particular concepts, however, can straightforwardly represent particular objects, like |Arthur Conan Doyle| or |Sherlock Holmes|, whether they exist or not as specified by their existential property (previously discussed in Chapter 3: *Concepts & Existence*). But if a universal characteristic like redness is abstract, then how can it participate in the concreteness of a particular physical object like my neighbor's car?[145] It would intuitively seem that a particular object should be defined by its particular characteristics. The suggestion here is to adopt a form of Trope[146] Nominalism (Williams, 1997), and make all characteristics of particular concepts particular themselves (as opposed to universal characteristics), so the redness of this particular apple is a particular redness distinct from the redness of any other red thing. All conceptual representation then cleanly divides between particular concepts with only particular characteristics, and universal concepts with at least one universal characteristic (thereby enabling universal concepts to have particular instances satisfying their universal characteristics).

This way of looking at the foundations of conceptual representation has consequences for how we use and interpret concepts. The basic decision of whether to use a particular concept or a universal concept to model or talk about something, will then depend on whether we think about that something as an individual particular thing, or as a category of things. At first glance these two options may seem incompatible for representing a single object, but it appears they can sometimes blend together. This kind of issue has come up in biology, where there has been some equivocation regarding the interpretation of the concept of species.

The controversy over whether to treat species as an abstract category or as a concrete individual thing has been going on for some time in evolutionary biology. The standard interpretation has favored thinking of a species as a kind or class (ie, abstract category), where the members of the species are particular individual instances of the class. But another way to conceive of a species is as a concrete (distributed) population of organisms regarded as parts of a whole, which together constitute the species. Stephen J Gould has compiled a list of criteria for deciding if something is a spatiotemporal biological individual vs a class of individuals (Gould, 2002). In an "individualistic" (ie, particularized) interpretation of a species, his central idea is that organisms are parts of or constituents of (not elements of or instances of or members of) that dispersed individual:[147]

---

[145] However, a realist will claim that the universal redness is not, in fact, abstract, but real. Yet it is then claimed that the single real redness is located in multiple places with each red object at the same time, making it a rather unusual kind of real thing.

[146] In metaphysics, a trope (Daly, 1997; Mellor & Oliver, 1997) is an elementary particular property of a complex particular object. For a physical object, its tropes are also physical, giving the physical object its causal propensities. Unlike a universal, each particular trope is unique for each particular object.

[147] Whatever a species is, it cannot be a simple *set* of individuals. A set is a different set if it changes membership. However, a species remains the same species in spite of births and deaths.

- An individual has a discrete and definable beginning. However, the precise starting point may be indeterminate.
- An individual has a discrete and definable ending.
- An individual has sufficient stability to merit continuous recognition as the same individual.
- A prototypical individual maintains a coherence of substance over time.
- A prototypical individual maintains a constancy of form over time.
- An individual may undergo change, but not so much so as to encourage redefinition.
- An individual's parts do not readily merge with parts of another individual.
- An individual maintains distinguished boundaries with its environment and other individuals of the same kind.
- An individual maintains a continuity of existence such that it does not fade in/out of existence over its lifetime.
- An individual has a continuous historical continuity.
- An individual's parts work together to produce an overall functioning/coherence. However, a thing which fails to do anything, or play any significant role, could still be an individual.

The intuitive concept of a species would seem to fit his criteria quite well, particularly if some causal criteria are added. These kinds of concrete properties, however, are difficult to reconcile with the traditional abstract class concept of species, because abstractions traditionally cannot be concrete. Using concept net notation, interpreting each species as a particular means that |platypus| is an instance of [species], where [species] is a standard level-1 universal concept and a kind of [biological group] (ie, a collective). (Recall in the standard universal interpretation of species, [platypus] is an instance of [[species]], a level-2 metauniversal concept and a kind of [[biological kind]].) Furthermore, |Polonius| being a particular platypus, is then a member of (rather than an instance of) the species |platypus|, represented as a particular group. In this account |platypus| and [platypus] represent different concepts, and which one best models the notion of species is a scientific/empirical matter. Spatiotemporal characteristics are a clue that the concept may not be abstract. In this case, since we want to talk about the species platypus as having a beginning perhaps 500,000 years ago, or having a range at this time somewhere in Australia, a particular interpretation seems appropriate.

However, we also want to talk sometimes about the kind [platypus] as an abstract entity possessing abstract characteristics which its concrete instances can inherit. By default |Polonius| and others of his kind lay eggs, have soft pliable bills, etc, because we understand that Polonius' characteristics are inherited from the abstract characteristics attached to the abstract concept [platypus]. Inheritance flows from abstract types to their instances, as the species [platypus] itself does not lay eggs, have a soft pliable bill, etc. Inheritance does not flow from a whole particular individual to its parts (or from a group to its members), as would be required if the species platypus were a particular.

There are at least two confusing issues here. First, the use of terms like "concrete" and "abstract", instead of "particular" and "universal", introduces real-world material properties into a discussion about mental representation. The distinction is akin to the earlier distinction between issues of metaphysics and issues of conceptualization, or between a physical representation of a concept in our brain and what the representation represents, which may or may not be physical. Concrete objects are usually taken to be those that exist, or could exist, in spacetime (or that are subject to causality); while prototypical abstract entities, like numbers and sets of objects, are not. However, some concepts like "platypus" (particular individual or kind) or "these three rabbits" (particular group or set) are at least plausibly both abstract and concrete. The distinction between a particular concept and a universal concept, on the other hand, is a distinction between being an individual thing and being a type (or category) of individual thing, whether it is spatiotemporal or not—a more fundamental epistemological difference. In discussing the epistemology of conceptual content, it seems better to focus on particulars and universals, rather than on the more empirical determinations of concreteness and abstractness.

Second, the conflict between the two uses of "platypus" above may only be apparent, since the characteristics attached to a universal concept can come in two forms. Both can be accommodated in a concept net.[148] The first form characterizes the universal concept as beliefs about the category directly, while the second form characterizes the concept's instances as beliefs to be inherited by those instances. A concept like [platypus] possesses both forms, and therefore harbors both particular characteristics (about the category) and universal characteristics (about its instances). But the two coexist peacefully. In this case [platypus] carries universal characteristics like [1[lays eggs]], and makes them available to its particular instances like |Polonius|. But [platypus] also inherits its own particular characteristics like |1|species' range|2| from a metauniversal like [[species]], which carries the universal characteristic [1[species' range]2] for all species. The point here is that in a concept net, a species concept like [platypus] can include both viewpoints at the same time with no contradictions. A separate concept of |platypus|, as a particular concept, is not needed, as long as such universals as [platypus] can possess some particular characteristics (even including spatiotemporal, material, causal, etc, characteristics) along with their other universal characteristics. Additional details of concept net organization to support concepts like [platypus] are shown in the following figure. This "blended" way of representing and understanding universal concepts in a concept net avoids the massive duplication of concepts (eg, [platypus] and |platypus|) that would otherwise be required, since virtually all universal concepts will have both particular (category) as well as universal (instance) characteristics.

[148] However, see Lowe for an alternate metaphysical position (Lowe, 1997). He believes it is a mistake to conflate the two concepts, since, for him, universals cannot have *any* particular characteristics.

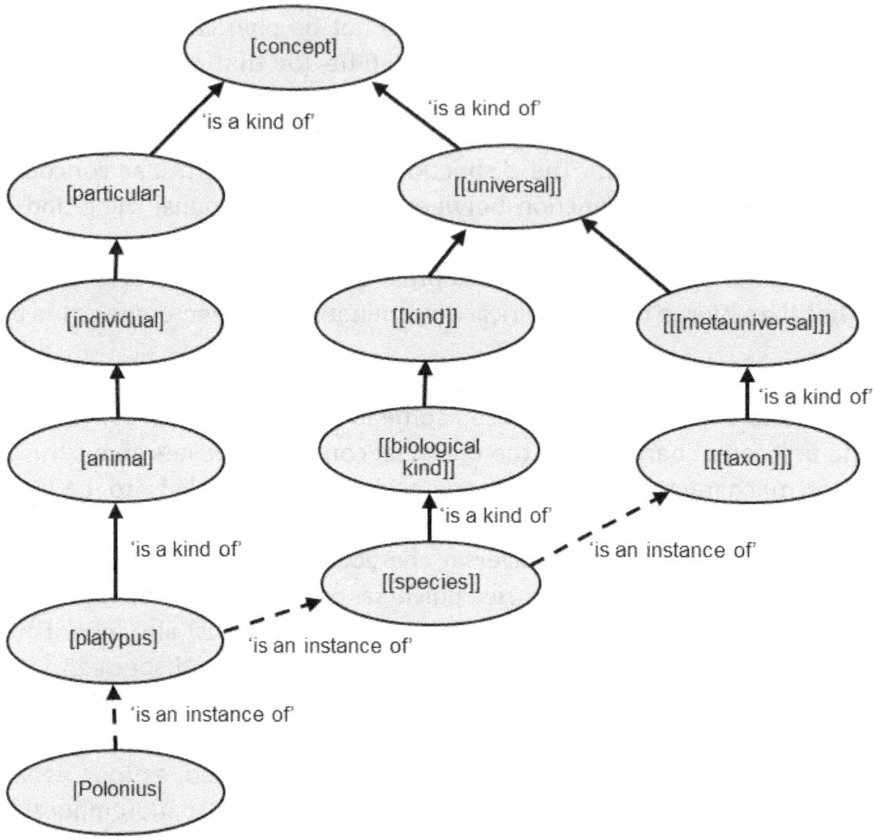

Figure 25: Representing the Species Concept [platypus]

Relations

In addition to the more-or-less "vertical" organization of concepts in the kind hierarchy, knowledge is also organized "horizontally" via characterizing *relations* between the universal concepts in the hierarchy. A characterizing relation describes a potential state-of-affairs that can hold among the instances of those universal concepts. It bridges universal concepts in the hierarchy by linking them to itself via associations serving as arguments to the relation. Placing a relation node like [[platypus][has part][platypus' bill]] in a concept net (often represented as a small dot in a diagram), joins that node with other nodes in the hierarchy—namely [platypus], [platypus' bill] and [has part] in this case. The characterizing relation node itself is part of the hierarchy too, since it is a concept; it is an instance of a [[relation]].[149]

[149] This definition of a relation as a universal concept mirrors the technical definition of a generalized relation in mathematics as a set of n-tuples. In logic, however, a relation is usually identified with a predicate. In this text the terms "predicate" and "relation" have different meanings: a predicate is a component part of a situation (see Chapter

A relation itself may have instances, which are broadly termed *situations*. Situations are the particular states-of-affairs that are authorized by the relation, where all variable arguments of the parent relation have been instantiated. Thus ||Polonius| |has part| |Polonius' bill|| is the situation concept modeling the particular connection between Polonius and his bill, authorized by the universal relation [[platypus][has part][platypus' bill]]. In this account a situation concept corresponds to an *atomic proposition*, as a logical form. The corresponding proposition is true or false depending on whether the underlying situation holds or not in the actual world.[150]

All binary relations (including relations representing primitive associations) come in pairs, such that one is the *converse* of the other. Whenever 1R2 is the forward relation from concept C_1 to C_2, then the converse 2R'1 is the reverse relation from C_2 to C_1. For example using shorthand notation, [2[has kind]1] is the converse of [1[is a kind of]2], [2[has instance]1] is the converse of [1[is an instance of]2], and [2[is loved by]1] is the converse of [1[loves]2]. Thus, if "[collie] 'is a kind of' [dog]" is believed, then so is "[dog] 'has kind' [collie]". Both mean the same thing, and usually only one or the other is represented—the other can be derived if needed. Care is sometimes needed to identify converses. Typically, the converse of a relation will have "of" appended, although not always—eg, the converse of [2[is greater than]1] is [1[is less than]2], and the converse of [1[has converse]2] is [2[has converse]1], since it is a symmetric relation.

Each relation also has an *obverse*, which is the negation of the original. Thus, [1[has kind]2] and [1[~has kind]2] are obverses, as are [1[is a kind of]2] and [1[~is a kind of]2]. Negation is occasionally needed to explicitly assert when some relationship does not hold: eg, "[whale] '~is a kind of' [fish]". It is useful to include obverse relations in the basic relational categories. For example, [1[+is an instance of]2] and [1[~is an instance of]2] together form a relational category notationally depicted as just [1[is an instance of]2]. [1[is an instance of]2] authorizes both positive and negative uses of the relation.

Relations in a concept net can also be standardly typed—eg, as reflexive, symmetric, transitive. Such specifications act as global constraints on how relations are used, and allow the automatic detection of logical and selective restriction errors during consistency checking of a concept net. Some of the standard relation specifications would include reflexive, irreflexive, symmetric, antisymmetric, transitive, and functional (indicating the relation takes only one value).[151]

Relations like [[[universal]][is a kind of][[universal]]] and [[concept][is an instance of][[universal]]], which were seen to form part of the kernel, are distinctive in that they are used to structure concept nets. Most relations, however, have nothing to do with the organization of knowledge itself. Compare with

---

3: *Generalized Situations*), while a relation is a concept where at least one argument is a free (unquantified) variable (ie, a universal)—with allowance made for relations with no arguments at all.

[150] The logical interpretation of a proposition as a well-formed formula seems consistent with its role here in characterizing generalized situations (ie, facts or counterfactual situations). However, this kind of standard role has been questioned lately. Other proposals argue that propositions may be event types (Soames, 2010) or relations with no arguments (Zalta, 1988).

[151] These most useful relation types are implemented in the OWL Protégé 4 program for building ontologies (Horridge, 2009).

the domain relation [[person][loves][person]], interpreted as the relation of loving between two persons, instantiated in "Drea loves Nynne". Universal relations and their particular instances (namely, situations) are used to encode knowledge about virtually any topic of interest.[152]

In a concept net the existence of a relation concept is interpreted as a specification for how the relation can be used. First, each relation concept is associated with zero or more universal concepts serving as arguments. A binary relation like [1[loves]2] (using shorthand notation) takes two arguments (the lover and the loved), [1[gives]23] takes three arguments (the giver, the gift, and the beneficiary), [1[laughs]] takes one argument (the person who laughs), and [rains] as in "it rains" or "it is raining" takes zero (no arguments).[153] Moreover, the relation specification authorizes certain uses and excludes others—eg, the loving relation above authorizes relation instances like "Drea loves Nynne", but excludes "The Crab Nebula loves Joe" because The Crab Nebula is a celestial object, not a person. Each argument associated with a relation incorporates such selectional restrictions, represented as concept categories, like [person], in [[person][loves][person]], specifying the loving relation between persons. Another way to think about a relation node is as expressing a *disposition* governing its instances. Thus [[salt][dissolves in][water]] expresses the disposition for particular samples of salt to dissolve in particular samples of water under normal circumstances, even though under "abnormal" circumstances a particular sample will not so dissolve—eg, if the water is already saturated. Default behavioral dispositions can be represented similarly—eg, as [[robin][flys]] indicating that robins normally fly. The dispositional interpretation of a relation in a concept net allows the straightforward representation of scientific laws as well as general default reasoning, since particular instances of the relation, while presumed, may or may not actually hold.

The remaining formal element of a relation typically occupies the second position, which indicates a *relationship*. A relationship is interpreted as a *truth function* for the relation. In the example the truth function is simply the concept labeled [loves]. A relation's truth function specifies the truth conditions that determine whether various instances of the relation hold or not. The truth function for many relations may not be accurately known, but we seem to learn such cognitive functions as a matter of course, based on numerous intuitive, but fairly reliable cues. For example, we can routinely tell whether two strangers are in love just by observing them together. A truth function has only two possible instances, assuming standard two-valued logic is modeled in the concept net—eg, "loves" or "~loves", corresponding to whether the modeled situation is true or not. In this case, the situation modeled conceptually is either "Drea loves Nynne" or "Drea ~loves Nynne". If there is insufficient evidence for either, the situation is not modeled at all in the concept net. The assertion of a statement like "[person][loves][person]" into the concept network adds the appropriate nodes and links to specify the relation.

---

[152] "Drea loves Nynne" is shorthand for ||Drea| |loves| |Nynne||, a particular situation concept made up of three particulars. A discussion of generalized situations is put off to the next chapter.

[153] A zero argument relation like "it is raining" can be instantiated in a particular situation specifying the time and place of the event.

To see the specifics better, consider the effects of asserting "[concept][is an instance of][[universal]]" from the prior discussion of the kernel. There are two arguments, [concept] and [[universal]], and a relationship [is an instance of], all of which are components of the relation. The figure below diagrams the details of this relation containing three new association links: 'has argument1', 'has argument2', and 'has relationship'.

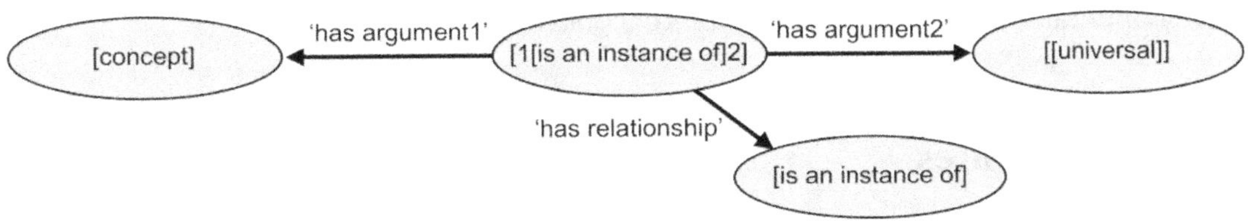

Figure 26: Representation of the relation "is an instance of"

This network fragment is abbreviated to the following, when the relation has a standard interpretation:

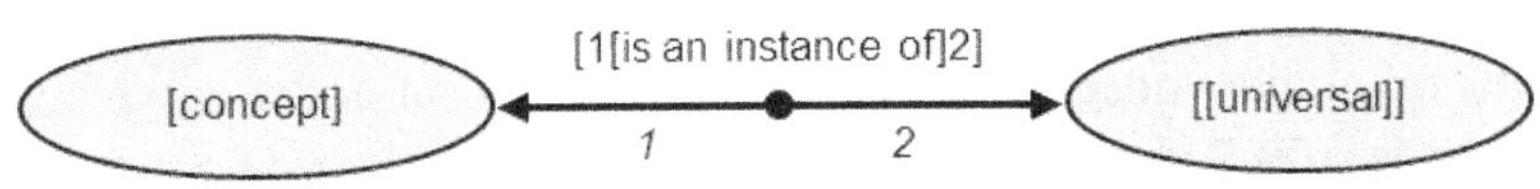

or even more succinctly, when the direction is clear:

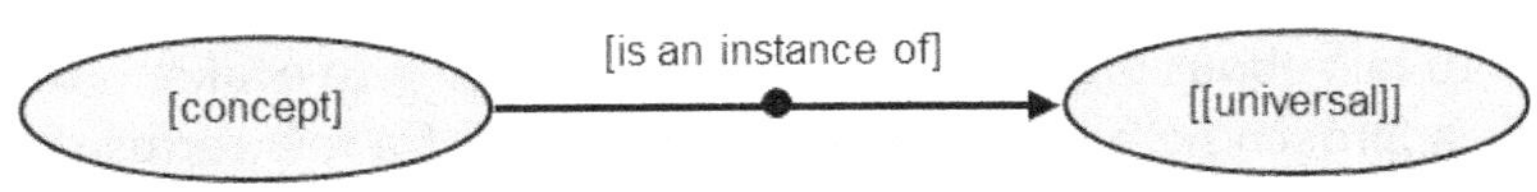

Figure 27: Diagrammatic Abbreviation of a Relation

The earlier kernel statement "[concept][is an instance of][[universal]]" thus expands into the primitive statements:

"[1[is an instance of]2] 'has argument1' [concept]",
"[1[is an instance of]2] 'has argument2' [[universal]]",
"[1[is an instance of]2] 'has relationship' [is an instance of]",

plus the associations that tie the relation and relationship into the existing network:

"[1[is an instance of]2] 'is an instance of' [[relation]]",
"[is an instance of] 'is an instance of' [[relationship]]",

plus the statements explicitly adding the three new associations as relations:

"[[relation]][has argument1][concept]",
"[[relation]][has argument2][concept]",
"[[relation]][has relationship][[relation]]",

each of which themselves need to be expanded into primitive associations! Thank God computers thrive on keeping track of this sort of thing. Actually, our own neural networks seem to keep track of a dizzying array of details, of which we are completely oblivious.

Properties & Attributes

In a concept net each concept becomes characterized in terms of its links to other concepts. Those links can be thought of, loosely, as expressing the intrinsic properties and extrinsic relations the concept carries—its total semantic characteristics. While a concept's relations describe its connections to other concepts in a more-or-less unrestricted fashion—eg, in asserting "robins eat worms" the concept of a [robin] becomes linked to the concept of a [worm]—a concept's properties usually characterize a more self-contained quantity or quality describing the concept itself. For example, when we say "normal robin body temperature is 109.7 degF" the concept [109.7 degF], as a kind of quantity, is used to intrinsically characterize the concept of [normal robin's body temperature]; and in "male robin breast color is reddish-brown" the concept [reddish-brown], as a kind of quality, intrinsically characterizes the concept [male robin's breast color].

This basic idea of intrinsic properties is relatively uncontroversial in philosophical studies, in spite of a wide divergence of views on exactly what properties are, extending all the way back to Plato. However the terminology used to talk about them is a mess. Without trying to evaluate the many terminological nuances involved, the approach here is to define the meaning of a few terms related to properties, so we can talk unambiguously about the relevant semantic distinctions.

Probably the most generic use of the term "property" has already been mentioned: it contrasts with the term "relation" in describing an intrinsic vs extrinsic characteristic of something. According to this approach, a property is generally regarded as an intrinsic characteristic of one thing, while a relation is an extrinsic characteristic involving more than one thing, exemplifying the relationship among those things—contrast one person laughing (interpreted as a temporary property of a person) with two people holding hands (interpreted as a relation between people). Alternately, in a widest possible interpretation, a property simply corresponds to the predicate phrase in any sentence describing some subject term, in accordance with the universal subject/predicate structure of language.[154] However, a property may also imply some scale of values by means of which something is quantitatively measured or qualitatively assessed. To say that "Phoebe is 14 months old" is to associate her with a particular value on a quantitative scale of ages, while to say that "Phoebe is beautiful" is to associate her with a particular (but imprecise) value on a qualitative scale of beauty. In these sentences either age and

[154] But identifying properties with unrestricted linguistic predicates is known to produce logical paradoxes, the simplest of which involve predicates that refer to themselves.

beauty, or 14 months old and beautiful might be regarded as properties (among yet other proposals) depending on point-of-view.

As seems evident, these various intuitive meanings of "property" are at cross-purposes. To settle on one sense here, properties will be grouped with relations, signifying that properties just *are* relations that take only one argument. This definition is consistent with the characterization of relations in the previous section, where a generalized relation can take any number of arguments—even zero arguments. A property, then, is simply a relation involving one kind of thing. As a relation, it is a universal (ie, not a particular thing), whose one argument is a universal. Thus, "men are mortal" expresses the property [[man][is mortal]] characterizing the universal kind [man]. Moreover, a property's instances are particular situations characterizing particular things. "Socrates is mortal" expresses the property instance ||Socrates| |is mortal|| which characterizes the particular |Socrates|. A property and an instance of a property (too easily conflated) have a different cognitive significance, because properties are universal abstractions, and instances of a property are particular situations. The same ambiguity is present for relations, in general, requiring some careful, if tedious, attention to terminology. That Maia and Phoebe are sisters, therefore, does not signify an abstract relation between them, but rather a particular situation involving them. That Maia is musical and that Phoebe is athletic likewise represent particular situations, which are instances of properties specifying that persons, in general, can be musical and athletic.

Another common, but confusing, meaning of the term "property" involves the use of, what will here be called, *attributes*. An attribute typically relates some kind of object to some kind of value. In saying that "normal robin body temperature is 109.7 degF", the attribute is the universal [temperature], where the concept of [temperature] is an instance of [[attribute]]. Similarly, for the sentence "stop signs are colored red", where the attribute is [color]. However, observe that attributes are often latent in colloquial expression if they are readily inferred from context, as in "stop signs are red", where the attribute [color] is implied. The explicit representation of implied attributes in a concept net is optional, but may enable a deeper analysis of the information provided in a statement by allowing comparisons with other attribute values. Given that an attribute like [color] is a universal concept, it can have conceptual instances like |the color of this particular stop sign|. Such instances of attributes have sometimes also been called properties, and they will be given the technical name of *attribution*, to avoid ambiguity. (Attribute instances, as unique particular properties, are commonly called "tropes" in the philosophical literature to distinguish them from attributes, as sharable universal properties.) Note that attribute concepts like [temperature], [color], etc, are then various kinds of [attribution], which all are instances of [[attribute]].

As mentioned previously, properties are also frequently associated with a quantitative or qualitative scale or set of *valuations,* commonly associated with attributes. But this use of the term "property" again conflicts with a property being a kind of relation. Valuation concepts like [109.7 degF] or [black] suggest selections from a quantitative scale like [Fahrenheit temperature value], or qualitative set of values like [color value], rather than structured relations with arguments. The universal concepts [109.7 degF] or [black] represent approximate, as opposed to precise, valuations. They are conceptual descriptions intended to cover a range of precise particular *values* like |109.749201... degF| or |Inky's exact shade of blackness|, where a particular value is an instance of a [value]. Under this rubric a universal concept like [109.7 degF] is a kind of [Fahrenheit temperature value], and [black] is a kind of [color value], where all four concepts are kinds of [value], and instances of [[valuation]]. To review the rather subtle high-level conceptual organization required to support quantitative and qualitative properties, a concept like [Fahrenheit temperature value] is an instance of a [[valuation]] but a kind of [value] that can have instances like |109.749201... degF|. A concept like [temperature] is an instance of an [[attribute]], but a kind of [attribution] that can have instances like |the body temperature of this particular robin|. Attributes and valuations are linked to each other through *relations* like [[temperature][has value][Fahrenheit temperature value]]; while attributions and values are linked through *situations* like ||the body temperature of this particular robin||has value||109.8 degF||. Additional details of the attribute-valuation and attribution-value characterizations are discussed by way of a more complete example in Chapter 5: *Attribute Values*.

We seem to intuitively appreciate that each particular physical object requires its own particular unique characterization, since no two physical objects are ever exactly the same. The color of this apple is necessarily slightly different from the color of that apple, even though both are characterized as red. The core body temperature of this person is slightly different from that person, although they may both be characterized as 98.6°F. The particular attributions of an object provide the means of uniquely characterizing each object. Thus, a particular like |Deven| will have only one body temperature attribution, namely the concept |Deven's temperature| or equivalently |the temperature of Deven|. However, this one attribution can be described in many ways—eg, as having a value of |98.6 degF|, as having a value of |37.0 degC| (both of which could be linked by the *equality association*),[155] and perhaps also |normal| (indicating a qualitative assessment). The following diagram represents this example of Deven's temperature.

As a nuanced point of philosophical clarification, a concept like |Deven's temperature| is not to be interpreted as a physical factor *causing* Deven to be the particular way he is. It simply represents a particular attribution which Deven is deemed to have. Attributions are conceptually independent of the particular objects which they characterize, rather than being integral features on which objects depend for their way of being. Other terms of philosophical art related to what here are called attributions, include tropes (Williams, 1997), abstract particulars (Campbell, 1997), and modes (Lowe, 2006). However, these terms come with excess baggage, typically implying a kind of causal power to effect change, or a role in natural physical laws, or at least a physical existence of some kind. The overall theory of concepts put forward in this text is more instrumental in flavor, casting concepts as useful rather than as veridical in some realistic sense. Imbuing Deven with such concrete elements playing causal roles in Deven's being, is a scientific hypothesis requiring further empirical support. If the theory becomes scientifically supported, specific trope, abstract particular, or mode concepts can be integrated into concept nets, along with relations, situations, attributes, and attributions, as described thus far.

[155] Two values are equal, if they are values characterizing the same attribution.

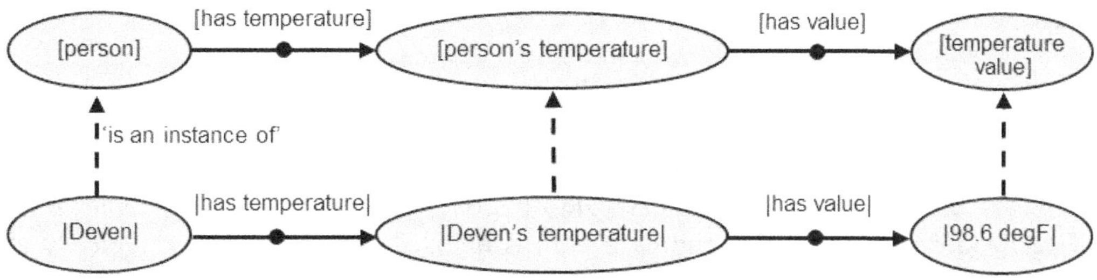

which, when the attributive meaning is clear, is more succinctly expressed as:

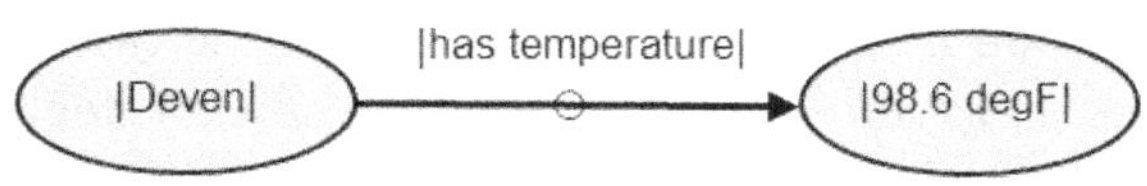

Figure 28: Representing (Object, Attribute, Value) Relationships

Intrinsic characteristics can also chain off of each other, such that one characteristic can itself have characteristics. For example, the physical attribute of [velocity] (of objects) has the further attributes of [speed] and [direction], typically called the components of velocity. Similarly, the psychological attribute of [color] is now known to consist of three subattributes: [hue], [saturation], and [lightness] (or [brightness]). Other complex attributes like [wisdom] are likely grounded in components like [experience], [knowledge], and [judgment], although exactly which components are relevant to a complex attribute may be controversial. These are all cases of an attribute having attributes of its own. This kind of componential analysis of attributes is common in classical physics, where vectors are analyzed in terms of other vector and scalar components. In physics a few primitive scalars (namely, length, mass, duration, temperature, electric current, amount of a substance, and luminous intensity) form the basis for all the more complex derived attributes. Borrowing from the insights of physics then, the concept of [attribution] can be conveniently subdivided into [scalar] and [vector] attributes.

Finally, it is common for some of the intrinsic characteristics for an object to become linked together, due to recognized correlations among the characteristics. We detect many such correlations unconsciously, and our most useful and informative basic-level domain categories represent clusters of correlated characteristics (Rosch, 1975). In particular, causal correlations help us make predictions and understand the underlying nature of things by supporting rational explanation, as when we notice that the hollow bones, wings, and feathers of birds enable flight. The general strategy of characterizing things in various domains of knowledge through attributes and attribute values seems to be a fundamental principle of knowledge organization. It encourages the organization of concepts into type hierarchies, defined principally by intrinsic characteristics.

Instantiation vs Characterization

Predicates in standard predicate logic are normally interpreted as relations, where each predicate takes some number of arguments. Examples of basic predications in logic include expressions like dog(Fido) and furry(Fido). Both expressions have the same syntax. But the linguistic roles of the two predicates, dog and furry, are different. In the first case Fido is being classified as a dog, typically expressed in English as "Fido is a dog". In the second case Fido is being described as furry, typically expressed in English as "Fido is furry". English makes the distinction plain: dog is a noun, while furry is an adjective. But logically, both are predicates. The distinction does not make a logical difference, but it can make an epistemological difference in the way knowledge is organized.

In a concept net the foregoing classification/description distinction is maintained. Classes and individuals (aka universals and particulars) are arranged in a classification hierarchy, while descriptions (aka properties, relations) can be attached to any concept in the hierarchy. The term "instantiation" applies to universal concepts and their instances, as in the case of dog(Fido): Fido is an instance of a dog.[156] The term "characterization" applies to any concept and its description, as in the case of furry(Fido): Fido is characterized as furry. Note that with this understanding, it is a misapplication of terms to say that Fido is an instance of furry (things), or that Fido is characterized as a dog. Classification is kept separate from description to facilitate the organization of concepts according to their characteristics.

The equivalent concept net representations are "|Fido| 'is-an-instance-of' [dog]" and "|Fido||is furry|". The first is a simple association between the two concepts |Fido| and [dog] expressing an instantiation of [dog]. The second is a situation (ie, an instance of a one-place relation whose argument is |Fido| and predicate is |is furry|) expressing a characterization of |Fido|. The different expression formats reflect the different roles these kinds of concepts play in a concept network. However, selecting the appropriate role for some concepts will still depend on the representational requirements for specific tasks. For example, whether to represent the color red (1) as a category (eg, [red]) admitting particular instances of red, or (2) as a predicate (eg, |is red|) for describing red objects, or (3) as both, will depend on requirements.

[156] This use of "instantiation", as a relation between concepts, is to be distinguished from another common usage in metaphysics as a relation between a predicate and reality. In a metaphysical interpretation a predicate is typically said to be instantiated iff there exists some object in the world that satisfies the predicate. This question of whether a concept corresponds to something in the world is a separate issue in a concept net. (See Ch 3: Concepts & Existence)

5. CONCEPT NET REPRESENTATIONS

Having outlined the basic theoretical issues related to developing concept networks in the last chapter, this chapter looks at specific strategies for actually representing the various concepts appearing in those networks. As will become apparent, there are many subtleties to consider in how best to represent the meaning inherent in a concept. Building effective knowledge representations will accordingly always be a work-in-progress. But the theoretical framework already established does provide enough guidance to begin extending the baseline toward a consistent and useful content covering some major core domain concepts.

The Concept of the Material Object

The notion of a cohesive three-dimensional medium-sized bounded whole material object (cat, stone, chair, book, etc) is one of our most familiar and deeply ingrained concepts. Most of our daily interactions with the environment involve perceiving and acting on such objects, and we have developed rather sophisticated strategies for dealing with them. Our object concept representations reflect this sophistication. Technically, this kind of object is often called a *sortal*, signifying a numerically distinct persisting kind of entity whose identity can be reliably tracked over its lifetime.[157]

Our intuitive grasp of the nature of material objects is complex and seems to span multiple levels of innate, perceptual, and conceptual representation. The ability to segment and track a perceptual figure multimodally over time (using mainly the visual, aural, or haptic senses), is certainly a deeply held aptitude, probably underlying the more advanced notions of object permanence and numerical distinctness from other objects. Put differently, successful tracking necessarily presupposes some degree of permanence and at least intermittent individuation of a perceptual figure (see Appendix B: *Early Perception* for discussion of perceptual figures). Experiments with infants show these faculties being expressed very early at 2-5 months (Aguiar & Baillargeon, 2002; Wynn, 1992). However explicit memory traces, developing later, are required to *conceptually* reidentify objects over extended periods of time.[158] The information derived from segmentation and tracking has been generally classified as *spatiotemporal information* by early childhood researchers, and is thought to deliver our more instinctive material object characterizations such as that two objects cannot occupy the same space at the same time, that one object cannot be in two places at the same time, that (solid) objects cannot pass through each other, that objects require support, that objects do not spontaneously break apart or

[157] We can conceptualize any given material object in any number of ways depending on our purposes or point of view. The very same object can be thought of as a statue, a doorstop, a lump of clay, or a totem, where each concept representation exhibits different characteristics. For example, shape is salient for [statue], [doorstop], and [totem], but not for [lump of clay]; animal or plant symbolism is salient for [totem]; etc. Also, note that the term "sortal" denotes a kind of object like [dog], not a particular object like |Fido| or a relation like [1[has part]2].

[158] Learning to group repeated occurrences of an experience (eg, Mama experiences) through conditioning is not the same as explicitly representing Mama as an enduring object (Quine, 1960).

merge, and that objects do not influence each other at a distance—eg, as agents do (Spelke, 1990). Spatiotemporal continuity thus underlies all of our generic object concepts derived from waking experience. However, spatiotemporal aspects of objects do not usually become explicitly represented in the conceptual system until later in childhood, when they take on more of an abstract theoretical functionality.[159] Some examples of underlying core domain spatial characteristics of objects, described from a conceptual viewpoint, would include:

- position (ie, where the object appears in an egocentric, and later, allocentric reference frame)
- direction (ie, the main axial direction of the object)
- distance (eg, between objects)
- spatial region (ie, the place or space occupied by the object—the basis for shape and size assessments)
- dimensional characteristics (eg, 2D surfaces of objects, 1D edges, etc).

Analogous core domain intuitions support the representation of temporal characteristics of objects (such as time of occurrence of changes and time intervals), and of combined spatiotemporal motion characteristics (such as object speed, heading, and rotation).

An appreciation for the *perceptual qualities* of objects (color, shape, etc) seems to develop alongside spatiotemporal intuitions, perhaps stabilizing at around 12 months, when perceptual observations eventually become able to override our more deeply held spatiotemporal intuitions in object classifications (Xu & Carey, 1996). Perceptual qualities summarize a broad range of useful information about everyday objects, as is evident from the following partial list, where each quality is typically a complex organization of interrelated facets:

- size (eg, approximate circumscribed volume, length, width)
- shape (ie, boundary including symmetry of form)
- location (eg, with respect to other reference objects, center of gravity)
- color (ie, values of hue, saturation and lightness over surface regions)
- texture (both color pattern and haptic rough/smooth texture of surface)
- orientation (eg, with respect to gravity)
- weight (including inertia and force dynamics)
- parts (ie, as elements of a whole plus their spatial arrangement and orientations)
- features (eg, the two ends of a stick or the dent in a car)
- motion (ie, speed, direction, and acceleration of movement plus path)
- sounds (eg, train's whistle including pitch, loudness, timbre and dynamics)
- smell (eg, aroma of a rose)
- temperature (eg, the coldness of an ice cube)

---

[159] Closely related to the spatiotemporal properties of an object, and probably derived from them, are its causal properties (Carey, 2009). Causal properties seem to characterize situational and relational aspects of objects based on their actions and effects on other objects.

Perceptual qualities are more readily translated into discrete conceptual characteristics than spatiotemporal intuitions, but the perceptual richness of qualities is of course lost in a characterizing description.

Given that perceptual qualities can only provide information about the appearance of objects, we can also add virtually unlimited *conceptual characteristics* for identifying the *kind* of material object before us, characteristics which convey additional information about what the object is. For example, we use such characteristics to distinguish rigid objects (like a chair) from articulated objects (like scissors), gelatinous objects (like Jello), deformable objects (like a beach ball), malleable objects (like a play-dough bracelet), and living objects (like a cat). Our appreciation for kinds of objects, as opposed to purely spatiotemporal objects (eg, regions) or perceptual objects distinguished by qualities, also begins appearing at about 12 months of age (Carey, 2009). Typically, kinds possess latent characteristics requiring some degree of conceptual sophistication to grasp, although some perceptual qualities (particularly shape) often play a role in characterizing various kinds of material object. Here is a short list of such latent characteristics that probably develop mainly as conceptual representations:

- composition (ie, what substances the object is made of)
- solidity (ie, the relative constancy of shape and impenetrability of boundary)
- origin (ie, where the object came from)
- hidden parts (eg, that a cat has a stomach)
- function (eg, tools and parts)
- classification in general (all manner of other distinguishing characteristics—eg, age for animals, pressure for closed containers, or resolution for a computer monitor)

Almost all of the material objects we regularly encounter become automatically encoded in all three kinds of representation—viz, spatiotemporal, perceptual, and conceptual. The concept net representations discussed in this chapter express knowledge in terms of beliefs about many different kinds of objects, where the relevant characteristics vary from concept to concept. For example, the characteristics of events (discussed later) will differ substantially from those of material objects.

The Psychological Basis of Material Object Representation

As discussed above, higher-level conceptual representations of material objects are built on the underlying perceptual object schemas and intuitive core domain models already present in early childhood. Although the physiological details of how the new conceptual representations grow out of the more primal representations remains somewhat clouded, a certain conceptual structuring is clearly evident in our everyday thinking about material objects. Material object representations are psychologically well-structured, well-organized, and well-adapted to classifying ordinary environmental objects in specific ways according to our needs. As a result, our object judgments are generally quite rapid, appropriate, and accurate in most circumstances. This is not to say that we cannot think about objects in a variety of different ways, but our basic understanding of material objects is, by default, organized in a particular way honed by evolutionary selection. Such organization processes are at work when we conceive of objects as belonging to various kinds (the type/token distinction), or when we think of objects as having characteristics (the object/property distinction) or parts (the whole/part distinction). It is hard to "think" about something that is not a kind of something, or about something which has no characteristics or parts, because these organizational principles are built into our way of looking at the world. But there is an even more primitive basis for conceptualization (and thought in general) that gets at the fundamental meaning of what it is to *be* a material object.

The psychological foundation for conceiving of a material object seems to revolve around the pure spatiotemporal location of a perceptual figure, divorced from any characterizing properties other than its perceived position in mental spatial and temporal frames of reference. The spatiotemporal perceptual processes, in essence, disregard other ancillary perceptual object information to extract a focal point-like representation of the figure. This "naked particular" subsequently serves as the perceptual object coat tree on which the figure's perceptual object characteristics are hung. The corresponding conceptual representation of the object, incorporating the point-like representation, then becomes the source of our intuition that an object can undergo sometimes dramatic changes, but still remain the same object. The object's naked particularity does not change. We normally don't notice the distinction between the naked particular and the whole perceptual object representation, because they are constructed together. However, psychological experiments are able to separate the two representations during multiple object tracking and object search tasks (Pylyshyn, 2003) and during tests of object behavior expectations in infants (Carey & Xu, 2001). These experiments correlate nicely with the physiologically well-established "where" (spatiotemporal) and "what" (property) information processing pathways in visual and auditory cortex.

From a strictly logical point of view, the naked particular directly represents the material object, to which we can assign a name or point to via a variable in a quantified proposition. For example if I say "Circe is a black cat", formally symbolized as the proposition *black(Circe) & cat(Circe)*, "Circe" is the name (ie, the representation) I use to signify that thing to which the characteristics black and cat are attached. Similarly in "Some cat is black", formally symbolized as $\exists x\ black(x)\ \&\ cat(x)$, the variable x represents that unnamed thing. According to standard logic, therefore, raw existence of an object is not a property of the object, but simply *is* the object of which properties are predicated. Each particular object and its properties are represented separately in any proposition. Moreover, the fundamental individual vs predicate distinction in logic, around which all propositions are organized, manifests in the primitive operation of *predication*—ie, where an individual (ie, a naked particular) is combined with a predicate (ie, a characteristic) to form an atomic proposition. Predication assumes this fundamental dichotomy between particulars and their characteristics. The same basic organization is universally expressed in all natural languages where some characteristic (expressed in a verb phrase) is predicated of some subject (expressed in a noun phrase) to form a sentence.

While this dual method of separately particularizing and characterizing material objects obviously works quite well, it is considerably less effective for things we want to talk about that have no spatiotemporal location, such as substances and abstract objects. It is composition, rather than location, which is relevant for interacting with substances, since samples can be spatiotemporally distributed in arbitrary ways. And abstract objects, like numbers, do not even have spatiotemporal properties. Nevertheless, we are constrained by logic and language to represent these kinds of things in the same manner as material objects. Such constraints can lead to philosophical conundrums, as when we try to reason about large noncontiguous objects distributed unevenly over space, intermittent objects distributed over time, or try to infer the particularized "reference" of intangibles like waves, fault lines, and personalities from a fully realist perspective. Reasoning consistently about nonmaterial objects can be awkward, given our tunnel vision on material objects.

Object Parts

In accordance with the fundamental strategy of describing objects in terms of their properties, there is one kind of property that holds a special place in our efforts to analyze and interact with material objects: namely, their parts. The relationship between a whole and its parts takes many different forms when talking about a material object. We can talk about a more-or-less integral *object* part (an engine is part of a car, the engine being a relatively independent component; or a sidewall is part of a tire, where the sidewall is conceptualized as a piece or segment of the whole tire). We can also distinguish *substances* as parts (ice cream is part of a root beer float), *features* (the surface is part of a table), *regions* (The White House is part of The District of Columbia), and *collectives* (a tree is part of a forest). We can even think about material objects (as well as many other things) as being parts of a *situation* (ie, a state, event, or process as in Zac is part of a parade). Finally, it is common knowledge that objects are made up of molecular and atomic *particles* (copper atoms are a part of a bronze statue), where a particle is the smallest structural unit relevant to a particular context.[160] All of these different concepts can enter into our understanding of a material object.[161]

It may not be readily apparent, but the word "part" is being used in two very different senses above. The first is exemplified by "an engine is part of a car", where one object (the car) is said to have another object (the engine) as a modular component part. This kind of relation will be called a "sub-" relation, because it can always be glossed as "x is a sub- of y" (eg, "an engine is a subpart of a car").[162] It applies to object fragments as well (eg, a sidewall is a subpart of a tire). The "sub-" relation generates a hierarchy of inclusive concepts, such that concepts lower in the hierarchy are "subs" of those higher up. For example, the subkind relation (eg, "collie is a subkind of dog") generates the kind hierarchy skeleton of a concept network. The subset relation of set theory is another example. With regard to the various subpart concepts mentioned above, it can generate a hierarchy for each of the concepts (ie, a hierarchy

---

[160] Other minimal structural units may be relevant for different objects—eg, cells for an organism.

[161] Left out of this hopefully comprehensive list of parts are four others mentioned by The Concise Oxford Dictionary. They are exemplified by "one-fourth of a pie is a part of a pie" (*fraction*), "the alto is part of a vocal score" (*voice*), "an afternoon is part of a day" (*time*), and "a chapter is part of a book" (*division*). Fractional parts of material objects are subsumed under *substances* since they represent an amount of something. Voice parts, (sub)divisions, and temporal (sub)parts of intervals are all subsumed under *object* parts as generic components. However, note that *particles* are *not* being interpreted as components of an object, as explained next.

[162] More formally, it is an inclusion relation, which is reflexive, antisymmetric, transitive, and supplementive (ie, every proper inclusion must be supplemented by another proper inclusion which is disjoint from the first).

of subobjects, subsubstances, subfeatures, subregions, subcollectives, subsituations, and subparticles), depending on the nature of each.

The second use of "part" is perhaps most clearly exemplified by "a tree is part of a forest". Here, a tree does not constitute a subforest within a forest; the relation is rather one of membership. Similarly, in "ice cream is part of a root beer float", the ice cream is more perspicuously an ingredient, in the sense that the root beer float is partially composed of ice cream. In fact, all of the other uses of "part" in the second paragraph above, except the first, fall under this second kind of usage. The difference is similar to the difference between a concept being a *subkind of* vs an *instance of* another concept; or a set being a *subset of* vs an *element of* another set.[163] Applying this second reading to the use of "part" in the examples, yields the following reinterpretations for the relations: "ice cream is an ingredient of a root beer float", "the surface is an immateriality[164] of a table", "The White House is located in The District of Columbia", "a tree is a member of a forest", "Zac is a participant in a parade", "copper atoms are microstructures of a bronze statue". (Note that particles are not themselves conceived of as objects with smaller parts in the context of normal use. However, a particle can be reconceptualized at a different level of understanding, as for example, when an atom is reconceptualized as having a nucleus and electrons. But then these subparticles are not normally thought of as "part" of the original object in the second sense of the word.[165]) To summarize, it is apparent that the original concepts endorse slightly different interpretations of the whole/part relation. It is this second sense of "part" that distinguishes the primary relations of the material object, since virtually anything can be decomposed into subparts in the first sense.

Having clarified the primary semantic relationships among concepts covered under the whole/part rubric, it becomes possible to outline the underlying conceptual structure. Laying out the concepts and relations described above, in the figure below, reveals the dependencies among the different concepts relevant to the second meaning of "part". The concept of an object is central, since an object can be a part of a collective, region, or situation, and can have parts which are features, substances, or particles:

[163] This second kind of parthood relation is irreflexive, asymmetric, and intransitive.

[164] The term "immateriality" is a generalization of "immaterial body" from (Casati & Varzi, 1994). An immaterial body designates a superficial discontinuity in a material—like a hole, boundary, or other disturbance (such as a wave or crack). See the section *Immaterial Objects & Information Content* later in this chapter for a discussion of immaterial objects.

[165] because the microstructure relation is not transitive. Similarly, a front door is a (sub)part of a house, and a doorknob is a (sub)part of a door, but a doorknob is not normally conceived of as a (sub)part of a house.

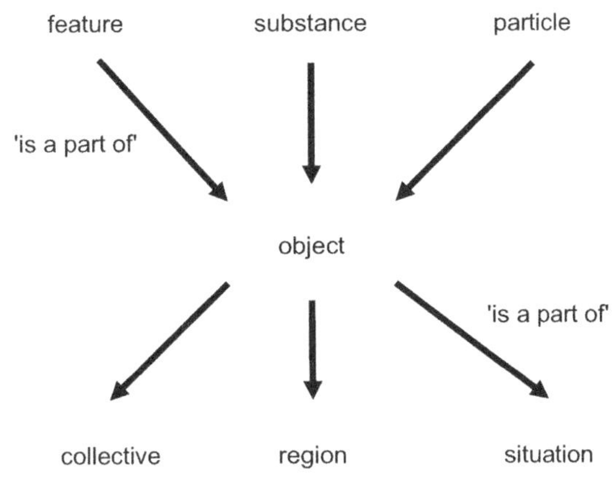

Figure 29: Dependencies Among Wholes & Parts for an Object

Note, however, a problem with the "object is a part of region" schema, exemplified by "The White House is located in The District of Columbia". Thinking of an object as a part of a region requires a different construal from all of the other parthood relations, more easily seen by listing them for comparison:

X is a part of Y
feature is an immateriality of object
substance is an ingredient of object
particle is a microstructure of object
object is a member of collective
object is a participant of situation
*object has location region

Y has part X (converse)
object has immateriality feature
object has ingredient substance
object has microstructure particle
collective has member object
situation has participant object
*region is a location of object

As is evident, the pattern for the region parthood relation is exactly backwards relative to the others. Its meaning is idiosyncratic, probably deriving from a different underlying semantics, and therefore represents a different sense of the parthood relation. In fact, the locative meaning of "part" does not directly appear in The Concise Oxford Dictionary, except when tied to the plural "parts", as in "she's not from these parts". After removing *regions* from further consideration (but only with regard to parthood), the relationships between the remaining concepts can be redrawn in preparation for investigating further hierarchical structure.

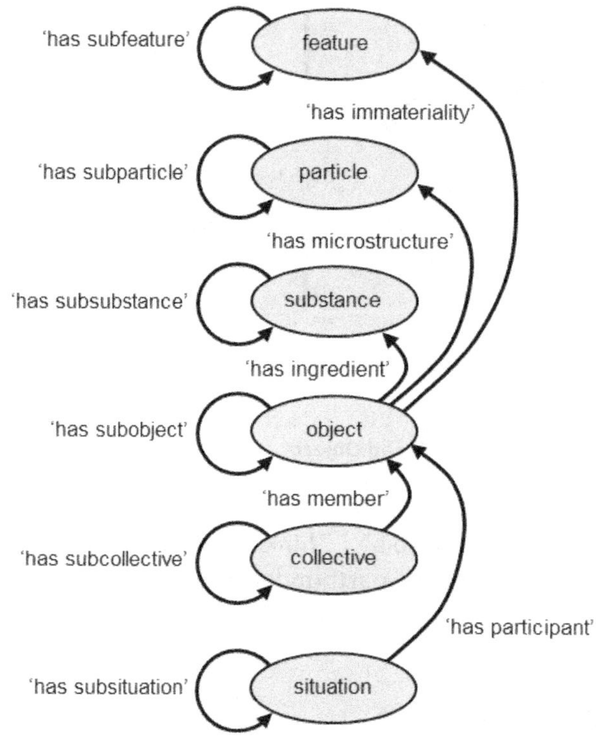

Figure 30: Object Part-Whole Relations

In looking at the nonobject concepts (features, substances, etc), discovering additional latent relationships is relatively straightforward. While an object can have a feature, substances can also have features, as for example when we talk about the holes in Swiss cheese or a crack in marble. English may not normally sanction using the word "part" in these cases (holes are a part of Swiss cheese?), but the relationship still describes a feature of a substance. Likewise, we sometimes view collectives as having features (eg, the Missing Man Formation flown by the Blue Angels, honoring the loss of a pilot). The idea of a substance being composed of a particle may seem strange at first, but particles are a form of matter in chemistry and physics, and reflect the atomic structure of all substances and objects composed of substances. Carrying the idea to situations and collectives, it becomes apparent that a situation can have any of the other concept instances as participants (eg, feature: the surface of this table shines like the sun), and that a collective can have any concept instance above it as a member (eg, substance: a herd of cows viewed as so much beef). The final picture that emerges is of a succession of whole/part relationships among these concepts, such that any concept can have parts of various kinds as long as they appear at the same level or above on the ladder of whole/part concepts shown below.

This kind of simplification has been found to apply to all concepts, not just material object concepts, and is the basis for the global organization and decomposition of concepts discussed in Chapter 6: *Concept*

Net Organization Principles. Global coverage, however, requires generalizing the parthood relationship to the broader notion of *constituency*, which extends the domain of material object concepts into the realm of all concepts, both concrete and abstract.

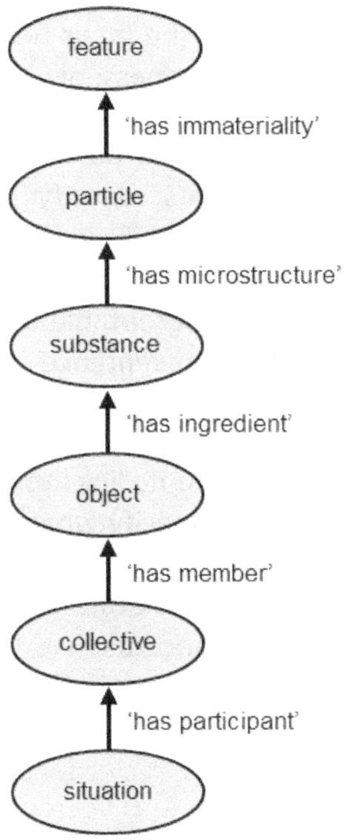

Figure 31: Hierarchy of Parthood Relations

The Stereotypical Meaning & Representation of the Noun "Lemon"

In 1975, Putnam published a landmark article entitled "The Meaning of 'Meaning'" (Putnam, 1975). In it, he proposed that the core meaning of a word is commonly understood in terms of its stereotype (but includes other aspects as well). The stereotype for a word (or more precisely for a word sense, since a word typically has multiple senses) consists of what a user's community generally expects any ordinary person to know about the word as used in normal circumstances. For example, if you don't know that a tiger is an animal, that it normally has stripes, is fierce, and eats meat, you don't know the core meaning of the word. Knowledge that a tiger is a mammal, although formally true, is not part of the stereotype. Stereotypes are what dictionaries strive foremost to express, but usually include other special information to guide usage in nonstandard contexts, as well. Although Putnam's understanding of core meaning as stereotypical is very different from the ontological interpretation proposed in this book, his discussion serves as a jumping off point for attempting to represent conceptual content.

The following discussion aims at translating the dictionary definition of "lemon" (one of Putnam's examples) into the concept net framework as an experiment in concept representation. Generally speaking, the meaning of a term in a concept net has a wider construal than just a stereotype, being dependent on extensive additional background knowledge plus context. However, standard use of the term "lemon" seems relatively straightforward, so the focus here is on basic representational issues, as opposed to higher-level issues like abstraction, vagueness, or use.

The Concise Oxford Dictionary defines "lemon" as "a pale yellow thick-skinned oval citrus fruit with acidic juice". Translating the definition involves identifying the embedded concepts and their relationships. Most of this definition seems to identify various attributes that a stereotypical lemon possesses. It is probably a stretch to assume most people know all of them (at least explicitly), but partial knowledge is likely good enough for many acceptable uses of the term. In any event, when the word "lemon" is used in the context of an occasion sentence, as in "this lemon is sour", the context is highlighting only some of the stereotype (ie, the acidic juice).

Any new concept ultimately takes its meaning from its links to other concepts in the concept net. The two principal links for any new universal concept specify what kind of thing it is, and what the concept itself is an instance of. In this example the definition states that a lemon is a kind of citrus fruit. Moreover, the concept [lemon] is an instance of a natural-kind concept, which is distinguished from artifactual concepts, like [spoon], and attribute concepts, like [color], (among many others). Thus far, then, the concept [lemon] can be linked to two other, assumed preexisting, concepts in the network:

[lemon]
    'is an instance of' [[natural kind]]
    'is a kind of' [citrus fruit]

The concept [[natural kind]] is a metauniversal concept because its instances are concepts like [lemon], rather than worldly objects like the instances of a [citrus fruit]. The links 'is an instance of' and 'is a kind of' are basic associations.

The remainder of the definition presents various characteristics of lemons. The first characteristic referenced is "pale yellow" which implicitly refers to the attribute of color. In a concept net an object's characteristic is always represented as a unique explicit concept, which in this case will be labeled [lemon's color]. "Pale" and "yellow" thus are subattributes. Now, psychological research over the last 50 years or so has established that the subjective experience of color is made up of three separate psychological components, so the attribute of color is structured (unlike the attribute of temperature which can be expressed in terms of a single value). The three subjective color components are usually called hue, lightness, and saturation; and "pale yellow" is taken as referring to values of these components. "Yellow" is clearly a hue, while "pale" is probably a combined characterization of a certain degree of lightness and saturation (somewhat toward white on the lightness scale, and slightly away from the richness of pure primary yellow on the saturation scale). We could invent a new scale, say called [color tone], of which "pale" is one value, but this approach ignores the underlying component organization of color. This component structure is typically not conscious, but is grounded in the way our visual system processes color. A lot of research supports the three component psychological theory of color, so it makes more sense to define "pale" by way of the more established color components, than create a new *ad hoc* component. The complete color characterization is then more complex, but is easily added to the previous stereotypical fragment of the concept of [lemon]:

```
[lemon]
    [1[has color]2]  [lemon's color]
                        'is a kind of'      [color]
                        [1[has hue]2]       [lemon color's hue]
                                               [1[has value]2]  [yellow]
                        [1[has lightness]2] [lemon color's lightness]
                                               [1[has value]2]  [light]
                        [1[has saturation]2][lemon color's saturation]
                                               [1[has value]2]  [rich]
```

The characteristic chaining effect is evident above in the characterization from the object [lemon], to [lemon's color], to [lemon color's hue], to the attribute [yellow]. Colloquial English usually short-circuits the full semantic chaining between the object and its attributes, as when we simply say "lemons are yellow".

The relation between the concept [lemon] and its color characterization [lemon's color] is mediated by the relation [1[has color]2], which is a concept in its own right (not one of the basic associations). Also, the color characteristic itself has characteristics (a lemon's hue, lightness and saturation), each of which expresses a certain attribute. The concept [yellow] is a kind of [hue], while [light] is a kind of [lightness] somewhat less light than pure [white]; and [rich] is a kind of [saturation] somewhat less saturated than [full saturation]. The 'is an instance of' relations for the attribute concepts are technical and not included, although they are necessary for a complete representation. The basic attribute scales of hue, lightness and saturation are part of the background knowledge of the concept net, and provide the perceptual grounding for using and comparing color terms. For example, [yellow], [green], [red], etc, represent value intervals on the scale of [hue]. Working out the details of such psychological scales counts as a new scientific discovery—in this case, the dual process theory of color vision.

The second characteristic mentioned in the definition is "thick-skinned". Chaining is again evident, since this is shorthand for the fact that a lemon has a skin, that the skin has a thickness, and that its thickness is typically characterized as thick. While thickness is an attribute, skin is a part of a lemon. Parts of objects generally are objects themselves, so a subpart can have object characteristics too. The full semantic representation of "thick-skinned" is therefore:

[lemon]
    [1[has part]2] [lemon's skin]
                    'is a kind of' [fruit's skin]
                    [1[has thickness]2] [lemon skin's thickness]
                                        [1[has value]2] [thick]

The third characteristic is "oval", referring to the shape of a lemon. Here, the reference is obviously to a 3D, as opposed to a 2D, shape, but the cognitive underpinnings of our experience of shape is not as well understood as for color (Palmer, 1999). As a first approximation to a reasonable representation, it is clear that "oval" is referring to the external boundary of the lemon. It would seem that there are other potential components to shape as well, but these are deemed irrelevant. (For example, one other major component might include a characterization of the internal boundary. Such a boundary would become relevant, in addition to the external boundary, if we were talking about a slab of Swiss cheese.) The representation for shape is a characteristic just like the other characteristics discussed above:

[lemon]
    [1[has shape]2] [lemon's shape]
                    'is a kind of' [3D shape]
                    [1[has external boundary]2] [lemon's external boundary]
                                                [1[has value]2] [oval]

The final characteristic mentioned is "acidic juice". And again, juice can be straightforwardly regarded as a (liquid) part of a lemon. The relevant attribute for the juice is acidity, with a valuation of acidic:

[lemon]
    [1[has part]2] [lemon's juice]
                    'is a kind of' [fruit's juice]
                    [1[has acidity]2] [lemon juice's acidity]
                                      [1[has value]2] [acidic]

This completes the concept net representation of the stereotypical meaning of "lemon", which includes all of the above characterizations, plus a few other links to concepts already existing in the background concept net. Most people's knowledge of lemons goes considerably beyond the stereotype, but it does furnish a useful basis to which additional information can be added.

The Concept of the Temporal Object

Like the basic concept of the material object (emphasizing the tangible material characteristics of an object such as a table), or the spatial object (emphasizing the spatial characteristics of an intangible object such as a triangle), the basic temporal object concept emphasizes characteristics related to the unfolding of a situation or happening over time. Most everyday objects exhibit all three kinds of characteristics (among others), but common *events* are prototypical temporal "objects", because they are fundamentally described in terms of temporal characteristics like action, duration, and time of occurrence. The event of driving a car to work in the morning is therefore a temporal object, in the same spirit that the car is a material object. Both are conceptual representations we can call up and use for various purposes.

Temporal objects also possess temporal parts, roughly in the same way material objects possess spatial parts, where "part" is being used in the first sense (the subpart sense) previously described above for material objects. (The structural parts of a temporal object corresponding to the second sense of "part" are dealt with in the next section.) A temporal part of an event or situation extending over time is simply a briefer temporal subpart of that event or situation. For the event of driving a car to work in the morning, a selection of subevents might include finding the keys, stopping off for coffee, and parking at work. A number of subevents, laid out in sequence with no temporal gaps, then make up the whole event; and these subparts of the larger event are themselves conceptually identifiable events, just like the subparts of a material object are themselves objects—and so on to smaller and smaller subparts.

Temporal object representations figure importantly in our ability to mentally simulate evolving situations and plan our activities before committing to action. This ability complements our more static capacity for imagining the configuration of material and spatial objects. Simulating situations, however, involves lining up subsituations in temporal sequence and "running" the sequence, in addition to mentally generating the spatial layout of material object subparts. The ability to run mental simulations enables us to build dynamic conceptual models of the world and even to experience the activation of the associated perceptual schemas (Barsalou, 2003; Evans, 2009). At the level of propositional representations, it allows us to chain inferences together in sequence to explore the consequences of "what if" questions.

Generalized Situations

The representation of generalized situations, as expressed by propositions in language,[166] is a major extension of the kernel concept network discussed previously. The kernel contains the skeletal framework for representing network structural knowledge using simple (binary) associations between concepts, but the capacity for representing propositional knowledge requires a new category: a [situation] represented as an instance of a relation with bound arguments plus ancillary arguments such as time, place, agent, theme, etc. The simple kinds of situations we talk about every day properly include events, actions, states, and processes. One conjecture about what we are doing when we use language to communicate is that we are providing cues to help others build mental structures corresponding to situations we are thinking about—eg, see (Fauconnier, 1997).

In a concept net a *situation* is a conceptualized segment of particular experience (also sometimes called a state-of-affairs, event, occurrence, circumstance, or happening), which is represented as an instance

[166] or well-formed-formulas in predicate logic—viz, those with no free variables.

of a relation.[167] For example, given the relation [[person][wears][clothing]] specifying one use of the relationship [wears], an instance of this relation might be ||Eli||wears||Eli's coat||, expressing the particular situation concept that Eli is wearing his coat (ignoring verb tense issues). |Eli|, |Eli's coat|, and the predicate |wears| are particular participants in the situation. The figure below diagrams the rudiments of the concept net representing this situation whose structure will be elaborated shortly. Note that a relation like [[person][wears][clothing]] is a kind of (not an instance of) [situation].

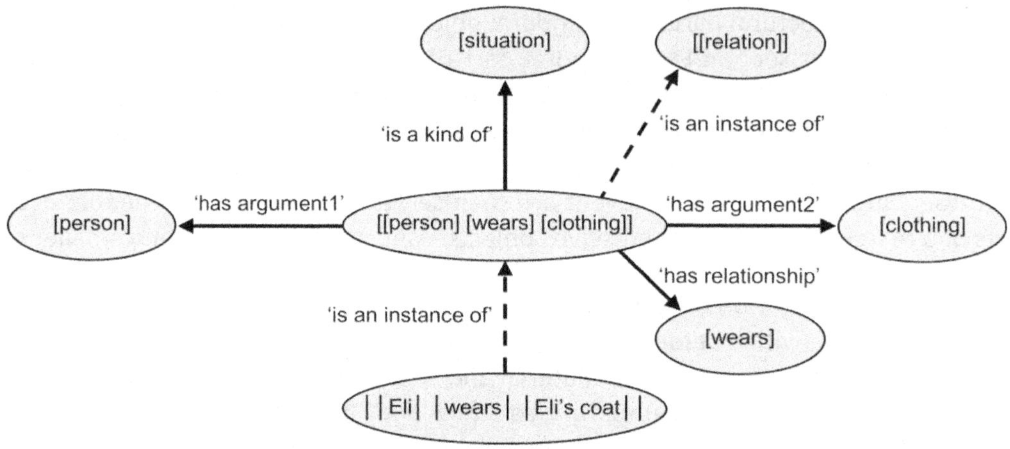

Figure 32: Representation of a Situation

As discussed previously, the alternative shorthand way of diagramming the same network is to collapse the relation node to a dot, abbreviate its arguments, and drop its other recoverable associations (like 'has relationship, 'is a kind of', etc):

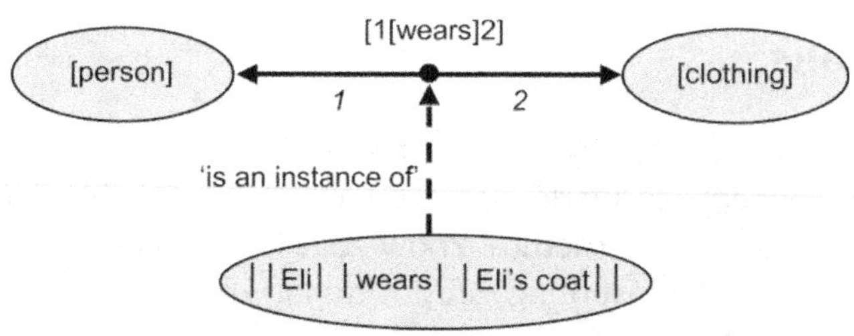

Figure 33: Abbreviated Representation of a Situation

Note that whole propositions serve as labels for situation concepts. Using propositions as names for situations is generally consistent with Parsons' account of *Events in the Semantics of English*, in which he concludes that the typical linguistic statement posits quantification over events (Parsons, 1990). That is, a statement like "Brutus stabbed Caesar" says that there is an unnamed stabbing event (e), such that the agent of e is Brutus, the theme of e is Caesar, and e has culminated. This is different from the standard

[167] Certain amodal conceptual situations may also link to modal perceptual impressions in episodic memory, providing an experience of the situation.

logic interpretation, where the sentence expresses a simple predication.[168] Positing an event structure to sentences (but preferring the generic term "situation" to "event" in the concept net framework) goes a long way toward explaining how we come to understand utterances via many low-level inferences related to the different roles participants play in the utterance.

The formal difference between a situation and a relation concerns the form of the participants involved in it. In a simple situation participants are simple particulars (like |Eli|), quantified particulars (like "every man"), or definite descriptions (like |Eli's father|). See the later section on *Quantification* for more about quantified propositions. However, in a relation one or more arguments are universals, which makes the relation itself a universal. A relation, therefore, is used in a modal sense to express a range of possibilities for instantiation as a situation. In the above example the relation [[person][wears][clothing]] specifies that it is possible for a person to wear a piece of clothing. This is part of the meaning of "wears" that authorizes its use with certain kinds of arguments, namely its selectional restrictions. Alternately, a situation like "Eli is wearing his coat" is an instance of that relation containing particulars, and in this case expressing an atomic proposition that conforms to the parent relation's authorizations (Eli is a person, his coat is a piece of clothing, and is wearing is an instance of wearing).

The concept [situation] fits into the concept hierarchy under [occurrent] as shown here:

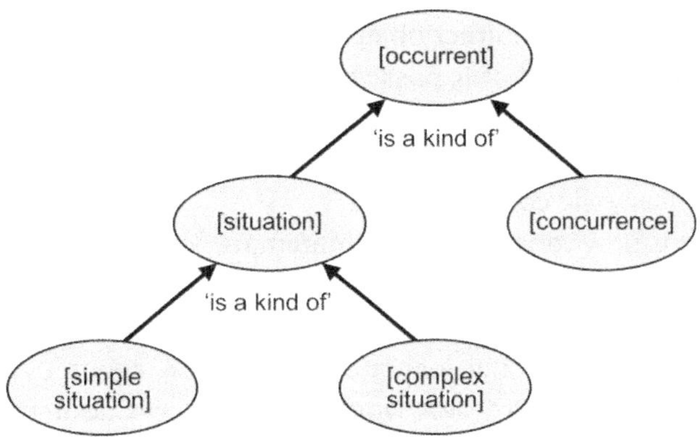

Figure 34: Situation Concepts as Occurrents

The two kinds of situation reflect the simple/complex distinction between simple situations that have no embedded situations, and complex situations that can have other situations as participants. Complex situations are entities such as propositional attitudes (eg, believing, desiring, promising, etc), while simple situations are simpler entities like general and specific facts and circumstances expressed by a single predicate. Most situations have spatiotemporal properties like time and place of occurrence

168 In predicate calculus the former translates into ∃e stabbing(e) & agent(e, Brutus) & theme(e, Caesar) & cul(e), and the latter into stabbed(Brutus, Caesar).

(although timeless situations like "Socrates is mortal" usually do not), and an assessed correspondence with reality (ie, the situation is either actual or nonactual).[169]

The participants in a situation play various roles that contribute to its semantic content. In the coat wearing situation, |Eli| fills the informal role of wearer and |Eli's coat| fills the informal role of the worn. The role that |wears| fills will be called a *predicate*.[170] Different predicates typically bring different roles to the situation.

Linguists (probably beginning with Fillmore) who have analyzed extensive bodies of text, have come up with ten or so semantic (case) role associations that seem to cover virtually any usage for all situations. For example, the obligatory role association links in a loving situation are 'agentive' and 'affected'. The 'agentive' role includes lovers and other animate beings instigating something happening; and the 'affected' role includes people who are loved, or other things that are directly involved in the action expressed by the predicate. The specific roles included in the concept network for situations are summarized below--from (Quirk, Greenbaum, Leech, & Svartvik, 1985):[171]

- *agentive*: the animate being instigating or causing the happening denoted by the verb (eg, **Angus** scratched Lynne.); typically the subject
- *affected*: a participant, animate or inanimate, which does not cause the happening denoted by the verb, but is directly involved in some other way (eg, Lynne lost **her glasses**.); typically the direct object, the subject, the indirect object, or the subject of copular verbs
- *recipient*: the animate being that is passively implicated by the happening or state denoted by the verb (eg, Dave gave **Jennie** $5); typically the indirect object, or the subject
- *attributive*: a participant that identifies or characterizes the subject or object (eg, My favorite color is **yellow**); typically the complement
- *external causer*: the unwitting, generally inanimate, cause of an event (eg, **Lightning** can fry your computer); typically the subject
- *cognate*: an event indicated by the verb (eg, Jennie sang **a song**)
- *instrument*: the entity, generally inanimate, which an agent uses to perform an action or instigate a process (eg, **This hard disk** would make a great doorstop.); typically the subject, or direct object
- *positioner*: a controlling participant that undergoes no change during the happening (eg, **Paul** lives in Houston.); typically the subject
- *locative*: designates the place of the state or action (eg, Mimi lives in **Houston**, too); typically the subject, or the direct object
- *temporal*: designates the time of the state or action (eg, I took Angus to the vet **yesterday**.); typically the subject

[169] The truth-value (ie, True or False) of the proposition labeling a situation is normally interpreted as an indicator of whether or not the situation holds in some specified (physical or nonphysical) world.

[170] The formal idiom used in predicate calculus is *predicate constant*, but this is a little unwieldy in diagrams.

[171] Helbig provides a similar useful analysis of situation roles in *Knowledge Representation and the Semantics of Natural Language* (Helbig, 2006).

- *eventive*: a nominalized verbal (eg, **The destruction of the world** will occur at midnight on the 23rd of October, 2020.); typically the subject, or the direct object
- *resultant*: an object whose referent exists only by virtue of the activity indicated by the verb (eg, Jennie baked **brownies**.); typically the direct object.

All formal roles depict associations in a concept net. A role, such as 'has agentive', links the situation node with an agentive participant, where "has-" is prefixed to signify its use as an association in a concept net. A participant can be any instance of [concept], since no specifiable topic is theoretically taboo. However, roles will impose special restrictions that delimit their applicability—eg, agentive takes an animate participant, and attributive takes a property or attribute of some sort. The predicate is also regarded as a role, since it participates in the meaning of the proposition. The graphical representation of a situation looks something like the hub of a wheel with various roles emanating like spokes. The figure below expands the previous diagram depicting Eli wearing his coat to show the association roles attached to the situation, as well as the authorizing relations for the new roles.

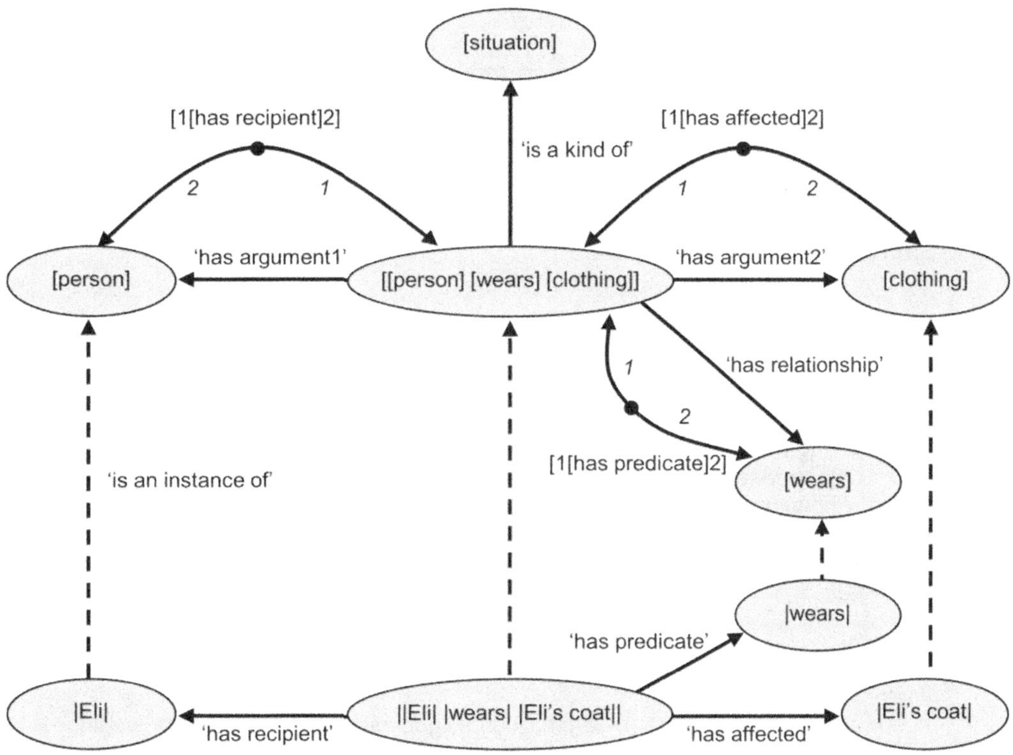

Figure 35: A Situation as an Instance of a Relation

Now that situations have been described, the relationship between associations, relations, and situations becomes clearer. An association expresses knowledge by directly linking two concepts in a concept net. It expresses the most elementary of beliefs. But you cannot explicitly talk about that association unless you reify it into a concept. To additionally represent the association as a concept in its own right (eg, for the purpose of describing it), turn it into a particular situation concept node. That situation is then an instance of a relation. The difference is diagrammed below, where the "use" of the association "[concept] 'is an instance of' [[universal]]" appears under the "mention" of it in a node:

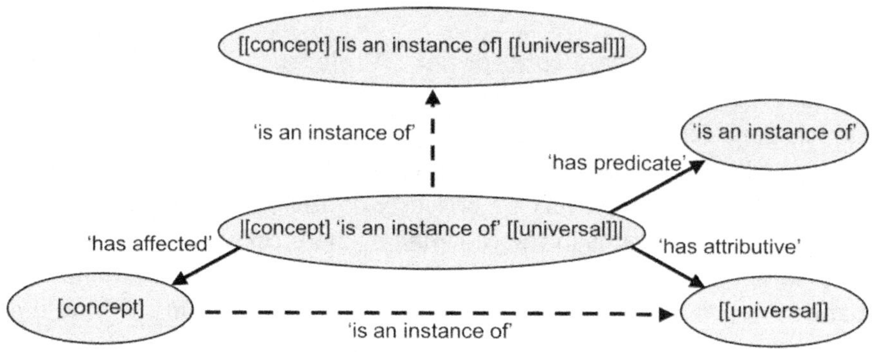

Figure 36: Representing an Association as an Explicit Situation

A wide variety of generalized situations is expressible in language via declarative sentences. One practical way of classifying situations is as indicated below. This breakdown attempts to provide for the orderly inheritance of roles, according to the kind of action indicated by a predicate, with the more specific kinds of situation lower in the hierarchy inheriting more roles from the more general kinds of situation. For the justification for this particular approach to decomposition, see Chapter 6: *The Organization of Situation (ie, Event) Concepts.*

- [simple-situation]
  - [state]
    - [condition]
      - [being]
        - [existing-as-something]
          - [permanent-existence] — eg: is a mammal, is a disease, is a time
          - [impermanent-existence] — eg: is a student, is a bachelor
        - [being-someway] — eg: is tall, is sick, is charming, is greater than 3
      - [having-something] — eg: has a dog, has measles, has a tail, has charm
    - [continuance] — eg: lives in LA, rain, boil, hunt, talk, hurt, see
  - [change]
    - [progression] — eg: improve, grow, eat, ripen
    - [status-change]
      - [instancy]
        - [transitional-change] — eg: turn off a light, catch a ball, arrive, die, begin
        - [momentary-change] — eg: flash, shoot a gun, hit
      - [repetition] — eg: beat a rug, repeat, iterate

Complex Predicates

Atomic propositions in a concept net consist of a predicate and some number (0 or more) of arguments, in accordance with standard logical form. For example, in the proposition |Lynne loves Dave| the predicate is |loves| and |Lynne| and |Dave| are arguments, where each constituent is a node in the concept net, with association links tying the constituents together.

But another way to represent the proposition is as a noun phrase followed by a verb phrase (ie, S -> NP VP), in accordance with standard sentential form. In sentential form, the proposition consists of a complex VP along with a single subject argument, the NP. For the proposition |Lynne loves Dave| the NP argument is |Lynne|, and the complex VP is |loves Dave|. In concept net terminology, a complex VP, represented as an individual concept, is called a "complex predicate". A complex predicate is just a complex VP that incorporates a simple predicate along with some number of arguments.

Representing a complex predicate in a concept net as a separate node (over and above the node representing the whole proposition), is a concise, oftentimes convenient, way of representing a complex proposition. It follows the intuitive idea that something can be predicated of any object whatsoever. Since the complex predicate from |Lynne loves Dave| is simply |loves Dave|, the only nodes required for an initially adequate representation are |Lynne|, |loves Dave|, and |Lynne loves Dave|. It may be most useful for quickly loading propositions into a concept net without worrying about how to precisely interpret potentially complex verb phrases. The advantage is better seen in more complex examples such as |Lynne likes the way Dave combs his hair|, where the generalized predicate is |likes the way Dave combs his hair|. Later, the generalized predicate can be expanded into its final, more useful, form with all propositional arguments represented. The following figure illustrates the supporting details of the move from a complex predicate representation to a full semantic (propositional) representation.

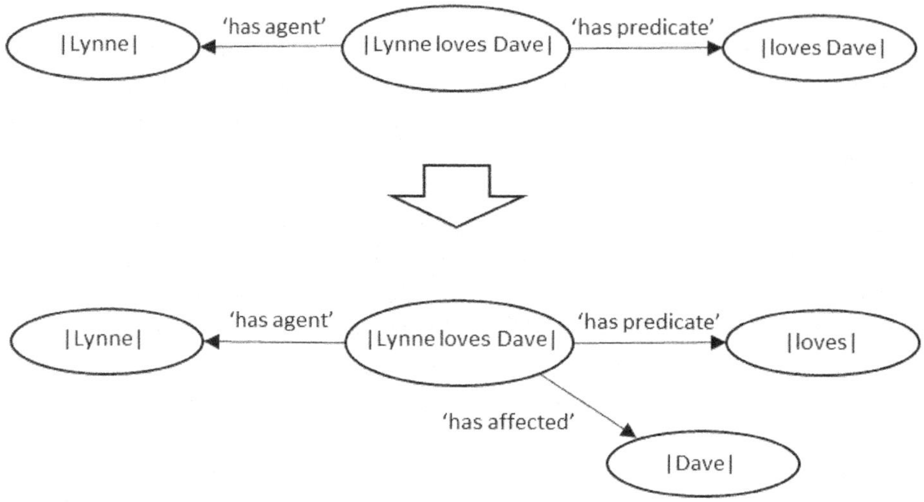

Figure 37: Representing a Complex Predicate and its Expansion

From this perspective, a complex predicate is one constituent of a proposition, where the complex predicate could characterize any of one of the participants (ie, arguments) in the proposition. For example, in the proposition |John told Mary a lie|, John is associated with the complex predicate of having told Mary a lie, Mary with the complex predicate of having been told a lie by John, and the particular lie is associated with the complex predicate of having been told to Mary by John. Taken together, there are then three potentially different complex predicates for the proposition. Yet these complex predicates add nothing extra over and above the generalized situation that is expressed by the proposition |John told Mary a lie|. Any one, or none, may be included in a concept net as a component of a proposition. Complex predicates only provide a convenient linguistic means to focus talk on one of the participants involved in the (arbitrarily complex) situation, but makes it difficult to reason about the other participants independently.

The main conclusion is that complex predicates probably should not be formally represented as specific separable concepts in a concept net (except perhaps as a precursor to further analysis). The proposal instead is to assimilate them into the representations of generalized situations. That is, it is more useful to think of situations as describing the interaction among several objects, than to think of each of those objects as being described by a complex predicate. This way of thinking may seem somewhat counterintuitive, since we so naturally conceive of objects in terms of their characterizing properties in language. But the intuition may turn out to rely on a limiting perspective—a perspective that is efficient for reasoning about simple characterizations, but more problematic for complex characterizations. It is a relatively straightforward matter to reason from 'All men are mortal' and 'Socrates is a man' to 'Socrates is mortal' using Aristotle's basic categorical logic (Kreeft, 2010), where the sentential forms are limited to unstructured predicates describing objects and categories. But our everyday reasoning typically exhibits much more complex inference processes, particularly when dealing with social situations.[172] Modern symbolic logic allows reasoning about much more than categories (namely propositions), but even these potentially complex relational and modal forms are child's play relative to our comparatively sophisticated innate reasoning abilities.

If generalized situation representations are, in fact, more useful for reasoning than (complex) property representations, our everyday practical conceptual knowledge is probably not *primarily* organized according to lists of semantic properties attached to individual concepts. (But see Chapter 6 for the suggestion that our epistemological knowledge—ie, high-level knowledge about our lower-level domain knowledge—is so organized.) An overall organization by properties would be expected to lead to a deep inheritance hierarchy of concepts, with knowledge about basic-level concepts (eg, dog, fish, bear, bird) tightly bound to knowledge about superordinate concepts (eg, animals, in general) and subordinate concepts (eg, collie, poodle, mastiff). But such wide-ranging hierarchical relationships have not been observed in neurological or linguistic studies of general folk knowledge.[173] Our observed semantic hierarchies tend to be rather shallow, and cover very gross semantic classifications such as creatures, nonliving things, fruits/vegetables/foods, and perhaps, tools (Cree & McRae, 2003). Instead, we seem to organize our working knowledge into a variety of *knowledge domains* consisting of propositions, where a diversity of concepts that are used together become interrelated, something like the organization of material in an educational course. Organization by propositional knowledge domains would appear to provide direct support for domain-specific inferencing processes (ie, potentially unique to each operational domain) based on representations of generalized situations. Secondary, cross-cutting organizations of domain objects and categories by properties could then add useful intradomain classification schemes for efficiency. Some of these issues are discussed further in Chapter 6.

An Alternate View of Situation Types

The preceding action-oriented approach to propositions has proven useful in representing situations in concept nets. However, there are other theoretical perspectives to organizing situations based on the static/dynamic dimension of verbs in general, since the identity conditions of situations also seem to have something to do with deeper temporal properties. The classification of situation types provided in (Quirk, Greenbaum, Leech, & Svartvik, 1985) provides an alternate viewpoint based on a dimensional analysis of verb phrase semantics (see table below). The salient feature dimensions they identified are stative vs dynamic, durative vs punctual, conclusive vs nonconclusive, and agentive vs nonagentive.

[172] For example, (Wierzbicka, 2015) analyzes the concept of prototypical lying as the following semantic explication: If someone X lied to someone Y, then someone X said something to someone else Y, and this someone knew that it was not true, and this someone said it because he/she wanted this other someone to think that it was true, and people think that it is bad if someone does something like this.

[173] However, exactly this kind of taxonomic organization would be expected of highly abstract ontological concepts like [[universal]], [[kind]], [[relation]], etc, whose representation is the subject of much of this book.

- When used in a sentence to express a situation, verbs have a tense-related feature, whose basic three values indicate whether the situation has: ended (past), is continuing (present), or is predicted (future). All situations of every type reflect this property.
- Usage examples below are in the simple past tense, since this is subject to the least complicated temporal interpretation.
- All verbs express a core situation type, which is instantiated in a particular situation when used in a sentence. Stative verbs often express a "being" or "having" state; dynamic verbs reflect other more complex activities. The activity may be very short (eg, in transitional events), but still is assumed to have some duration.
- Verb types differ according to their features—specifically to what degree they exhibit stative vs dynamic characteristics, whether they express durative or punctual occurrences, whether the occurrence is conclusive or nonconclusive, and whether they take an agent or not.

While this salient feature approach to verb classification does generate a relatively complete and coherent set of verb categories, it lacks a deeper explanatory organizing principle, such as the one that underlies the former analysis of situation types. Instead of organizing situations according to four independent features, the former analysis organizes according to the underlying kind of "action" indicated by the verb. This action-oriented approach to situation kinds is consistent with the broader approach to all other concept kind hierarchies developed in this text. Chapter 6: *The Organization of Situation (ie, Event) Concepts* outlines the various kinds of verbal actions in more detail.

Table 2: Alternative Classification of Situation Types (t = transitive) (Quirk, Greenbaum, Leech, & Svartvik, 1985)

VERB	USAGE	CORE ACTIVITY
Quality		
Be tall	A was tall	Being tall
Have 2 legs	A had 2 legs	Having 2 legs
Be a student	A was a student	Being a student
State		
Be angry	A was angry	Being angry
Be ill	A was ill	Being ill
Love(t)	A loved B	Loving
Resemble(t)	A resembled B	Resembling
Think (that)	A thought that B	Thinking
Own (t)	A owned B	Owning
Stance		
Live	A lived near B	Living
Stand	A stood near B	Standing
Lie	A laid near B	Lying
Sit	A sat in B	Sitting
Goings-on		
Rain	It rained on Wed	Raining
Snow	It snowed in A	Snowing
Boil	A boiled at 100°	Boiling
Shine	A shone brightly	Shining
Glow	A glowed dimly	Glowing
Activity		
Drink	A drank yesterday	Drinking
Sew	A sewed yesterday	Sewing
Write	A wrote yesterday	Writing
Hunt	A hunted on Tue	Hunting
Play (t)	A played the piano	Playing
Talk	A talks a lot	Talking
Process		
Ripen	A ripened nicely	Ripening
Grow up	A grew up quickly	Growing up
Improve	A improved today	Improving
Separate	A separated from B	Separating
Turn red	A turned red	Turning red

VERB	USAGE	CORE ACTIVITY
Accomplishment		
Write (t)	A wrote a novel	Writing
Eat (t)	A ate meat	Eating
Drink (t)	A drank alcohol	Drinking
Fill up (t)	A filled up the tub	Filling up
Discover (t)	A discovered B	Discovering
Momentary Event		
Sneeze	A sneezed	Sneezing
Explode	A exploded	Exploding
Blink	A blinked	Blinking
Flash	A flashed slowly	Flashing
Bounce	A bounced high	Bouncing
Momentary Act		
Tap (t)	A tapped the table	Tapping
Nod (t)	A nodded off	Nodding
Fire	A fired too quickly	Firing
Kick (t)	A kicked B	Kicking
Transitional Event		
Drop	A dropped B	Dropping
Receive (t)	A received B	Receiving
Catch (t)	A caught a cold	Catching
Take off	A took off	Taking off
Arrive	A arrived	Arriving
Die	A died quietly	Dying
Transitional Act		
Sit down	A sat down slowly	Sitting down
Catch	A caught a ball	Catching
Shoot (t)	A shot a rifle	Shooting
Begin (t)	A began a project	Beginning
Stop (t)	A stopped the car	Stopping

Meanings & Representation of the Verb "Paint"

In a concept network every concept lives in a neighborhood of other concepts that contribute in different ways to its potential meanings on various occasions. Since utterances containing verbs like "to paint" typically express semantic situations (ie, denote instances of relation concepts), as opposed to object nouns like "lemon" (denoting kind concepts), their neighborhood consists mainly of participants in the situation. The situation specifies the semantic roles (eg, agentive, benefactive, instrument, locative, etc) that the various sentence participants play in instances of the relation, where the participants are fillers for the arguments of the parent relation.

The listing below captures a range of uses of the verb "to paint" along with its possible relational arguments, where each argument (in brackets) is associated with the preceding participant's role in the sentence.

Max [agentive] paints. (Max fills an agentive role for painting activities referenced in this sentence)
Max painted the office [affective]. (the office is affected by the painting activity)
Max painted in Covelo [locative].
Max paints for his father [benefactive].
Max painted it with a camelhair brush [instrumentive].
Max painted it green [color]. (green fills an attributive color role for "it")
Max painted iodine [instrumentive] on her wound. (the iodine is used to do the painting)
Jen paints in oils [instrumentive].
Jen painted a portrait [resultant]. (the portrait results from the action of painting)
Jen painted a flower [resultant]. (more perspicuously, Jen painted a picture of a flower)
Jen painted spots [resultant] on the egg.
Jen painted yesterday [temporal]. (when the painting took place)
Jen painted it quickly [manner]. (how the painting occurred)
Jen paints well [evaluative].
Jen paints a little [degree].

Other roles for "paint" likely exist, but this list covers most of them. Also, there are two other uses of "paint" which are usually considered similar (ie, polysemous) to the above meanings, but which require separate treatment in a concept net:

Zoe painted a dismal tale of woe. (ie, described as if verbally painting a picture)
Zoe painted the town red. (ie, an idiomatic usage meaning Zoe celebrated)

What seems to unite all exemplars of the first group is the meaning schema of an agent applying a painting material to a surface of some object. The identity of the agent, material, surface, and object vary considerably from sentence to sentence, and may not even be directly mentioned at all, but all four participants seem necessary to the meaning of "paint". They are core roles associated with "paint" because their presence can always be inferred from any such use of the word. Even in the highly figurative use "Jen painted a flower", a surface and a resultant picture object are implied. The remaining arguments of "paint" are optional (eg, [temporal]), and are used merely to provide additional information about a particular painting situation. The second group of meanings is distinct from the first in that they express different categories of activity. While the first group falls under the theme of applying activities (along with applying bandages, applying fertilizer, etc), the second group falls under describing and celebrating activities, respectively. The core meaning of "paint" is categorically different for each of these three kinds of activity.

The proposed representation of the concept of painting is given below. The first two associations in any concept specification tie the concept into the background concept net via its 'is an instance of' and 'is a kind of' associations. Here, the concept of painting is a relation—namely [1[paints]234], taking four core arguments, and denotes a kind of activity in which someone applies something to the surface of something else:

[1[paints]234]
  'is an instance of' [[relation]]
  'is a kind of' [1[apply]2]
  [1[has agentive]2]* [painter]
  [1[has instrumentive]2]* [painting material]
  [1[has locative]2]* [painted surface]
    [1[is a part of]2]* [painted object]
    [1[has color]2] [painted surface's color]
  [1[has instrument]2] [painting instrument]
  [1[has resultant]2] [picture + figure]
    'is a kind of' [product]

The possible arguments characterizing the activity of [1[paints]234]are listed next—viz, [1[has agentive]2] ... [1[has resultant]2]. The painting relation inherits the agentive, instrumentive, and locative arguments from the parent applying relation, adds new selectional restrictions to them (most notably to the locative argument making it into a surface of some sort), and further entails that the surface is part of some object. The starred arguments are necessary to any instance of painting, while the unstarred are optional—ie, every instance of painting will involve at least an agent, a painting material, some location where the material is applied (in this case the surface of some painted object), even if they are not explicitly mentioned in a sentence.

The agentive role for [1[paints]234]is filled by the concept of a [painter]. Most activity verbs commonly conventionalize the agent role by adding "-er" to the root form of the verb (hunter, robber, sweeper, etc). The instrumentive role is filled by the substance that is applied during the process. The last necessary role is filled by the painted surface, which is conceptualized as a part of some painted object.[174]

The remaining optional roles include the color of the painted surface (as in "Max painted the trim green"), the instrument used to paint with, and the resultant characterization of what was painted (ie, a picture or figure of some kind). The concept [picture + figure] encompasses the joining of two normally independent concepts into one concept, but it is important to note that the join is an exclusive or. The resulting creation can have characteristics of either a picture (eg, landscape, nude) or a figure (eg, triangle, stripe), but not both simultaneously. Perhaps other kinds of painting resultants could be added in a more thorough study. The list of arguments excludes the other generic optional arguments (like

[174] more precisely, a feature of some painted object (see previous section *Primary Relations of the Material Object*)

temporal, evaluative, manner, etc) because they are directly inherited with no change from the most general kinds of relation concepts in the hierarchy.

Particulars as Unique Universals

A proper name, like "Becky" or "Empire State Building", is normally used in a sentence to denote a unique individual thing (ie, a particular). Similarly, a definite description, like "Andreas' gecko", "the father of Eli" or "Lynne's weight", is typically used to denote a unique individual thing, in this case a particular gecko, person or weight. There are still more, including possessives ("my cat"), demonstratives ("that woman"), and even some indefinite descriptions ("a dog bit her", denoting a single unspecified dog). All of these descriptions cite unique individuals.

The most straightforward way to represent the denotation of a unique individual in a concept net is as a particular. For example, in "Socrates is a man", Socrates is standardly interpreted as a named particular, and the proposition is expressed as: "|Socrates| 'is an instance of' [man]", or graphically as:

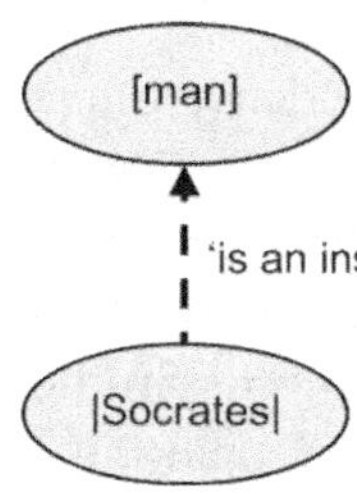

Figure 38: Representing a Unique Particular

However, another way to represent the same proposition is to treat Socrates as a unique universal (ie, a category which has exactly one instance, indicated by a unit superscript): "[Socrates]^1 'is a kind of' [man]", or:

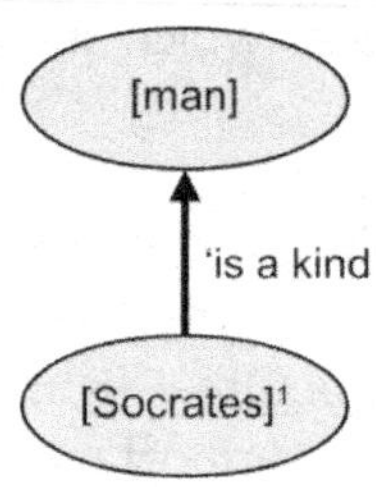

Figure 39: Representing a Unique Universal

The representation of Socrates is now implicit, since he is represented conceptually as a category of one, and the predicate is accordingly changed to 'is a kind of' rather than 'is an instance of' to signify a relationship between two universals. Seen in this light, names and other unique descriptions in language can sometimes be thought about as talking indirectly about particulars—ie, as predicates. The distinction between a particular concept like |Socrates| and its unique universal counterpart [Socrates][1] also is sometimes technically useful, because it allows talk about the representation of Socrates as a concept as well as Socrates himself (eg, to explicitly represent that he is a unique individual). A complete representation may include both:

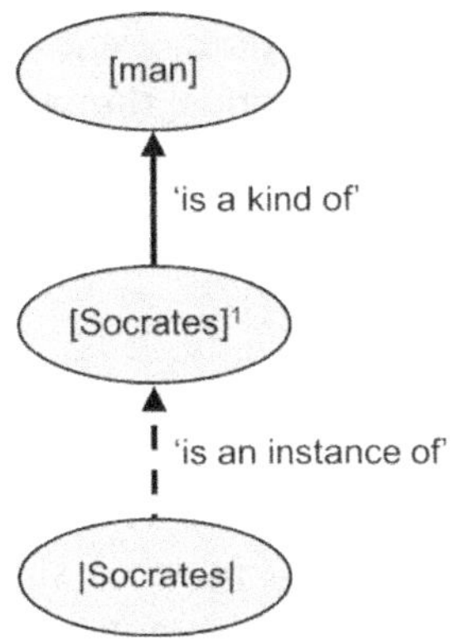

Figure 40: Relationship Between a Unique Particular and a Unique Universal

Such unique universals are sometimes awkwardly referred to as "abstract particulars" when it is convenient to emphasize their distinction from other universals (Campbell, 1997).

But the justification for including unique universals, along with particulars, in concept nets goes beyond technical considerations. We seem to intuitively appreciate the difference between the various occasion-like presentations of an object (eg, at different times and places for physical objects), and the summary of those presentations we regard as the enduring individual object. This intuitive difference may correspond to the well-researched representational difference between exemplar concepts and prototype concepts, as discussed in Chapter 2: *Three Basic Varieties of Concept Representation*. Under this line of reasoning, exemplar concepts are conceptual models of perceptual encounters representing particulars like |Zachary #1|, |Zachary #2|, etc, on different occasions; while prototype concepts are conceptual models of the persisting individual, here represented as [Zachary][1].[175] In other words, our normal understanding of an individual object more insightfully corresponds to a unique universal, rather than to a particular instance of a generic category. Although we make use of both conceptual forms for different purposes as explained in Chapter 2, our core experience of a world populated by individual objects is based on an abstraction covering a range of nonabstract presentations. Note, in passing, that a particular presentation, like |Zachary #1|, is a continuant, and not an occurrent concept (eg, an event), even though the concept arises out of an occurrent experience of Zachary at a certain time and place. An occurrent has participants in an action, but Zachary himself is being represented here, rather than the occurrent experience of Zachary.

---

[175] Prototype concepts are normally thought of as modeling universal categories like [toddler] (summarizing the characteristics of typical toddlers), but the prototype theme is extended here to include unique universals like [Zachary][1] (summarizing the characteristics of many presentations of Zachary).

The particular vs unique universal distinction in content also brings out a distinction in representation. The representation of a particular thing would seem to require representing the particular values of its intrinsic properties (and particular participants in the situations of which it is a part). There is no variability or change inherent in the particular—it just is exactly what it is, typically at some time and place. But a universal, especially a unique universal, can summarize a number of instances by specifying the range of values (or the statistical distribution) exemplified by its particulars. The universal represents a category precisely because it specifies the acceptable variability or potential change distributed across its instances. Thus while |Zachary #34| may include a characterization of his height as 2'8", [Zachary][1] may specify any height in the interval 1'8"—2'10", assuming he is now 2'10" tall, where the interval covers all of his previous heights.

Unique universals may also allow us to more easily share concepts with each other. We each probably acquire our own set of particular exemplars from personal experience, and these exemplars will generally vary from person to person for any given object. But the distribution parameters we extract from those exemplars usually will be quite similar (given a sufficient number of exemplars) due to the laws of statistical smoothing. Representing the relevant distribution measures in a unique universal produces an explicit concept, which can serve as a convenient basis for talking about roughly the same thing. In normal dialog, what is communicated from one person to another is the propositional content derived from a common unique universal representation of the proposition.

In a concept net, universal concepts that represent a unique (ie, single) individual exhibit the formal property of *uniqueness*. Names and other object descriptions, as mentioned above, denote unique concepts, and are contrasted with concepts like [man] or [weight], which can represent an arbitrary number of individuals. Furthermore, each unique concept requires a unique label or index in a concept net, which distinguishes it from other concepts. If Lynne and Becky both weigh 137 lb, then [Lynne][1] and [Becky][1] (as individuals) will both be associated with [137 lb][1], since they are characterized as weighing the same, even though their weight values are not precisely the same. Particular instances of [Lynne][1] and [Becky][1]—eg, |Lynne #1| and |Becky #1|—however can express the precise difference, if needed—eg, as |137 lb #1| and |137 lb #2|, respectively. Contrast these measurement values with quality values like |female| and |male|, where occurrences of |Lynne #1| and |Becky #1| are both associated with the same identical attribute value |female|, since, colloquially at least, sex does not admit scalar values.[176] As a final point of clarification, a unique universal concept, even though it represents a single individual, may have any number of presentation instances. If the multiple instances are conceptualized as alternative presentations of a single object, then they become unified in the unique universal concept. At least for perceptual objects, a unique universal implies a *clustering* of its particular exemplars, where only one cluster is present. Other universal concepts, like [Becky's

[176] The sex characteristic "|Lynne #1| |has sex| |female|" could alternately be represented in terms of a situation like "|Lynne #1| |is female|" if desired, but the property representation is generally more informative for analytical or quantitative inference—see subsequent section on *Attribute Values*.

children], may exhibit more than one cluster, each cluster circumscribing exemplars of each child, in this case three.[177]

The discussion above has focused on the representation of continuants as unique universals, but the treatment for occurrents is the same. Explicit particular situation concepts, like ||Eli| |wears| |Eli's coat||, are also amenable to implicit representation—viz, [[Eli]^1 [wears]^1 [Eli's coat]^1]^1, where each former participant becomes a unique argument in a relation. The relational representation implies the situational representation, but expresses it as an implicit particular—ie, a unique universal. In this case, the instances of the unique universal are the various particular presentations of Eli wearing his coat, normally recorded as exemplar situation episodes in our autobiographical memories. The graphic portrayal of the implicit situation is shown below—contrast with the explicit diagram shown in the prior section: *Generalized Situations*.

This same explicit to implicit approach can be followed for all other particular objects, events, collectives, etc, as well. In fact, Quine has shown that in any logical system, any individual (ie, particular) can be turned into a relation (ie, universal) without changing the resulting system's truth conditions. In essence, what we choose to call a particular depends on our perspective on the world, and no privileged set of particulars (scientific ones included) has an absolute claim to reality.[178] This result is in keeping with the basic principle that representing the world is a fundamentally indeterminate matter—there are no concepts that refer directly to particular "objects" in the world, just useful mental representations. For concept net representations, it is perhaps possible to excise particulars altogether, representing them instead as unique universals, but this tack seems foreign to the natural ways we think about particular individuals as being fundamentally distinct from the abstract categories we use to classify them. The current approach embraces both ways of representing individuals, allowing the most perspicacious choice in each circumstance.

---

[177] But for collectives like [Becky's children], the clustering is technically among the *members* of the collective, not its instances.

[178] The particulars in scientific theories, however, tend to achieve a higher level of internal & observational consistency than folk theories.

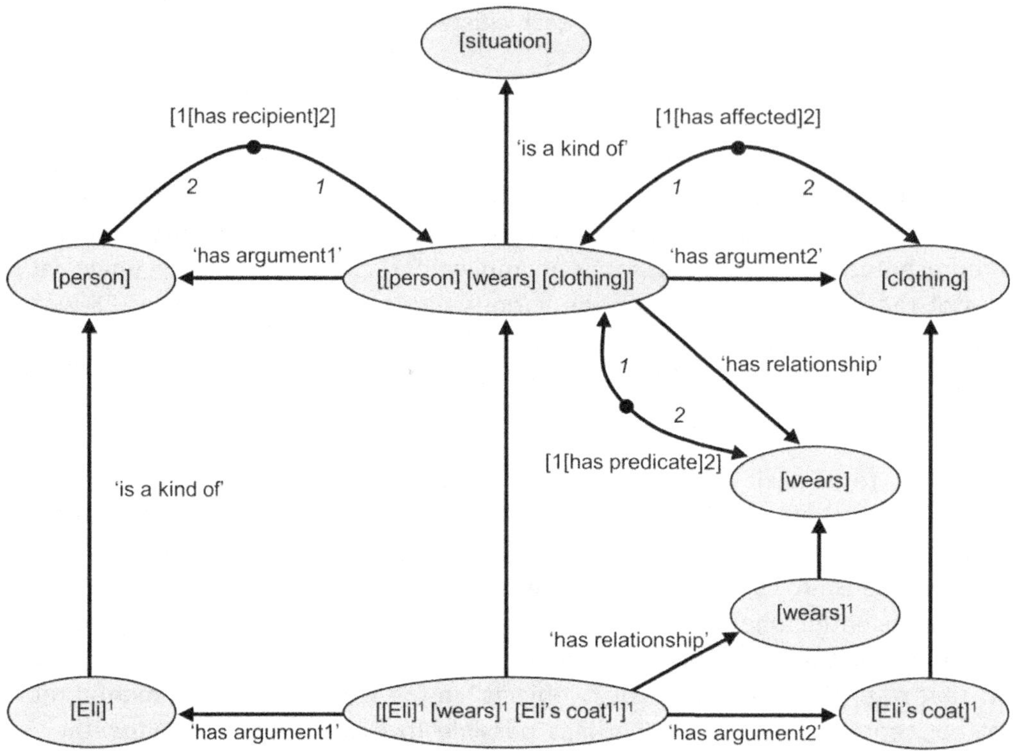

Figure 41: Representing a Particular Situation as a Relation with One Instance

Finally, an abstract representation for particulars permits a useful generalization of the notion of a unique universal. If the concept [Jennie's father]^1 is a universal representing only one individual, then a concept like [Jennie's parent]^2 can represent two individuals, and so on for any superscript index. In fact, concepts like [Jennie's atomic number]^0 or [round square]^0 which are meaningful, but impossible to instantiate, may still become topics for representation in a concept net. The superscript notation on concepts will be said to express the *tally* of the concept, where tally is a property possessed by all universals.[179] Tally, thus, replaces uniqueness as a fundamental property of universal concepts. Most universal concepts like [dog] will have an unspecified tally value, but others like those mentioned above will have numerical values.

From this point forward, the unique universal notation (eg, [Lynne]^1) and the particular instance notation (eg, |Lynne|) will be used interchangeably to represent a unique individual object, which is what we intend talk about most of the time. [Lynne]^1, as a universal, is the technically correct representation for a unique individual object, but |Lynne|, as a particular instance of a universal, is more in line with our everyday simplistic way of thinking about enduring objects and seems easier to

[179] The tally for a concept (representing a category of individuals) is analogous to the cardinality of a set (representing a plurality of elements).

comprehend in complex diagrams. A true particular (eg, |Lynne #1| with an index) will continue to represent a single presentation of the enduring object.

Sets & Pluralities

Singular and plural terms can occur in nearly any grammatical English sentence (and in most other Indo-European languages). They are used to express the fundamental semantic distinction between one and more than one item. This conceptual distinction develops very early in infancy at 15-18 months (Carey, 2009). Furthermore, every verb is required to agree in number with its subject. In some other, principally Eastern languages like Inuktitut or Sanskrit, the syntactic distinction exhibits a three-way number division: singular, dual, and more than two. (Actually, there are also a few duals in English that denote exactly two things, like "both" and "pair".) In languages like Chinese, where number is only optional, context normally provides disambiguation when needed.

The most straightforward concept of multiple items is the *set*. A set represents multiple elements either by directly specifying an enumeration of those elements—eg, {|Bush|, |Clinton|, |Obama|, |Trump|}—or a rule for picking out those elements—eg, {the 21st century US presidents}. The curly brackets are standard for representing sets. Every set has defining characteristics, mainly *cardinality* and *element,* which uniquely distinguish the set. A set's elements cannot change (and still remain the same set), but its elements can be arbitrary particular items, including other sets. The following diagram illustrates the concept net representation of a two-element set.

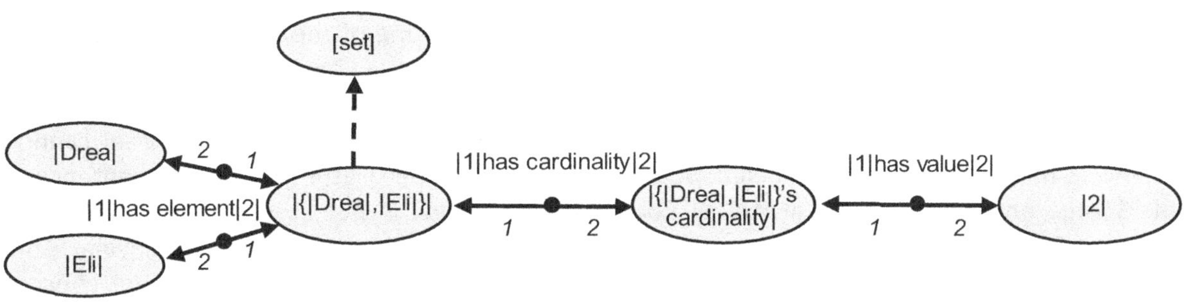

Figure 42: Combining Arbitrary Particulars into a Particular Set

In a concept net, a set always represents a particular thing, in this case indicated by the vertical brackets surrounding the two elements which are its particular elements: |{|Drea|,|Eli|}|. A set is *not* an abstract universal thing like [boy], which is designated by square brackets. But a set is also not a physical (concrete, material, spatiotemporal) thing like the boy |Drea|. A set of physical boys like |{|Drea|, |Eli|}| is still a nonphysical particular thing, since the set is conceptually distinct from its elements. In general, a set can have elements which are arbitrary particulars, including other sets, but cannot have universals as elements. For example, the abstract universal [dog] cannot be a member of a set, but the particular concept of dogs, namely |[dog]|, can be. The distinction is subtle, but maintains the fundamental meaning and dichotomy between particular concepts and universal concepts. This line of thought places the concept of a set near the top of the concept hierarchy:

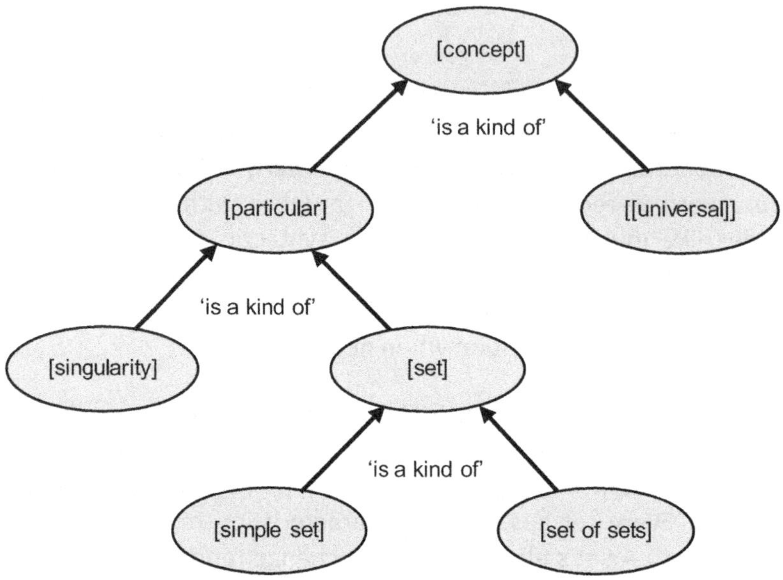

Figure 43: Locating Sets in the Concept Hierarchy

In certain limited contexts, a set interpretation is also appropriate for terms like "The Beatles" — rendered as |{The Beatles}|—when talking about the set consisting of the fab four. The concept of a set, however, should not be confused with the concept of a *plurality.* The *group* (ie, the band) labeled |The Beatles| in a concept net, is implied in a sentence like "The Beatles played *I Want to Hold Your Hand*", since it is the unified group as a whole doing the playing. It is not a set, since a set cannot play anything. Similarly, the plurality labeled |John, Paul, George, Ringo| as implied in the sentence "John, Paul, George and Ringo played at Abbey Road Studios today", is not a set. It is a *collective* with characteristics that can be applied to any of the individuals in the collective (eg, playing a musical instrument). While the concept of a set is contrasted with that of a singularity in the concept net hierarchy, the concept of a plurality is contrasted with that of an individual:

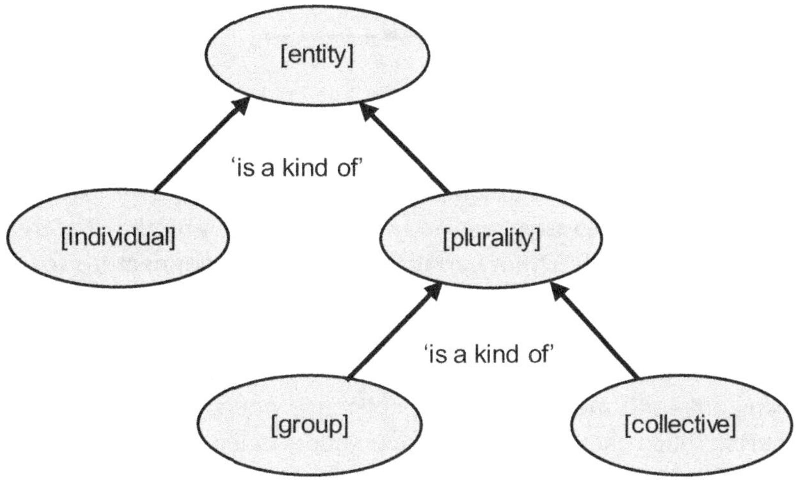

Figure 44: Locating Pluralities in the Concept Hierarchy

Consider the sentence "Drea and Eli like Nynne", and its representation in a concept net. A straightforward interpretation produces a sentential conjunction, being equivalent to the two sentences "Drea likes Nynne" and "Eli likes Nynne". Therefore, the plural subject "Drea and Eli", along with the plural verb form "like", is simply a shorthand way of talking about two individuals and two acts of liking separately. The separate representations would look like:

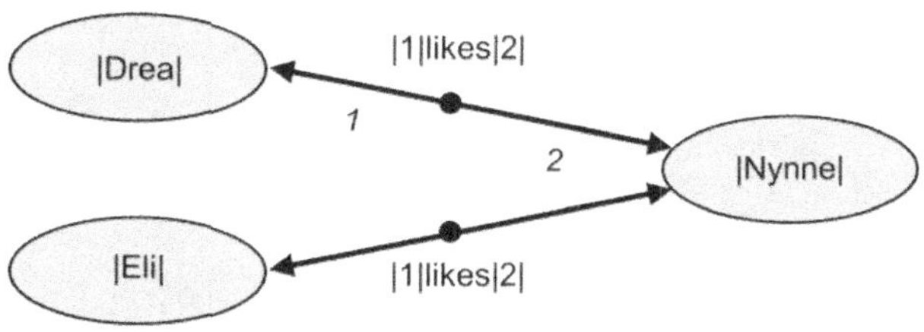

Figure 45: Representing Parallel Relations

Contrast this use of the plural with "Drea and Eli lifted Nynne", where it takes their combined strength to do the lifting. In this case the meaning is not equivalent to "Drea lifted Nynne" and/or "Eli lifted Nynne" independently, even though how the lifting was actually done is left unspecified. Under a standard interpretation only one lifting is implied, and it was accomplished jointly. The combination represents a *collective* concept, where there is a contextual connection between the collective's *members,* analogous to a set's *elements*. Drea and Eli are now conceptually combined into one particular collective:

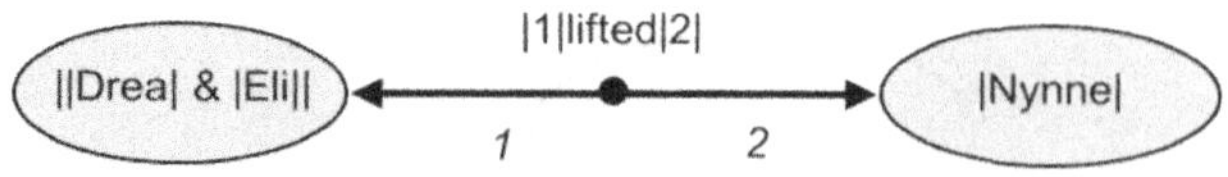

Figure 46: Forming a Collective with Members

The range of collectives expressed in language, however, goes beyond simple conjunction. Another is disjunction (both inclusive and exclusive or) as in "Drea or Eli ate the ice cream", represented as ||Drea + Eli| |ate| |the ice cream||. Other variations on the interconnected "togetherness" implied for some actions include open-ended collectives like "those crooks robbed the bank": ||those crooks| |robbed| |the bank||, the plural form of the particular concept indicating togetherness. The following diagram illustrates some relevant concepts and relations for the simple collective concept |my arms|. My left and right arms are both members of the collective and are also "parts" of |my body|, in the sense of being subparts, which themselves can have subparts like hands and fingers (see *The Primary Relations of the Material Object* above for a discussion of subparts):

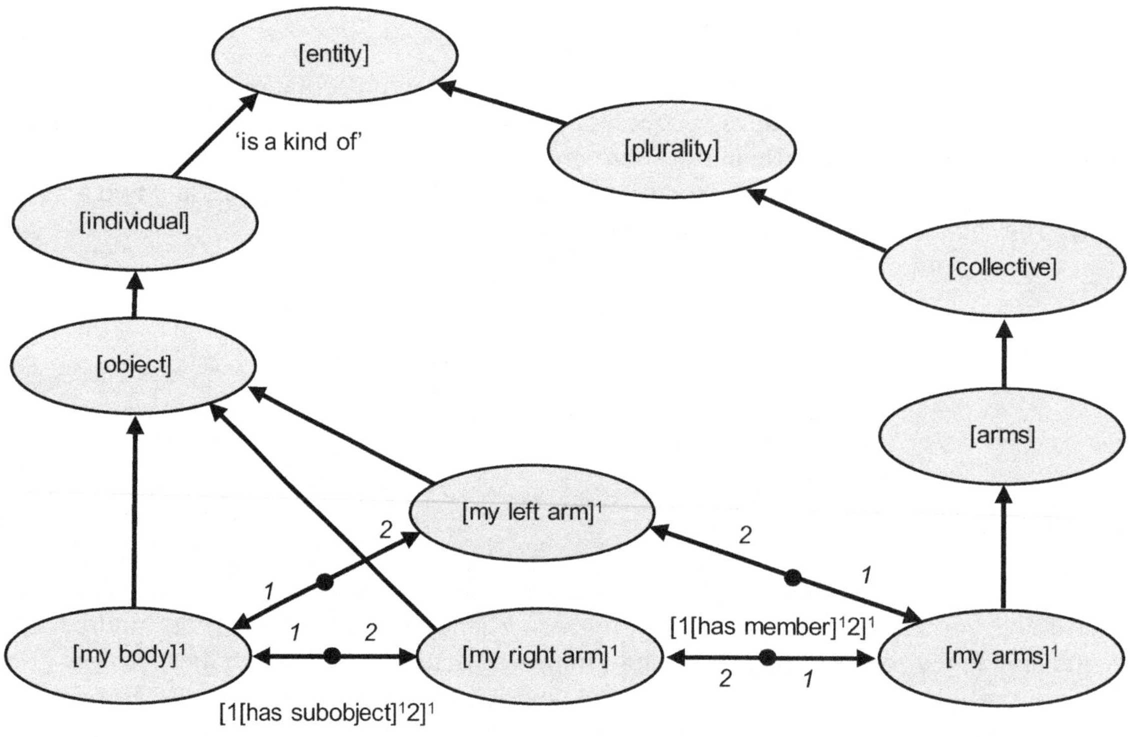

Figure 47: Representation of a Collective

So-called collective nouns, however, usually signal a group concept—eg, "a flock of seagulls terrorized the school", where |a flock of seagulls| is interpreted as a cohesive group. Here, note that "flock" is a singular term; it expresses one thing (a group) which is made up of a number of other things (its members). The same idea is likely expressed by "seagulls terrorized the school", but now the plural concept |seagulls| expresses a collective explicitly (and probably somewhat more loosely). If there is a choice between interpreting a singular noun as a plurality vs an individual, interpreting it as a plurality leaves open the possibility of adding information later about potential members, while an individual interpretation closes out that possibility. Likewise, interpreting it as an individual, allows adding information about potential parts, rather than members. Pluralities (in analogy with sets) can even be compounded to produce complex pluralities like ||the Montagues| & |the Capulets||.

At a deeper semantic level the logical form of many simple pluralistic sentences is standardly expressed through the use of quantifiers like $\exists x_D\, F(x)$ "there exists at least one D which has characteristic F", and $\forall x_D\, F(x)$ "every D has characteristic F", combined in various ways. Logical forms afford one way of reasoning about knowledge, and quantifiers generally indicate how many things (some, all, and other derived quantities such as exactly two) are talked about by the predicates in the sentence. Extensions of logic to include set theory provide additional power to represent sets of elements, in addition to numbers of elements. It would appear there is something fundamentally innate associated with our ability to talk and think about various pluralities. Chomsky calls it "an elementary property ... which is exhibited in its purest form by the natural numbers 1, 2, 3, ... Children do not learn this property" (Chomsky, 2000).

Objects & Substances

The earlier discussion of wholes and parts hints at the common difficulty in separating out the underlying concepts associated with conceiving of one thing as being made up of many things. It haunts the metaphysics of what it means for an object to be composed of substances, as well. In fact, the distinction between an individual object and the substance which serves as its medium or substrate has been one of the most difficult to reconcile in philosophy (Cruse, 1986; Simons, 1987). The intuitive distinction is clear enough; we have no trouble talking about a porcelain statue or a baseball as being composed of string, glue, cowhide, etc. The main difficulty seems to revolve around the existential status of the substance: does it exist in the same sense as the object of which it is a part?

From a realist perspective, most of us would say the object clearly exists. Elaborate perceptual processes strive to deliver object representations of various kinds based on information in the environment, and those object representations provide the bedrock for our intuitions about what is real. While the ultimate nature of the object necessarily remains uncertain, we nevertheless assume "something" is there. However, our conceptual characterizations of that something are very different from the thing-in-itself, as Kant would say. Our concept of substance, as the material extent of an object, is simply one characterization of the thing-in-itself, much like the concept of its weight, as the inertial extent, or shape, as the spatial extent of the object. The extent to which we confer existence on concepts like an object's substance, weight, or shape only derives from the assumed existence of the thing-in-itself.

The substance of which an object is made, therefore, does not exist in the same way as the object itself. "This rubber ball" (object) and "the rubber in this ball" (substance) are two different concepts, and they have a different basis. Objects are fundamentally different from their compositional properties. This is why two rubber balls cannot be located at the same place at the same time, but a rubber ball and the rubber of which it is made can. The concept of location is being used to describe both the object and its substance without the conflict inherent in describing the overlapping location of two things-in-themselves. Similarly, any characteristic of a rubber ball (eg, its weight) is derivatively located at the same (possibly partial) place as the rubber ball at the same time. In this sense, substance is just a useful property concept we have developed for thinking about objects in general; it does not represent a separate thing-in-itself.

There are many ways to characterize objects and substances. The following observations attempt to clarify the differences between the concept of the prototypical object and the concept of prototypical substance:

- Individual objects are generally countable, while substances are not. You can count the total number of pennies in a jar, but not the total metal of which the pennies are made. However, countability is insufficient to characterize the difference alone, since many other things are uncountable as well (eg, color).
- Substance is usually characterized in terms of amount (eg, the amount of gold in a ring), where an amount is a continuous quantity of substance (eg, 0.5 oz of gold). But again, other things come in amounts, namely quantitative properties (eg, the amount of force exerted in a press). Assuming an amount is the same as a quantity, it does not by itself distinguish between the concepts of substance and quantitative properties.
- An individual object is usually characterized in terms of a shape or function, rather than amount. A strawberry normally has a distinctive heart-like shape and is edible, while a doorstop serves a distinctive function.
- Another distinguishing criterion is based on how a substance vs an object divides into subparts (or combines into wholes). A substance is usually thought of as dividing continuously, such that the subparts of a substance are themselves a substance. On the other hand, the subparts of an object divide discretely and are themselves objects, even if they are fragments. Of course, divisions of both substance and object eventually reach a minimal atomic subpart, so the distinction is purely conceptual, but it is useful in reasoning about everyday affairs. In making a chocolate topping, it may be simpler to add four ounces of chocolate (amount of substance) to a recipe as opposed to adding 16 small chocolate squares (individual objects).
- Another distinction is based on identity conditions. The identity of a substance seems to depend not only on the kind of substance it is, but also on it being that particular substance. For example, the gold in a ring remains the same gold if the ring is melted down and recast as a cross. The gold in the ring and the gold in the cross are different concepts, but the gold in each is identical. Moreover, if some of the gold in the original ring is even partially replaced with some other gold, the gold in the original ring is then not identical to the gold in the ring following the substitution. For the ring object however, melting it down and recasting it in

exactly the same form normally counts as a different ring. Alternately, simply substituting some of the ring's gold for other gold (or even another metal) does not normally make it a different ring. The spatiotemporal form of an object is more central to its identity, whereas the particular materiality is more central for substance.

- If division is combined with identity considerations, we can also say that any division of a substance remains the same *kind* of substance. But dividing an object arbitrarily will produce objects of a different kind. An object does not survive a significant arbitrary division (or a significant arbitrary addition), although an incremental loss or gain usually leaves it the same object.
- Substance can be reconceptualized as an object by applying an objectifying partitive operator, such as *piece of* or *lump of*, to the mass term. For example, the gold in a ring is substance, but a piece of the gold in a ring is an object. However, it is important not to confuse particular instances of a substance with particular instances of objects made of that substance. For example, the concept [gold] is a substance concept, and it has particular instances like |this gold|, |the gold in this ring|, or |the gold in this nugget of gold|. These particular instances are not particular objects; they are particular instances of a substance. The particular substance instances are what particular objects are composed of. Many common nouns can also be used to represent either substance or an object, where the operator is simply a definite or indefinite article (chicken vs a chicken, glass vs the glass, etc). The appropriate representation will, as usual, depend on the intended meaning.
- Finally, the classification of different kinds of substance is parallel to, but distinct from, the classification of different kinds of object. Substance, it is proposed, is conceptually subclassified into either homogeneous substance or heterogeneous substance, such that different homogenous amounts combine together to make a heterogeneous amount of substance. Copper and tin combine in this way to make bronze. In fact, the field of chemistry is, in some fundamental sense, a study of the properties of homogeneous and heterogeneous materials (see Chapter 6: *Organization of Chemistry Concepts*). Likewise, objects can be subclassified into homogeneous or heterogeneous objects, depending on whether their parts are viewed as being the same or mixed. A brick wall is a homogeneous object composed of bricks (assuming the mortar is irrelevant in the context), while a broom (with handle and bristles as subparts) is heterogeneous. Most everyday objects then come out as heterogeneous, with homogeneity reserved mostly for idealized objects. However, we can also readily reconceptualize a heterogeneous object, by thinking of it as having the same kind of particles as subparts, as when we think of an object as being made up of atoms.

Object Fragments & Fusions

The concept of an object *fragment* is distinct from the concept of a whole object, because a fragment is implicitly dependent on the object of which it is a fragment. Fragments include perceptual or conceptually motivated parts like my arm or your circulatory system, neither of which is normally thought of as an independently functioning whole, as well as "grammatically constructed" segments of various kinds of things like the crust of a pie, a lump of clay, or a page in a book. Fragments represent the partitioning of a whole individual object into other individual things (ruling out substances), with the restriction that none of the fragments are themselves thought of as whole objects. Thus Rosie represents a whole object, but Rosie's tail is a fragment. Fragments, however, may themselves be subdivided into additional subfragments (eg, a piece of a slice of a pie), as is the case with any concept.

A *fusion* is the formal name for a mereological sum of individual things, where the sum can be either arbitrary or motivated. For example, I can think about my cat Rosie as a whole object, or alternately as a sum of parts (head + tail + body + legs, etc), or perhaps as the sum of so many elementary particles (ie, molecules, atoms, etc). The whole vs fusion viewpoints are two distinct ways of thinking about her, even though they denote the same thing. Fusions are generalized sums, and so can represent an arbitrary sum of individual things, but the connection among them is much tighter than for a collective group. Whereas a collective (like an orchestra) has relatively loosely bound members, a fusion (like an electronic circuit containing embedded capacitors, resistors, etc) implies a more permanent integral configuration of components.

Immaterial Objects & Information

The range of immaterial objects we talk about in everyday speech rivals that of material objects. Denizens of this comparatively elusive category include things like holes, quantitative and qualitative values, surfaces, persons (in their agentive role, as distinct from human beings in their material object-body role), responsibilities, literary content, geometric figures, and information (as meaningful content of representations). Immaterial objects are particular things, and therefore are not abstract like universals. They often have spatial or temporal location, but are not a material phenomenon in the sense of having mass-energy.

Casati and Varzi discuss the relative advantages of a logical theory of immaterial bodies vis-à-vis theories that include only material bodies (Casati & Varzi, 1994), and the distinction is adopted here for concept net representations. Zalta, in his *Principia Metaphysica* (Zalta, 1999), has also developed a full-fledged logical theory distinguishing concrete from nonconcrete objects (although his “abstract object” terminology is somewhat different from the “immaterial object” terminology used here). Zalta’s formal theory demonstrates that immaterial objects can be consistently reasoned about just like normal material objects. The ontological category of immaterial objects thus supports a coherent and useful fine-grained analysis of intangible concepts to complement the familiar conceptual categories of ordinary tangible objects.

From one point of view, the ontology of *immateriality* seems to run more-or-less parallel to that of materiality (outlined previously in Chapter 5: *The Primary Relations of the Material Object*). That is, there are corresponding relations between an immateriality and its "parts". A triangle can have characteristics, like sides and angles, and a message can be thought of as being composed of information (similar to an object being composed of a substance). Our ability to grasp immaterial concepts is likely based on an ability to project from material concepts by analogy.

The "immaterialities" (more technically called *insubstantialities*) we tend to affiliate with ordinary material objects (eg, a hole in a bowling ball) seem to do with the way those objects manifest themselves. We can talk directly about the way something is, as if it is something which can be individuated from the object itself, in other words as if it is a *continuant*. The two most basic kinds of insubstantiality characterizing the way something is, are here termed *values* and *modality*. Values are simply qualitative or quantitative assessments of objects (eg, heavy or 240 lb), and are addressed subsequently in the section on *Attribute Values*. Values characterize objects in terms of attributes (eg, weight). A modality, however, pertains to a characteristic more directly inherent in the underlying object—namely, its mode of presentation or the *way* it is—and can be usefully thought of as a nonmaterial constituent of the object (eg, the information in a message or the pattern in wallpaper).

The ancient philosophical idea of a *form* may provide a familiar anchor for analyzing these kinds of modalities,[180] if we can contrast form with *content*. Assuming the form of an object refers to the overall configuration of the object's parts, rather than the parts themselves, the content derived from the form is described in terms of the information extracted when the form is interpreted. The form is the pattern, while the content is associated with the pattern's meaning. Forms are said to convey information content for agents smart enough to interpret them. In this way of looking at meaning, information becomes a *measure* of the interpreted content derived from a form. An informed interpretation makes more information available to an agent than a less informed interpretation. For example, if my background beliefs about bachelors simply characterize them as unmarried men, and your beliefs characterize them as never-married men, then your classification of an individual as a bachelor will provide you with more information about him than mine. In this case the additional information can be roughly quantified in propositional terms as an extra constraint on my bachelor concept. The content of more picture-like forms may be better quantified by nonpropositional information measures.

Many patterns (eg, the one on my kitchen tile floor) are not normally designed to convey information, but information is typically available in the pattern, if one knows how to exploit it. For example, a distinctive pattern may indicate which company manufactured the flooring. Such a form lacking a purposive information conveying function will simply be called a *configuration*. Other forms, usually called *representations*, are conventionally understood to convey information (eg, symbols, portraits, language, and other conventional signs). The information content we associate with forms in general results from the interpretive processes we apply to them. Both the form and any derived informational content are immaterial objects, tied to a material substrate, but conceptually distinct from it. Following this approach for analyzing intangibles produces a tentative hierarchy of immaterial concepts, as outlined in the following figure. (See Chapter 6: *Explanations for the Top-Level Epistemological Concepts* for further comments on the concepts in the figure.)

---

[180] The existence of form concepts for the purpose of representation is very different from the existence of forms themselves, which historically has proven to be highly controversial.

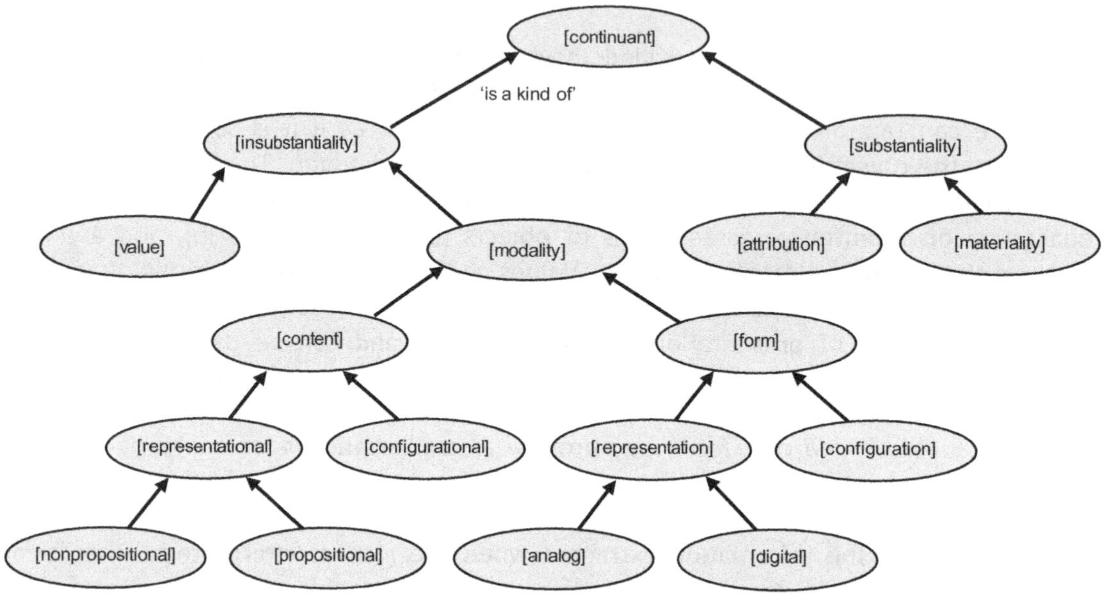

Figure 48: Hierarchy of Immaterial (Insubstantial) Concepts

The principal distinction between form and content is perhaps best understood in terms of a simple correlated interaction between two objects, one object hosting a form (ie, structural pattern) which consistently affects the second object in some way. The resulting change in the affected object constitutes the content (or significance) of the interaction, where the first object is the information producer and the second object is the information consumer (see Chapter 2: *Representations in General*). If the consumer is a sentient agent, it is said to actively "interpret" the form; but for lesser subpersonal consumers, the consumer is merely passively "affected by" the form.

A bear in the woods eating wild berries illustrates a passive interaction. Here, the berries constitute the producer, and the bear the consumer, where the ripeness configuration of the berries (expressed in their color and texture) is noticed by the bear. If the particular form of the berries is a toxic variety (expressed in their chemical structure), then the bear may become sick. The bear's sickness is then the content or meaningful significance resulting from its interaction with the berries. The "state" of the bear (more accurately, its *configuration*) has been affected by the form of the berries it ate. In a potential follow-on interpretation, the bear itself may become the producer for another information consumer, such that the form of the bear (speaking colloquially, its state of being sick) leads to further content for another consumer.

The same kind of correlated interactions seem to occur at many levels in and between organisms generally. In fact, the DNA in the nucleus of a cell might be viewed as an information producer, with other cell components serving as consumers. The information derived from the pattern of DNA nucleotides during the processes of transcription and translation eventually results in the production of proteins configured in specific ways—newly created forms which are useful to other consumer processes in the organism.

As previously stated, a concept is a mental representation physically realized in the neural substrate. When talked about this way, a concept is said to be explicitly *mentioned*. However, the concepts appearing in a concept net are tightly bound together through consequential associations with each other. When the concepts in an integrated concept net are implicitly *used* by interpretive processes to draw inferences or otherwise produce content, they are interpreted, producing belief or thought content. The subtle use/mention distinction is important for properly understanding representations in general.[181] As discussed previously in Chapter 4: *Basic Terminology*, when a concept is labeled |Fido| in a concept net, the label informally summarizes the concept's intended content, expressed by a convenient locution. The content associated with |Fido| may be derived from background knowledge about Fido stored in the concept net—eg, that he is a dog, is shaggy, has two ears, etc. The concept itself, as it appears in the concept net however, is not its label or content, but rather a pattern of neuronal connections. As such, it is a representation (of something else, in this case a particular dog) possessing explicit characterizations attached to a nexus in the concept net. Ordinarily there is no need to think about concepts as representations; we simply use them to create content on the fly. But, still, we do create some concepts to serve as explicit representations, when we want to build models of the world. We then turn around and use those model representations to develop predictive and explanatory content.

In order to explore the nuances involved with representing informational content, consider the example of tree rings. The growth rings observable in a cross section of a tree trunk are known to correlate with the age of the tree. One ring typically corresponds to one year of growth, so the total number of rings corresponds to the total number of years the tree has been growing. This relationship amounts to background knowledge that most people can use to infer the age of a given tree section.

The following diagram outlines the concepts which might be involved in the explicit representation of information about the rings in a tree trunk. The concept of a tree itself, along with the part of the tree which hosts the tree rings (ie, its trunk) are viewed as normal material objects, as discussed previously in Chapter 5: *Primary Relations of the Material Object*.[182] The tree ring pattern (as a general form), however, is an immaterial feature of the tree trunk. Thus, while the wood in the trunk is material, the

---

[181] In language, the use/mention distinction is somewhat easier to understand: If I say "Deven rocks", then I am *using* the words "Deven" and "rocks" (as linguistic representations) to say something about Deven, the dancing toddler. But if I say "'Deven' is bisyllabic", I am *mentioning* the word "Deven" (still as a linguistic representation) to say something about the word itself. The convention in language is to enquote a word or phrase to mention it, but in a concept net, the most straightforward way to mention a concept is probably to prefix the concept with the type of representation it expresses—eg, |the symbol π|, |the name Fido|, |the pattern of the tile floor|, |the x-ray of Bob's lungs|, etc.

[182] The tree trunk is simply glossed as a generic part of the tree, although "part", as previously noted, is highly ambiguous. According to the full ontology developed in Chapter 6, the trunk is better regarded as a *fragment* of the tree.

particular way the wood is arranged into a ring pattern is not. If the form remains uninterpreted, it simply expresses the concept of the ring pattern as an (uninterpreted) configuration of a tree trunk. But an agent with suitable experience analyzing tree rings is capable of extracting information from the pattern, for example by counting the number of rings. The number of rings constitutes the information extracted from the raw configuration—one piece of information perhaps among many others available in the pattern. In this case the ring pattern is being classified, and subsequently interpreted, as a representation (ie, as conveying information).

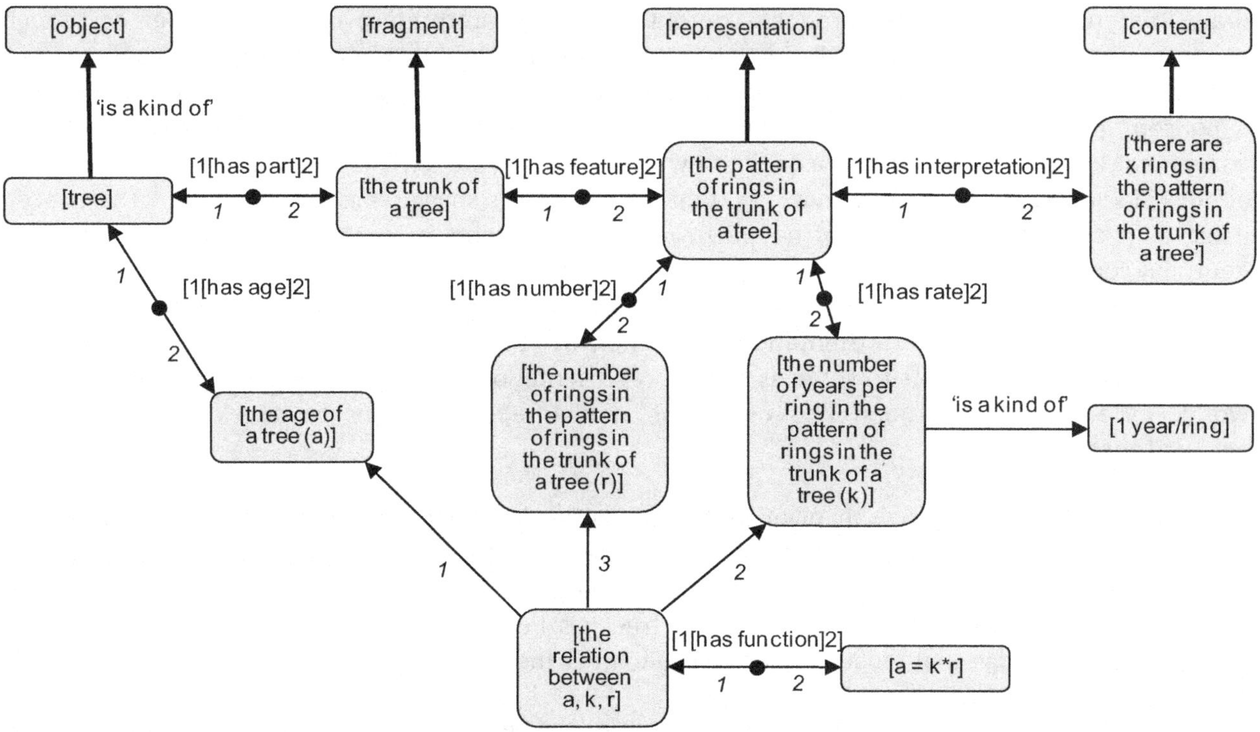

Figure 49: The Representation of Tree Ring Concepts

Once information is extracted from a representation, it may become generally available for inference operations (not shown in the diagram). Given a new belief that a particular pattern of tree rings contains 150 rings, and the background beliefs that each ring corresponds to one year of growth, and that annual growth is a measure of age (a known attribute of a living thing), the agent can infer that the tree's age is about 150 years.

More specifically from the diagram, background knowledge is assumed to include three attributes (ie, three instances of attributes, or equivalently, three kinds of properties): the age of a tree, the number of rings in the pattern of rings, and the rate of ring growth (specified as one year per ring).[183] Knowledge of the functional relation binding these three attributes together, then allows calculation of the

[183] Immaterial objects can have attributes just like material objects.

attribute value for either age (a) or the number of rings (r), given the other, for any particular tree. Note that the raw nonpropositional content derived by an agent reading the rings of a particular tree must first be translated into structured propositional content—ie, into an attribute value for the property |the number of rings in the pattern of rings in the trunk of this tree (r)|—before the information can be represented conceptually. In general, deriving informational content is only the first step in putting that information to use. In a broader learning sense, the conceptual structures involved with specifying the previous property for r, could only have been created in the first place by translating information content. That is, the agent was probably either told that the number of tree rings corresponds to the same number of years of tree growth, or came to this realization independently, perhaps by induction after many observations. Either way, our powers of inference (in which we process proposition-like information) are wide-ranging, and oversee the constant structuring and restructuring of our background concepts.

Finally, some additional clarification of the difficult notion of *semantic content*, from a more philosophical point-of-view, may be relevant—viz, how it is distinguished from the closely related notions of *reference*, *extension*, and *intension*, all of which are similarly bound up with the meaning of a concept. Such ideas have long held a central place in formal semantic theories (Caan, 1993). As already proposed, the content of a concept on some occasion of use is the information which is derived by an agent from the concept's embedded representation. In a wide sense, content amounts to the interpretation of current beliefs and thoughts plus potential inferences about them. Semantic content itself may or may not feed back into long-term memory as new propositions, thereby restructuring our beliefs and thoughts. But if it does, those new propositions are subsequently available to reasoning and other interpretive processes.

The notions of *reference* and *extension*, as standardly used in linguistics and reapplied to concepts, typically relate to the assumed things in the world that a concept represents. The concept |Fido| then *refers* to a worldly dog object, which constitutes the concept's extension. But if the concept |Fido| is a particular delimited arrangement of neurons in an agent's head, that arrangement is simply a passive pattern of connectivity until it is interpreted. It is the interpretation of that particular arrangement which gives it meaning as a representation of something. In other words, it is the *content* derived from the representation, rather than the passive arrangement of neurons, that "refers" to the worldly dog Fido. It is the content of a concept which has an extension, rather than the concept itself. The traditional use of "reference" thus bypasses content. In fact, the whole idea of reference, as a conceptual construct, can perhaps be dispensed with as discussed at the beginning of Chapter 2, if we view conceptualization as the mere attempt to make some (any) sense out of our raw experiences, rather than as discovering preexisting conditions or objects in the world.

The notion of *intension*, as used in linguistics, typically applies to whatever it is that gives a word its extension—ie, that *determines* how it refers to some things in the world and not to others. The intension of a word then is usually associated with the set of properties that pick out a word's extension (what it denotes). For the psychological/conceptual leaning advocated here, the intension of a concept is preferably thought of as the potential activation pattern expressed in all its physical connections with other concepts in the current context, since this pattern shapes the concept's interpretation.

Intentional Objects

Our mental experiences seem to consist of relatively long-term mental *states* (like being angry, loving someone, believing that Santa Claus exists, etc) and short-term mental *events* (like perceiving a pin prick, deciding whether a sentence is grammatical, thinking that I should wear green today, etc). In philosophical terms, all such states and events (and processes, happenings, states-of-affairs, etc) are called *occurrents*, because they constitute dynamic situations occurring at or over a period of time. They are characterized in terms of an action (eg, being angry) and the participants in that action (in this case, a person). The dynamic nature of states and events is typically expressed in language by verbals like being, loving, believing, perceiving, deciding, thinking, as in the examples above.

However, when we want to talk about things existing in time as static objects, we switch to nominal expressions like anger, love, belief, percept, decision, thought. Such objects are called *continuants*, because they are enduring entities, continuing to exist (wholly present at any moment) over time,[184] and are characterized by intrinsic properties. The difference between occurrents and continuants is important here, because intentional content will be analyzed as a property of a mental representation (a continuant). Much of the philosophical literature, alternately, seems to think of intentional content as one of the participants in or characteristics of a mental state or event (an occurrent) (Chalmers, 2006).

Mental representations will be said to have *content*, which is what those representations represent for the agent, what they are about. For example, my mental representation of my cat Rosie (ie, my concept of Rosie) has Rosie as its content. But its content is not the real cat Rosie, and not some other related representation of Rosie. It is rather the *information* that I have, or that I am currently entertaining, about Rosie. It is the intangible result of certain interpretive mental processes operating on, or disposed to operate on, my representation of Rosie in some context.[185] As discussed in the previous section, contents are technically immaterial, but nevertheless physical, being most usefully classified as some kind of informational element.[186] This interpreted content is then actionable, in the sense that it can produce physical mental effects. An analogous process occurs in a computer when the representation of an instruction like (+ 1 1) is interpreted, executed, and the result (the binary representation of 2), stored in a register.

Only some kinds of mental content are typically called *intentional*. The usual examples of intentional content include the contents of beliefs, knowledge, hopes, etc (associated with mental states like believing, knowing, hoping, etc) and thoughts, judgements, imaginings, etc (associated with mental events like thinking, judging, imagining, etc). There are surely other kinds of mental content which are

---

[184] The lifetime of a continuant or occurrent can be long or short. A belief normally exists for longer than a thought, for example; and similarly, for believing something and thinking something. The static vs dynamic distinction is about the object vs event distinction, not duration.

[185] Exactly what information *is*, is a controversial topic. The view here is that it is "embodied" in the way neurons are connected and activated together, forming (static) connection and (dynamic) activation patterns in the brain. Our disposition to interpret those patterns in some way is the information content.

[186] By way of contrast, the *concept* of content (as a representation, rather than what it represents), is a certain physical interconnectedness in the neural substrate of the brain, like any other concept.

not intentional, such as being in love or being in pain. But the present discussion is mainly about intentional content and the representations from which it is derived.

Intentional content is often introduced in English sentences by a 'that' clause (Burge, 1979) as in "I believe that God *is* love", where my belief is 'God *is* love'.[187] My physical mental representation of this belief is the basis for my personal interpretation of the belief (ie, its content). Intentional contents and representations are particular to each agent—not shared between agents. In a situation where two people are commonly said to share the same belief, they are each simply interpreting their own belief in the same way for the purpose at hand. Exactly what it means for two interpretations to be identical, or sufficiently alike in a given context, remains problematic; but in general, it means the respective interpretations convey the same information to both parties.

Those particular mental representations (ie, beliefs, thoughts, etc) that are associated with intentional content will collectively be called *propositions*. In the previous example, the belief 'God *is* love' is a proposition, labeled |<God *is* love>| in a concept net, indicating it is a particular instance of a proposition. The proposition is enclosed in angle brackets to distinguish it from a linguistic concept like |"God *is* love"|, which represents a particular instance (or use) of the *sentence* "God *is* love". The content of the proposition (ie, its interpretation on a particular occasion) is further distinguished as |that God *is* love|. Keeping these related concepts separate goes some way toward clarifying how intentionality works.

Propositions are regarded here as real mental entities, since they are representations; but they are personal, not directly shared between agents, as is sometimes thought. They are *logical* (or more likely *quasi-logical*) representations, used by the executive processes for the purposes of reasoning. For example, given the propositions |<Alan is taller than Bill>|, and |<Bill is taller than Charlie>|, we can infer |<Alan is taller than Charlie>|. In theory, many of the usual, so-called natural, inference rules are available to us as executive functions, although how they are so effectively (and often so ineffectively!) applied is notable.

Propositions also do not possess truth values directly. Whether a declarative statement is true or false depends on whether what the statement says corresponds with the way the world is (Kaplan, 1977 & 1989). Since propositions are only used for logical analysis, any proposition itself carries little or no contextual background linking it to the world. For this reason, truth values are better associated with the content derived from propositions. To achieve declarative content, a *context* must be added to the proposition during interpretation to say something about the actual world. For example, the propositions |<She is taller than the girl next door>| and |<The girl next door is taller than my sister>| allow the inference |<She is taller than my sister>|, even though 'she', 'the girl next door', and 'my sister' are left unspecified. Once they are identified in context (eg, as Alissa, Barbara, and Carol), then the content that Alissa is taller than Carol evaluates to true or false depending on the actual

[187] Somewhat perversely, I can also say that "I believe that I believe that God *is* love"! In this case, I also maintain a representation of the belief that 'I believe that God *is* love' in addition to the belief that 'God *is* love'.

circumstances. The diagram below offers a rough sketch of intentionality as outlined in the previous comments.

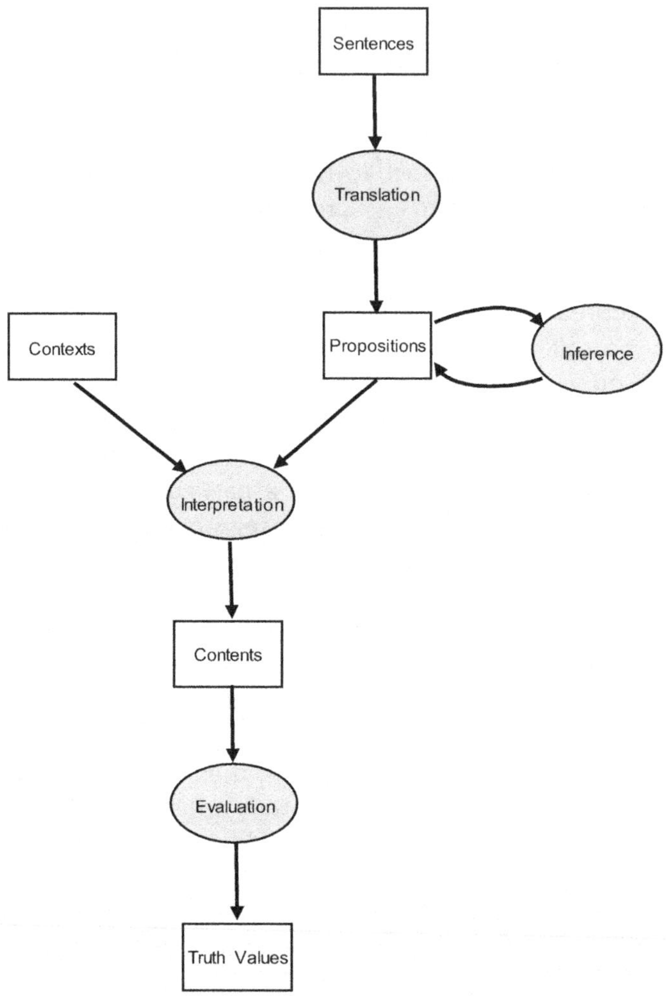

Figure 50: Sketch of Propositional Intentionality

To reflect on a speculative illustration of intentional processing, suppose I say to you, "My wife is a nurse". Presumably you would first build up a mental linguistic representation of my utterance, call it |"My wife is a nurse"|, to represent a particular instance of that spoken sentence. A full sentential representation would include connections to both syntactic and semantic concepts, which determines how you understand the literal sentence. Once fully represented, it is then amenable to translation into a quasi-logical form, call it |<My wife is a nurse>|, namely a proposition suitable for use in your reasoning about wives and nurses. But the syntactic and semantic properties of the propositional form at this stage probably remain relatively independent of each other to permit inference based on the rules of symbolic logic. In quasi-logical form, it can combine with other propositions, via inference, to produce new propositions (eg, my wife probably is a caring person, has been to nursing school, etc).

As we are taking part in a conversation, you would also be unconsciously sensitive to our conversational context. Assuming you mentally represent such conversational contexts, call it |the context of "My wife is a nurse"|, it would likely have characteristics such as {speaker=Dave, time=morning, month=March, place=LA, audience=me, ...}. Different contexts highlight different characteristics, probably triggered by word associations in the original sentence, as well as salient objects in the current environment.

Once propositional and contextual representations are in place, it then becomes possible to interpret the proposition by identifying references (eg, that 'my wife' signifies Dave's wife), establishing links to background emotions and knowledge (eg, that Dave's wife is Lynne, and that I like her), and entertaining new information (eg, that Lynne is a nurse). The result of your interpretation, call it |that Lynne is a nurse|, is tacitly embodied in the static connections and dynamic activations of concepts involved in the interpretation. It is the informational content reflecting the meaning of the sentence to you, what the sentence is actually saying about the way the world is.

Finally, you can optionally evaluate the truth of the statement. You can consult your background knowledge about Lynne's profession, gather additional knowledge about her, or appraise the reliability of the speaker (in this case her husband). If you deem this new information reliable, you can turn it into an explicit proposition like |<Lynne is a nurse>| and store it as one of your background beliefs. It may even cause some revision in your other beliefs, depending on how your network of beliefs is organized.

This somewhat tenuous discussion of intentionality, of course, leaves many questions unanswered. However, the present goal is simply to provide a tentative framework for more thorough investigations of how we mentally process information. The concepts of interest are now more clearly seen to revolve around the nature of—and interactions among—mental representations, information content, and the perceived world.

Quantification

A relation among several arguments imposes selectional restrictions on those arguments, restrictions that, apart from metaphorical use, are imposed on all situation instances of the relation. The epistemological function of a relation in a concept net, therefore, is to authorize legitimate usage, such that its situations express well-formed propositional knowledge about topics of interest. For the "speaks" relation, specified as [[person][speaks][language]], the first argument requires a person and the second argument requires a language of some kind. An instance of this relation—ie, a situation like ||Lynne||speaks||Spanish||—must conform to the selectional restrictions of [[person][speaks][language]]. The relational specification provides a global constraint on possible instances; namely, it simply authorizes usage conforming to "a person can speak a language", without regard to specific instances. Another way to think about the difference between the relation [[person][speaks][language]] and the situation ||Lynne||speaks||Spanish|| is that the relation contains two arguments with free variables (ie, [person] and [language]), whereas an instance of the relation contains no free variables. In fact, any relation will contain at least one free variable, making it an abstract universal.

As an example of a constrained relation, consider "every person speaks some (ie, any) language", which creates expectations on its participants beyond those imposed by simple selectional restrictions. To accommodate such propositional knowledge (with bound variables) requires the introduction of quantifiers into the concept net framework. The two standard quantifiers are "some" (or "a" or "at least one") and "every" (or "each" or "all") of predicate logic, symbolized by ∃ and ∀, respectively. The quantified assertion above represents a relation because it contains the variables, [person] and [language], bound by quantification (as expressed by "every" and "some"). Other nonstandard quantifiers like "most" can also be represented in a concept net, although there is no consensus on the best semantic interpretation, as there is for "some" and "every".

Given any particular relation like [[person][speaks][language]], taking a specified number of arguments (in this case two), we can quantify over its arguments in a variety of ways using ∃ and ∀. In the language of predicate calculus the options are the following for dyadic atomic formulas:

∀x∀y speaks(x,y)	-every person speaks every language
∃x∃y speaks(x,y)	-some person speaks some language
∀x∃y speaks(x,y)	-every person speaks some (any) language
∀y∃x speaks(x,y)	-every language is spoken by some (any) person
∃x∀y speaks(x,y)	-some (single) person speaks every language
∃y∀x speaks(x,y)	-some (single) language is spoken by every person.

Whenever the normal argument order x,y is reversed in the operators, it means the converse of the relation is being used—in this case expressed in the passive voice. The alternatives multiply for triadic and higher n-ary relations, but the pattern is the same—eg, "every boy gave some girl a gift" (at the party) is expressed as ∀x∃y∃z gave(x,y,z). For unary relations there are only two possibilities; for example, ∃x sleep(x) or ∀x sleep(x) for "someone/everyone slept".

We can translate the above statements into quantified (restricted) relations for assertion into a concept net as follows:

[[person]|∀1| [speaks] |∀2|[language]]
[[person]|∃1| [speaks] |∃2|[language]]
[[person]|∀1| [speaks] |∃2|[language]]
[[person]|∀2| [speaks] |∃1|[language]]
[[person]|∃1| [speaks] |∀2|[language]]
[[person]|∃2| [speaks] |∀1|[language]]

where each quantifier refers to an argument number, and the order of the quantifiers (from left to right) conforms to the prenex normal form of the proposition.

For example, to assert "every person speaks some (any) language" create the relation [[person][speaks][language]], and specify the order of the quantifiers.

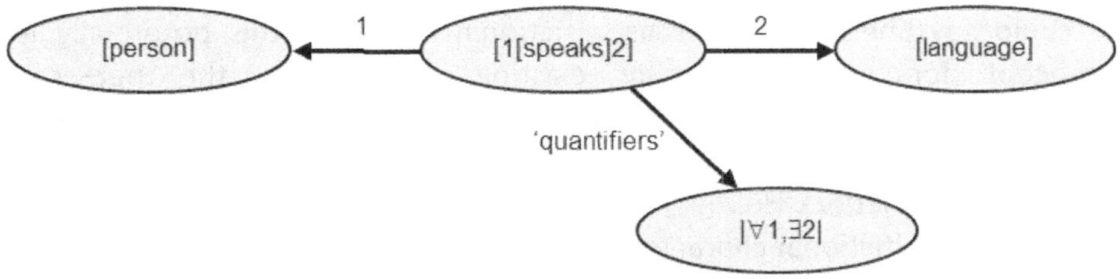

Figure 51: Representing Quantified Propositions in Prenex Form

Now all instances of a person speaking a language—such as Lynne speaking Spanish—must conform to the specified (constrained) relation.

The above concept net representation provides a relatively straightforward way to express propositional quantification—that is, by attaching the quantifiers for each argument of a relation to that relation in a list of quantifers. But this prenex normal form of a proposition is semantically awkward, because it separates each argument from its quantifier. A more natural representation would keep each argument and its (optional) quantifier together, in accordance with normal English phrasing. For example, to represent "some (single) language is spoken by every person", the existential quantifier naturally belongs with the first argument [language], the universal quantifier belongs with the second argument [person], and the passive voice reverses the argument order from the default active voice, namely a [person] speaks a [language]. The following diagram provides the more natural, but more complex, representation by moving the quantifiers from the relation to the individual arguments.

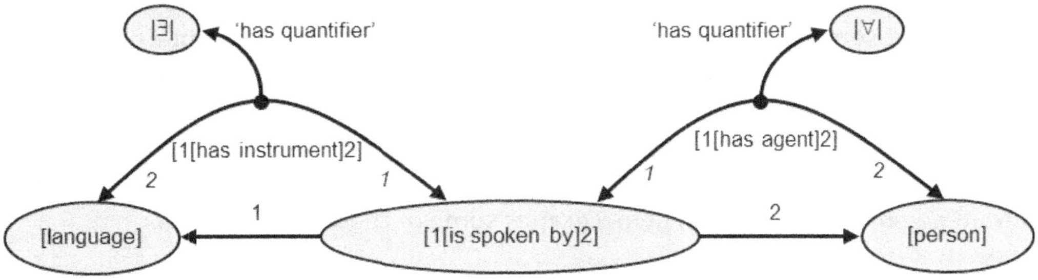

Figure 52: Representing Quantified Propositions in Natural Argument Form

Pseudo Concepts

An *imaginary* concept is a particular concept which does not represent anything in reality. It exists as a concept in a concept net, but is characterized as some variety of nonphysical object (eg, fictional, mathematical, functional). |Sherlock Holmes| is an example of an imaginary concept. It is an instance of [fictional character], from which it inherits real properties, namely its literary properties, such as being a character created by Sir Arthur Conan Doyle, and the real characteristic of being |fictional|. |Sherlock Holmes| also has imaginary properties, like being tall and thin. Normally these kinds of ascribed properties would be inherited from [man], but Sherlock Holmes is not really a man, as say Mark Twain is. We could say Sherlock Holmes is a fictional man, but this potentially leads to a dual, nonphysical concept for every physically existing concept in the net—rather inefficient representationally (but possibly justified for concepts like [fictional detective] since there are so many of them). It seems better to make imaginary concepts parasitic on real concepts: "|Mark Twain| 'is an instance of' [man]", but "|Sherlock Holmes| 'is a pseudo instance of' [man]" (in addition to "|Sherlock Holmes| 'is an instance of' [fictional character]"). The 'is a pseudo instance of' association, along with the 'is a pseudo kind of' association for universals, then permits inheritance of imaginary, in addition to real characteristics.

A *pseudo* concept is a concept that, like an imaginary concept, alludes to being something it is not. Its significance, however, extends beyond imaginary concepts. For example, a toy soldier is not really a soldier. (It is, however, a real toy.) Waxed fruit is not really fruit (it is a decoration), and counterfeit money is not really money (it is a forgery). Many pseudo concepts are real things in their own right, having real characteristics, but also include imaginary characteristics. The toy-related characteristics of a toy soldier are real, but most of the soldier-related characteristics are not. If a concept has any imaginary characteristics, then it is a pseudo concept.

A pseudo concept inherits its characteristics (real and imaginary) from its parent instance/pseudo-instance and kind/pseudo-kind concepts, respectively. The concept [toy soldier], for example, inherits its real characteristics (such as possibly having a current owner) from [toy], given that "[toy soldier] 'is a kind of' [toy]" and toys are generally owned by someone. It similarly inherits its imaginary characteristics (such as having a military rank) from [soldier], given that "[toy soldier] 'is a pseudo kind of' [soldier]". Inheritance flows along the 'is a pseudo kind of' association like it does for the basic 'is a kind of' association, but some of the real characteristics of [soldier] may be reinterpreted as imaginary characteristics of [toy soldier]. In particular, the rank characteristic of [toy soldier] will become an imaginary rank, since it does not express a real military rank with official recognition, responsibilities and privileges. Such imaginary characteristics of pseudo concepts will be called *ascriptions*[188] to distinguish them from real characteristics, in much the same way characters (in novels, mythology, etc) are distinguished from persons. The diagram below graphs some of these relationships.

[188] van Inwagen somewhat reluctantly uses the word "ascription", and contrasts it with "predication", to talk about the difference between ascribing something to a character in a novel, and predicating a property of a real thing (van Inwagen, 2001).

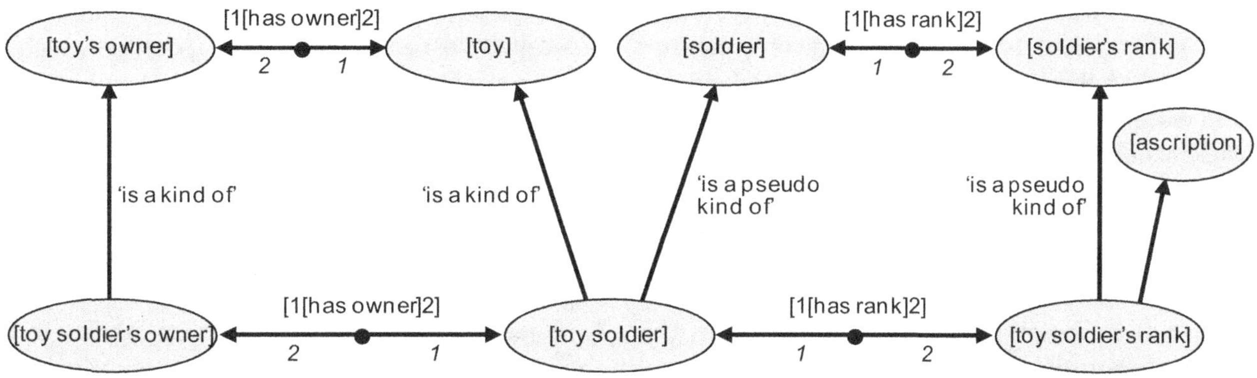

Figure 53: Representation of Pseudo Concepts

Some characteristics inherited via a pseudo association can also remain real. A toy soldier's general shape (but not size) is inherited from the default shape of a real soldier. A pseudo association enables an imaginary interpretation of characteristics, but does not require it. Typically, the pseudo concept will inherit via the pseudo association certain real perceptual characteristics, which superficially describe it and give it some resemblance to its real parent concept. A facsimile is an extreme example of superficial similarity—eg, [counterfeit coin], [mirage]—compared with the partial similarity of a replica—eg, [porcelain cat].

Generally speaking, all imaginary concepts are pseudo concepts (because they have imaginary ascriptions), but not all pseudo concepts are imaginary (because they may be actual). A pseudo concept always 'is a kind of' some (real) kind concept which specifies its identity characteristics (this is true even for imaginary concepts), but also 'is a pseudo kind of' some other (real) kind concept which typically adds imaginary normalcy or cue (ie, recognition) characteristics. Identity properties cannot be inherited via the pseudo association since they specify what something *is*, rather than how it appears.

Natural vs Nominal Kinds

Kinds and *relations* are the two fundamental building blocks for representing all generic knowledge in language or in concept nets. We use common nouns (eg, father, continent, dog, freedom) to talk about various kinds or classes of entities—kinds of persons, places, things, and ideas according to one dictionary. Correspondingly, we normally use verbs (eg, is tall, love, give, tell) to say something about or relate various kinds of thing to each other. Kinds and relations together make it possible to represent propositional knowledge.

In examining the wide variety of common nouns, many philosophers over the years have also proposed that there are two fundamental categories of kinds of things. These categories are usually called *natural kinds* and *nominal kinds*. The distinction has always been controversial, but here, and for the purpose of effectively structuring concept nets, a natural kind signifies a category built around some, usually hidden, naturally occurring phenomenon. Standard examples of natural kinds include tiger, water, oak, and gold, where each term conceivably harbors a so-called "essence" or intrinsic *nature* which is essential to determining instances of the kind. Tigers and oaks seem to possess unique DNA patterns, and water and gold are identified by their unique chemical microstructures. Alternately, a nominal kind like bachelor, boat, father, or lunch is built around conventional agreement about the use of the term, such that every nominal kind includes some form of semantic stipulation. Because of their conventional components, nominal kinds do not exhibit a cohesive underlying intrinsic nature that can be investigated and discovered, but instead often constitute useful groupings of things. Nominal kinds may possess identity characteristics just like natural kinds (eg, being an unmarried man is usually a necessary and sufficient condition for being a bachelor), but there is no naturalized "essence" of bachelorhood to be characterized over and above its semantic/linguistic meaning. Anything that meets the appropriate (usually prototypical) characteristics in the appropriate context may be classified as an instance of the nominal kind.

The idea of a natural kind concept harboring a hidden nature is a fuzzy notion, at best. Since our perceptions (as interpretations constructed by the mind) do not allow us to directly apprehend the world, "nature" will always remain a mystery to a substantial extent. (See the opening sections of Chapter 2 for additional discussion of indeterminacy.) However, we apparently can apprehend nature's effects on us, and these effects show up in our conceptual structures, particularly in our natural kind concepts. The following points are an attempt to further clarify and consolidate thoughts about the latent nature underlying a natural kind without attempting a comprehensive specification:

Natural kinds are generally thought of as "given" objectively to us, mostly independent of our subjective and oftentimes imaginative thought processes. The source of this objectivity presumably originates outside ourselves in the world, and relates to the nature or essence of the natural kind.

Science strives to analyze the nature of natural kinds and discover their attributes empirically. Technology successfully exploits those attributes to develop useful devices, methods and processes. A deeper understanding of natural kinds affords us a wider scope of application and explanatory power, since the universal laws of nature pertain to the attributes of natural kinds. Even though we do not apply to everyday beliefs the same rigorous standards as science does, we nevertheless intuitively assume that many kinds express an inner nature that makes instances of the kind what they are. The fact that we do not normally understand what that nature is, is usually irrelevant, but subject to further investigation.

The overt descriptive characteristics associated with a natural kind spring from its nature. For example, the fact that gold is yellow and malleable at room temperature presumably arise from the complex nature of its atomic and crystalline microstructures—ie, the natural physical processes that cause its observable attributes.

The underlying nature of a natural kind is presumed to have a *causal basis*. For example, silkworm disease became a provisional natural kind when Pasteur discovered that silkworm disease is caused by a bacterium. Likewise, what causes gold to be gold is the number of protons in the gold atom. But having a causal basis, by itself, is apparently not enough to warrant that a concept is a natural kind. The concept bachelor is not a natural kind even though a man can become a bachelor through divorce (ie, becoming a bachelor is caused by becoming unmarried). Causation seems to be a necessary but not sufficient condition for a natural kind.

A further condition on a natural kind posits a *naturalized causal mechanism* that systematically explains how the causal basis comes to be. While the causal basis mainly involves the identification of the source (A) of a natural kind's (B) nature (B is caused by A, expressed as a proposition), a naturalized causal mechanism will describe how A comes about through the systematic workings of a naturalized process. The reason why bachelor is not a natural kind is that in this case the causal basis simply involves a change of state brought about by a conventional process (eg, by divorce or by becoming an adult), not a naturalized process. The reason why silkworm disease and gold are regarded as natural kinds is that each is brought about by naturalized causal mechanisms, namely the molecular and cellular processes of infection, and the interplay of the binding forces and energy levels in the atom, respectively. Note however, that further investigation may show we were mistaken about a causal mechanism, and thereby undermine the status of the natural kinds it supports. A natural kind's scientific significance will always be questionable.

The operation of a naturalized causal mechanism explaining a cause and its effect is typically characterized in terms of the relevant objects, properties, and relations associated with the cause, plus their interactions which bring about the effect. Therefore, the explanation is generally regarded as achieving a deeper level of understanding than the initial causal basis. Achieving deeper and deeper levels of understanding about the naturalized causal mechanisms of a natural kind increases confidence in its standing as a natural kind (Thagard, 2012), but because there is currently no known lower limit to successive explanatory achievement, a natural kind's explanatory basis will always be asymptotic to an ideal.

The mental processes which register the nature of natural kinds are either subconceptual or intuitive. Many registrations of distinctions are innate and perceptual (eg, contact, containment, red, etc), but others are learned and abstract (eg, animate, number, agency, etc), developing within the first few years of life, and depend on exposure to appropriate environmental stimuli. Evidently the development of new registrations (eg, creative insights) can continue throughout life with immersion and study. Conceptual characteristics are built on the underlying subconceptual registrations.

The complete nature of a natural kind cannot be precisely described, or otherwise represented, either linguistically or conceptually or perceptually. The nature of a natural kind is therefore detectable, but ultimately inscrutable. It is usually assumed to have a determinate form, based on the detection of recurring patterns and regularities, but that form can only be approximately represented or modeled.

The most successful ways of describing natural phenomena are based on mathematics, where differential equations express the changing relationships among various attributes over variations in time or space. But even mathematics necessarily over simplifies and idealizes its models, in order to deal with the synergistic complexity of the natural world. The subsequent translation of mathematical equations and probabilities into explicit linguistic or conceptual characteristics still further muddies the waters, because different systems of representation are generally incommensurable to some extent. Evolution has given us a basic ability to sense the regularities exhibited by instances of natural kinds, but has knitted up a patchwork quilt of strategies for understanding them.

The nature of a natural kind is presumed to possess a certain internal unity with respect to the given natural phenomenon. And there is only one essential nature associated with each natural kind. A natural kind cuts the world at natural joints such that the natural kind is segregated in logically coherent ways from other natural kinds and takes its place in logical relations to them.

The nature of a natural kind is different from its so-called identity conditions, which are its assumed necessary, sufficient, dispositional, or potentiality characteristics (as described in Chapter 3: *Concept Characteristics*). Even identity conditions are subject to revision in the light of a revolutionary reinterpretation of knowledge, but the underlying nature is not. When Berzelius discovered that water was a chemical compound in 1826, we changed our identity characterization for water, but we did not assume that the inherent nature of water had changed. Likewise when whales were reclassified as mammals instead of fish.

Natural kinds, to the extent that their natures can be identified and understood, represent an optimal basis for consistent knowledge representation about the world. Such conceptual representations improve overall as knowledge about the nature of things expands and becomes more organized. The organization of subjective knowledge, however, is not well-suited to natural kind representation, because subjective knowledge is based on utility, as discussed next.

No less fuzzy is the idea of a nominal kind harboring some purely semantic component of meaning. The following discussion attempts to clarify this notion as well.

In much of the philosophical literature the idea of a nominal kind is tied to its dependence on human affairs for its meaning, leading to the independent vs dependent dichotomy for natural kinds and nominal kinds, respectively. The meaning associated with a dependent nominal kind then is based on semantic conventions (ie, consensus about the meaning associated with certain terms in a language community, and the way we humans represent knowledge), while the meaning associated with an independent natural kind is based on a causal theory related to the nature of some external natural phenomenon.

However, a pure independent/dependent distinction is difficult to maintain in practice due to the lack of a universal theory of either causality or semantic meaning. In addition, causal characteristics are almost always profoundly intertwined with semantic/conventional characteristics for almost all concepts. For a nominal kind concept like bachelor, the embedded quality of being male seems more like an independently occurring natural phenomenon than a conventional stipulation. Similarly, the grasp of a natural kind concept like water seems to rely more on convention (perhaps its look and feel and utility) than chemical composition, the structure of which most people do not even think is relevant to its everyday meaning.

A second proposal, argued by Devitt, casts the main distinction as one between a descriptive vs nondescriptive difference in meaning (Devitt, 2005). Nominal kinds emphasize descriptive meaning, whereas natural kinds emphasize nondescriptive meaning. The characteristics attached to any concept amount to its descriptive meaning, while for natural kinds only, the concept additionally counts as an indexical placeholder for its presumed nondescriptive meaning. The distinction remains philosophically problematic without saying what a nondescriptive meaning amounts to, but for the purposes of conceptual representation, it may not matter, because nondescriptive meaning cannot be represented conceptually.[189] When we interpret a concept's identity conditions as deriving from some ineffable nondescriptive component (eg, based on intuition), we then simply treat it as a natural kind. If we intuit meaning merely resulting from the combination of characteristics derived from descriptive components, we treat it as a nominal kind. We can always change our mind (usually with minimal impact on daily affairs) depending on how much emphasis we wish to give to our hunches, and how much consistency we endeavor to maintain with other like concepts. I don't believe in ghosts (they are nominal kinds for me), but some of my daughter's friends seem to treat them as natural kinds.

In spite of the difference in emphasis, these two views of the dichotomy (independent/dependent and nondescriptive/descriptive) are nevertheless complementary, and together provide some traction for the natural/nominal distinction. A nondescriptive meaning dovetails nicely with the inscrutability of nature (ie, it is impossible to fully describe the nature of a natural kind, like gold), and a descriptive meaning aligns well with the linguistic/conceptual understanding of a conventional nominal kind (ie, providing something closer to a definition for some terms, like bachelor as an unmarried man). Considerable confusion between natural and nominal kinds may arise because nondescriptive (or independent) meaning cannot be represented, either conceptually or linguistically, and thus cannot be straightforwardly analyzed.

For natural kind concepts the characteristics are normally construed as *deriving from* the nondescriptive component, as being fundamentally incomplete, and as being subject to major revisions as more knowledge about the nature of the underlying phenomenon surfaces. For nominal kind concepts the characteristics are normally construed as *determining* its extension (ie, what it applies to). Such characteristics can also change, but the source is different: any changes are due to changes in accepted usage.

All concepts necessarily have a descriptive component, and so are conventional to some extent. This is a basic principle of conceptualization, and ultimately ties the interpretation of all concepts to a first-person human conventional perspective. An objective third-person perspective is derived. Conferring a nondescriptive objective nature on some concepts is an effortful undertaking (with the aim of discovery), but often pays off in useful ways through the discovery of latent characteristics. However, the unavoidable component of conventional meaning in natural kind concepts always has the potential to interfere with the presumed integrated unity of the underlying nature, and can't help but turn it into

[189] It may alternately involve perceptual representations, representations of context, or perhaps even evaluative judgments (as in emotion), but none of these constitute descriptive meaning, as the term is used here.

a disjunctive concept. We can approach the understanding of natural kind concepts asymptotically, but they seem to be idealizations.

In keeping with their utilitarian uses, nominal kinds come in a rather eclectic variety of breeds, including roles, artifacts, states, and disjunctions. Perhaps the most straightforward nominal kinds are those intrinsically dependent kinds (often called case roles) that we consistently associate with relations. Each relation possesses some number of arguments, and from the point of view of the relation, each of its arguments fills a certain *role*. For example in the [[human being][loves][human being]] relation, exemplified in "Drea loves Nynne", there are two roles, namely a [lover] and a [loved]. These are conventional kinds, because they are dependent categories defined in terms of the "loves" relationship. In contrast, the natural kind [human being], which is used to *specify* the "loves" relationship, is independent of the relation, since human beings exist independently of their participation in loving situations. Concepts like [lover] and [loved] are specializations of the more-or-less established semantic role categories for all relations, namely [agentive], [affected] (sometimes called theme), [beneficiary], [attributive], [locative], etc—see Chapter 5: *Generalized Situations* for other case roles. Lower-level concepts like [lover] and [loved] are then kinds of these topmost roles, in this case [agentive] and [affected], respectively. The conventional role of concepts, like [hunter], can usually be made explicit by identifying the situation underlying the concept. Thus a hunter is a person who participates in hunting situations—ie "x hunts", where the hunter is x. A husband is a male person who participates in a marriage situation—ie "x is married to y", where the husband is either x or y. A bachelor is a man who is single—ie "x has marital status single".

Artifacts are another major category of nominal kind, distinguished mainly by their functional use. They are objects created or used for a purpose, and therefore are grouped into kinds according to their function. The underlying nature of an object is often relevant to the performance of a function, but for artifacts the function is the basis of for identifying the kind. Doorstop, chair, Europe, and minute are all artifactual and therefore nominal.

Some other common nominal categories include temporary states or circumstances that apply to pre-existing objects or persons. This includes occupational roles, like doctor or president; the so-called phased or stage sortals like caterpillar or ice[190]; states of being like student or bachelor; and various configurations of matter like constellation or orchestra.

Disjunctions of natural kinds can form an inclusive nominal kind too. Jade is a good example of a disjunction of the natural minerals nephrite or jadeite. A given sample of jade can be either. A simple disjunction of any number of different nominal kinds can also form a nominal kind. For example, a purely extensional concept of an English vowel is any one of the concepts: a, e, i, o, or u.

---

[190] A phased sortal (reflecting the temporary state of an object) is easily confused with the object itself. A tadpole, represented as a nominal concept, can grow into a frog, represented as another nominal concept. However, we also use the same word "frog" as a natural kind to talk about the enduring individual animal, possessing an intrinsic nature, which is undergoing growth changes from conception to death. These are two different polysemous uses of the word "frog", and express different concepts.

In a concept net all of the kind concepts represent the core interpreted content or meaning of categories, organized in a coherent hierarchy. Both natural and nominal kinds can appear in the hierarchy. What then is the significance of the distinction for the representation of knowledge? From one point of view, natural kinds would seem to have a claim to being more fundamental, since they are ideally independent of human perspective in aspiring to reflect universal laws. Consequently, it might be argued, such concepts should make up the core skeleton of the hierarchy, based on the so-called *rigid* use of such concepts to refer to a given natural essence (Guarino & Welty, 2000). Nominal kinds then are parasitic on natural kinds, and are defined in terms of them. For example, the nominal concept [bachelor] is parasitic on the natural concept [human being] (or more technically *Homo sapiens*). If a bachelor is characterized as an unmarried man, then a [bachelor] is a kind of [man], another nominal concept, containing as it does the conventional interpretation of being an adult male. Continuing this line of reasoning, since a man is usually characterized as an adult male human being, then also a [man] is a kind of [human being], completing the link to the core skeleton. Similarly, since a sample of [jade] can be either of two distinct natural minerals, it links to both, such that [jadeite] is a kind of [jade] and [nephrite] is a kind of [jade] with properties common only to both. The concept [jade] then is a kind of [mineral], which is the governing natural kind. Generally speaking, natural kinds, like [gold], have only one natural parent kind (eg, [element] specifying that gold is a kind of element), and tend to have mutually exclusive natural sibling concepts (eg, the other chemical elements). Nominal kinds, like [jade], often represent a disjunction or useful grouping of natural kinds, and these groupings may overlap among themselves or with other convenient groupings. In keeping with the established philosophical distinction, the relevant universal concepts in a concept net would then be labeled [[natural kind]] and [[nominal kind]], where each is a subkind of [[kind]], and an instance of a [[[metauniversal]]], and where [[natural kind]] carries the presumption of a latent nature.

The above strategy for representing skeletal natural and parasitic nominal kind concepts would seem to work well for much of the concept hierarchy, accounting for the vast majority of our practical knowledge. However, the uppermost levels of the knowledge hierarchy are clearly nominal kind concepts, having no perceptible inherent nature apart from how they are defined. The concept of a [[universal]] is simply anything which can have instances. The concept of a [[kind]] is anything which can have determinate classification characteristics. The concept of an [occurrent] (eg, an event) is anything that can have participants in some action. The concept of a [collective] is anything that can have members, and so on. In Chapter 6: *Three Tiers of Knowledge*, it is suggested that a concept net be subdivided into multiple layers according to the sort of knowledge being represented: the upper epistemological levels, the middle theoretical levels, and the lower domain levels. The significance of the natural/nominal distinction, it is proposed, then concerns the level of knowledge being addressed.

Under the tiered knowledge hypothesis, the natural/nominal distinction is important because its presumptions (mainly about a latent nature) apply best to concepts at different levels of the hierarchy, rather than to individual concepts in isolation. A concept like [bachelor] is a nominal concept parasitic on the natural concept [human being], located in the middle-lower levels of the concept net hierarchy; while [collective] is a nominal concept, located in the upper echelons of the hierarchy. Here, [human being] is a theoretical concept, [bachelor] is a technical domain concept, and [collective] is an epistemological concept. Given the emphasis on tiering, the distinction between natural and nominal kinds probably should not be directly reflected in the core knowledge hierarchy as universal concepts like [[natural kind]] or [[nominal kind]] as suggested previously, but rather understood as metaknowledge about how we conceptualize the hierarchy itself. Other philosophers have concluded that concepts like essence and natural kind have outlived their usefulness in scientific explanation (Brown, 2007); and that even biological concepts like species no longer support the idea of naturalized essences (Dupré, 1993). The concept of a natural kind then is just a convenient fiction to go along with the psychological essentialist bias, such that natural kinds are simply those kinds of things we intuitively take as relatively independent of our thought processes. Furthermore, while the structure of the core hierarchy at the upper (nominal) and middle (natural) levels will ideally exhibit a strict hierarchy with single inheritance, the lower level nominal concepts, being based on utility, will tend to form a tangled hierarchy with multiple inheritance.

To conclude this rather heterogeneous analysis of conceptual kinds, I will simply close with a tentative point-of-view regarding the significance of the natural/nominal kind distinction:

- The concept of a natural kind is probably best interpreted as a convenient philosophical idealization, in somewhat the same way that a point mass is a useful concept in physics. Strictly speaking, there are no such things as natural kinds or natural essences that determine what something really is. It is rather a useful asymptotic notion, which we effectively use to organize our knowledge, and which evolution has selected to structure conceptualization. It is the best solution devised so far for stabilizing our classifications of worldly objects and experiences.

- Our concept of a natural kind is probably built on our more intuitive grasp of causality. If we can discover some fundamental naturalized causal mechanism that systematically explains why an object has the properties it has, we are intuitively led to regard that mechanism as a privileged source of the properties of the object. We then interpret it as the root cause that makes the object into what it is, and why it is similar to other objects with the same root cause. But root causes are always subject to revision, since we also intuitively know that we could be mistaken and that root causes are themselves caused by other more fundamental causes, which we may not have uncovered yet.

- All kind concepts are ultimately nominal, assuming natural kinds are idealizations. However, we judge some kinds as more nominal than others, depending on our assessed balance between mind-dependent stipulation and mind-independent naturalism. Stipulation and naturalism then represent opposite poles on an intuitive implicit scale of nominal kinds.

- The backbone of our concept network hierarchy becomes organized around our best natural kind concepts. We use these concepts to represent our most stable and most consistent understanding of the world. Other nominal kind concepts become parasitic on natural kind concepts, in order to maintain a stable knowledge base that is both rational and useful at the same time.

Attribute Values

The representation of attribute values in a concept net draws on the simple observation that objects (more generally, things of any kind) can be characterized in terms of attributes, and that those attributes can have values. All three (object, attribute, value) interrelate in a concept net, contributing to what is often called the inherent nature of something—ie, its intrinsic characteristics. However, as discussed previously in Chapter 4: *Properties and Attributes*, we need to be careful about the terminology if the terms are to have consistent semantic significance.

To summarize the terminology of attribution established previously: a universal binary *relation*, like [[person][has weight][weight]], relates a *kind* (eg, [person]) to an *attribute* (eg, [weight]) via a *relationship* (eg [has weight]). And similarly, [[weight][has value][value]] relates an *attribute* to a *valuation*. Stringing these two relations together then enables the full representation of some object being characterized by some attribute, which in turn is characterized by some value. Correspondingly, the two particular binary *situations* attributing a particular *value* to a particular object would take the form of situation instances like ||Lynne||has weight||Lynne's weight|| and ||Lynne's weight||has value||137 lb||, where |Lynne's weight| is an *attribution* (ie, an instance of the attribute [weight]), and |137 lb| is a *value* describing that weight attribution (ie, an instance of a [weight value]), joined by the two *predicates* |has weight| and |has value|.[191] Furthermore, *values* (eg, quantities like |137 lb|) are themselves characterized in terms of *magnitudes* like |137| and *units* like |lb|, respectively. As it stands, Lynne has only one weight, but this weight might be described in many different qualitative and quantitative ways. The basic organization of concepts to express the sentence "Lynne weighs 137 lb" is outlined in the figure below.

Generally speaking, an attribute *value*, whether a *quantity* (like |137 lb|) or *quality* (like |heavy|), is selected from a prespecified collection of values, depending on the attribute. The value is a concept just like other concepts, but it exists along with other interrelated values that together comprise a *value set* or *scale* for an attribute. For example, |137 lb| is one particular value on a scale of other avoirdupois weight values ranging from 0 lb on up.[192] The values in this case comprise a totally ordered set, allowing their magnitudes to be compared, and the values are mutually exclusive with respect to each other. The representation of object qualities (like |heavy| and |light|, or |hot|, |warm|, |lukewarm|, |cool|, |cold|) is handled in the same way, except the values are discrete. All of these kinds of attributive situations, then, describe a state-of-affairs where an object becomes intrinsically characterized by a value in some independently existing attribute value space. Values are particulars, but their status as being existent in the physical world is left open. They simply exist as useful epistemological concepts in a concept net, useful for characterizing entities of all manner, shape, and form.

---

[191] In general, two physical things will never have the exact same weight, so each instance of any weight description could be labeled something like |137 lb #7|, indicating it is unique. However, for nonphysical things like numbers, attribute values may be shared (but only if they are regarded as particulars, rather than universals). Both 4 and 6 are even, but their "evenness" is not distinct.

[192] Note that two values like |137 lb| and |62.1 kg| are just two different concepts with the same denotation (since they are equal weights), in the same way that |Cicero| and |Tully| are two concepts representing the same person. They are represented as being equal to each other, assuming this information is known.

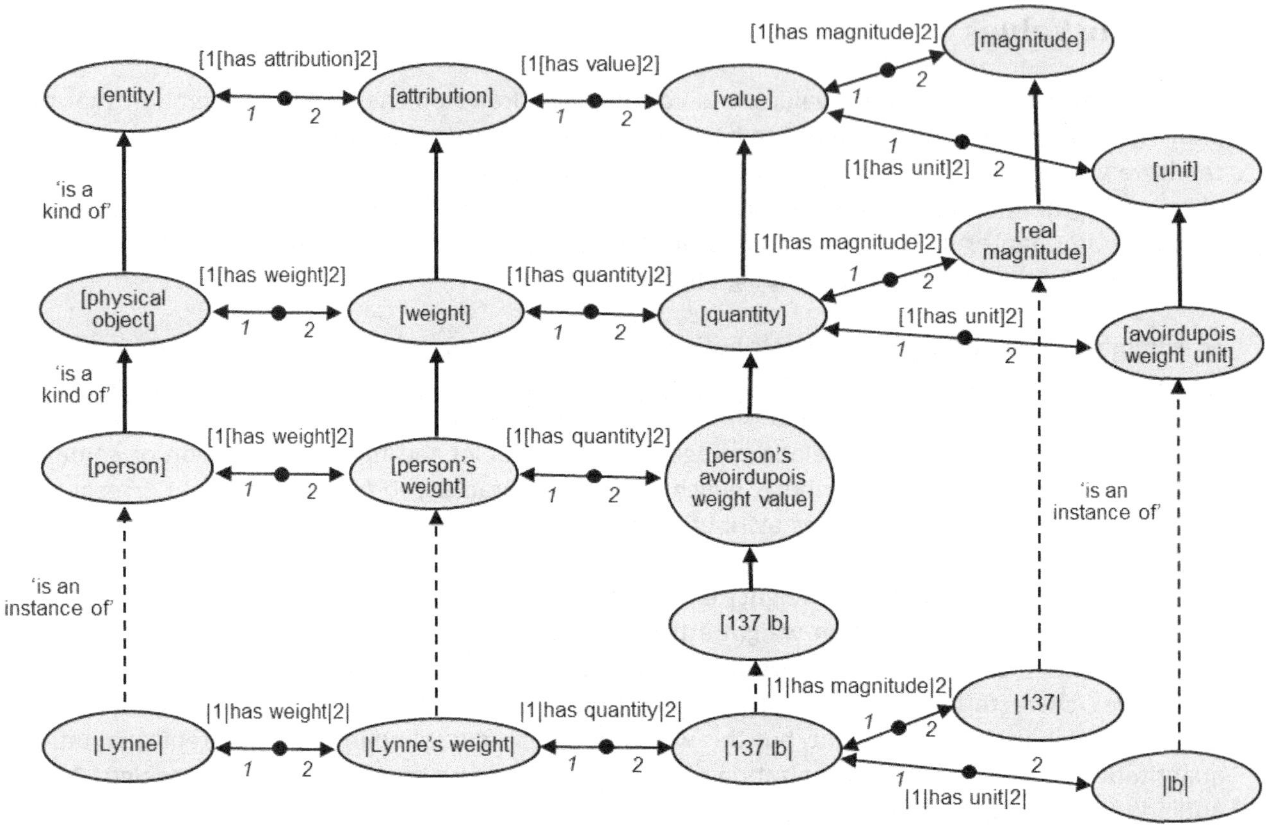

Figure 54: Representation of Attributes and Attribute Values

The attribution of a qualitative *value* to an object can be compared and contrasted with a simple *situation* characterizing an object. Several psychological experiments have shown that people regularly employ both attribute concepts (eg, [color]) and valuation concepts (eg, [red]) in their conceptual representations of objects, often covertly (Barsalou, 1992; Ross, Perkins, & Tenpenny, 1990). For example in a statement like "fire engines are red", the attribute of color is only implied, but nevertheless likely becomes encoded in the associated conceptual representation—more accurately rendered as "fire engines have color red" or "fire engines are colored red": [[fire engine][has color][fire engine's color]] and [[fire engine's color][has value][red]]. Language often economizes on expression when meaning can be readily inferred. Such shortcut expressions will normally translate into fully attributional conceptual structures encoding object/attribute/value, rather than a simple object/value association (Gentner, 1988). The simple object/value representation, however, can be used to express the simple predication when convenient (eg, for later elaboration), or when the value scale is unknown or nonexistent. The statement "fire engines are red" then straightforwardly translates into a relation: [[fire engine][is red]]. The straightforward relational (or situational) representation is also useful in cases where the contrast set of values is unclear: [[coffee][is aromatic]]. Although young children are typically unable to separate attributes from values, this innate ability matures with normal development (Shepp, 1978).

Considering values a little more thoroughly, there are several different kinds of attribute space which correspond to the kind of structuring imposed on the values making up the space (ie, its scale of values). The structure of a scale is primarily determined by what kind of ordering relation holds among the values. This relation is normally written "≤" (ie, the less than or equal sign), but it has a slightly different interpretation for each different kind of scale. For example the value sets [137 lb] and [400 lb] are related as [137 lb] ≤ [400 lb] on the weight scale; while in the military rank scale [lieutenant] ≤ [major]. On the weight scale the relation implies that the distance between two weights is meaningful, but for military rank only the relative ordering of values is meaningful.

The different types of value scales can be roughly organized according to the amount of information provided by the ordering relation. The stronger scales like avoirdupois weight allow absolute numerical comparisons among values, while weaker scales like military rank allow only relative comparisons. Six types of scale potentially useful for characterizing attribute values are listed here:

Nominal Scale. The weakest kind of scale simply contains a set of values with no ordering relation whatsoever. An attribute value (eg, a student's answer to a test question) might be selected from [true, false] or [a, b, c, d, e]. The shape of an object might have a value specified as [triangular, circular, square, egg-shaped, ...]. Nominal scales are probably good representations for attributes that have complementary values like [dead, alive], [inside, outside], or otherwise unordered values like for race or sex [male, female, hermaphrodite, asexual]. The above are examples of finite nominal scales (except for shape, which may be denumerable), but nondenumerable nominal scales, although rarely encountered, are also possible. Perhaps the orthographic names of all of the real numbers (ie, sequences of digits with a decimal point) form a nondenumerable nominal scale.

Cyclical Scale. Stronger than nominal because its values are at least ordered, but the ordering relation is very weak. A cyclical scale contains ordered values but there are no maximal or minimal values—the values form a cyclic order in which each points to one or more predecessors. The days of the week are a circular (linear cyclical) ordering. The subjective property of color is often described on a circular scale, moving from shades of red to yellow to green to blue and back to red. The ordering relation is reflexive and antisymmetric, but not transitive. Only adjacent values are comparable.

Partially Ordered Scale. The ordering relation is reflexive, antisymmetric, and transitive such that the values form a hierarchical tree or lattice, and some values may not be comparable (eg, those on the leaf of a tree) but others are (eg, a value's ancestors). An organism's species may be an example of a partially ordered value scale. The species values (eg, cheetah, koala, Homo sapiens, etc) are organized by evolution as a binary branching tree, where a parent species may branch into two descendant species. The structuring of objects into parts also organizes values into hierarchies and could define a partially ordered scale (eg, of body parts—arm, hand, finger, digit).

Ordinal Scale. In an ordinal scale, like military rank, all the values have meaning in relation to each other. All values are comparable and form a totally ordered set. In the most common ordinal scales, the scale forms a linear chain of values, like the five-point Likert scale of qualitative degree of agreement [strongly disagree, disagree, neither, agree, strongly agree], but the chain may be complex as well. The energy level of an electron in an atom may be an example of a property with a denumerably infinite ordinal scale.

Interval Scale. An interval scale is a linear ordering in which the distance between any two values has meaning and is unique. There is also some unit of measurement that can be used to express the distance. Typically the values can be put into correspondence with the natural or the real numbers, but unequally spaced values are possible too. The mean and standard deviation of a distribution of values also becomes meaningful. The Fahrenheit temperature scale is an example of an interval scale since each unit of temperature on the scale represents the same degree of temperature difference no matter where it is placed on the scale—ie, the difference between [10 degF] and [20 degF] is the same temperature difference as the difference between [90 degF] and [100 degF].

Ratio Scale. A ratio scale is an interval scale with a zero value that indicates the complete absence of a relevant attribute.[193] The previous Fahrenheit temperature scale is therefore not a ratio scale, since [0 degF] does not mean the complete absence of temperature. The Kelvin temperature scale, however, is a ratio scale because [0 degK] means absolute zero (defined as the complete absence of molecular motion). Furthermore, on a ratio scale proportions between values are the same over the entire scale. This means that it is meaningful to talk about one value being twice as much (or some other proportion) as another (due to the presence of an absolute zero value). A temperature of [100 degK] is twice as hot as [50 degK], but [100 degF] is not twice as hot as [50 degF] since on the Fahrenheit scale there is no absolute zero from which to measure proportional distances.

Translating Sentences into Concept Nets

Once the basic organizing structures of a concept network are in place, it becomes possible to add new knowledge to it. The ultimate ambition is to receive textual input in a "learning mode", automatically interpret the text using knowledge already in the network, and finally restructure the network to incorporate the new knowledge. But this kind of natural language processing is still a remote vision. This section, instead, outlines a simple manual procedure for application to descriptive text. Taking a complex input text, the end result is a list of elementary statements encoding knowledge for assertion into the concept net.

Semantic Analysis

Analyzing the meaning of a sentence is troublesome for a variety of reasons: complex constructions, abstract ideas, implicit references, nuances, and ambiguity, among others. This is a discussion of a technique for analyzing a sentence that helps work out the latent semantic relationships among its

[193] The value 0 is not the same as not applicable. An example of a nonapplicable property for an apple is atomic number, since atomic number is applicable only to atoms.

constituents. Once a discerning representation of the sentence is available, it is more amenable to in-depth logical analysis and assimilation into a larger structure of background knowledge.

The goal of the analysis is to reduce a complex sentence into a logically related collection of simple sentences[194], each of which consists of a predicate or relationship and zero or more participants. The reduction typically unfolds as two processes which are interleaved. One process reduces a complex sentence (or sentence fragment) into a collection of semantically related simple sentences. For example, the sentence "Jennie works, because she likes expensive clothes" is a complex sentence composed of two simpler sentences: "Jennie works"(X) and "Jennie likes expensive clothes"(Y). The logical relation between the sentences is expressed by "because": (X 'because' Y), which is a simple sentence. The other process reduces complex constituents. In the example, ('Jennie' 'works') is already an atomic sentence. However, Y contains a complex construction (ie, the phrase "expensive clothes"), and can be reduced to ('Jennie' 'likes' Z) and (Z 'are' 'clothes') and (Z 'are expensive'), all of which are now simple sentences. The analysis thus proceeds recursively, until all semantic information is expressed in simple sentences.

This approach skips the standard syntactic analysis of a sentence achieved via parsing into phrases. Formal parsing can accurately identify hidden sentence structures, but it does not directly support identifying the underlying semantic relations. While formal parsing could certainly aid the analysis, the informal techniques offered here are usually sufficient to produce a useful preliminary analysis.

The following descriptive sentence will be used to illustrate the process. It is taken from p.349 of *The Smithsonian Guide to Historic America* (Logan, 1989):

"Seattle's central business district, bounded by Yesler Way, Route 5, Stewart Street, and the waterfront, has among its sleek glass monoliths smaller buildings with gargoyles and other lively early-twentieth-century decorations that give it an interesting texture."

Since understanding and accurately representing the meaning of a sentence is as much art as science, the following notes are better viewed as guidelines rather than fixed procedures. Each sentence presents its own unique challenges. Although the following stages reflect a top-down style, any comparable order of analysis is sufficient, as long as it is complete.

---

194 Simple sentences in English typically adhere to the form (<subject> <predicate> <object> <complement>), where all participants except the predicate are optional and repeatable. Examples: "it is raining" is an atomic sentence consisting of a predicate 'rain' with no participants ("it" is a dummy participants); "Pegasus flies" contains the predicate 'fly' and one participant "Pegasus"; in "Jennie loves Max" the predicate is 'love' with two participants, "Jennie", the subject, and "Max", the object; in "A gave B to C" the predicate is 'give' with three participants, the last of which is a complement. Predicates with more than five obligatory participants are rare.

First-stage Analysis

Identify the individual clauses (component sentences) in the sample, if there is more than one. Each clause will likely be organized around a main verb—identify it to help identify the clause. Then establish the logical relationship between the clauses and express them as atomic sentences.

After studying the example it eventually becomes clear that there is only one clause and that the main verb is "has". A paraphrase of the sentence core is then something like "district has buildings" corresponding to the form (<subject> <verb> <object>). The first stage of the analysis then produces:

(A 'has' B)

where

A = 'Seattle's central business district': "Seattle's central business district, bounded by Yesler Way, Route 5, Stewart Street, and the waterfront",

B = 'A's smaller buildings': "among its (the district's) sleek glass monoliths smaller buildings with gargoyles and other lively early-twentieth-century decorations that give it (the district) an interesting texture".

Since there is only one clause in the original sentence, there are no logical (between sentence) relations such as 'and', 'or', 'if...then', or 'because'.

The clause (A 'has' B) effectively partitions the sentence into three parts, one each associated with A, 'has', and B. The sentence fragments are carried along with the complex concept labels to enable further analysis in the next stage.

A, B, and 'has' are labels for new concepts introduced by way of analyzing the sentence, where A and B are complex constituents and 'has' is a generic predicate. (Since pinpointing the intended sense of "has" is not required to develop a correct translation of the clause, it is not rendered as a complex constituent.) A is shorthand for the concept label 'Seattle's central business district', which adequately characterizes the underlying concept to the extent provided in the sentence. However, labels are only mnemonic devices, and any other distinguishing label would suffice. The information about the district being bounded by the indicated features could also be included in the concept's label, but the result is rather unwieldy. Similar labeling criteria apply to B.

The analysis removes the distinction between singular and plural references in the sentence: A denotes a single district; B denotes several buildings; and the predicate 'has' probably denotes multiple instances of having, namely of A having B_1, of A having B_2, etc, for each of the smaller buildings. (Interpreting 'has' as a plurality has the advantage that if one of the smaller buildings is subsequently talked about—eg, the bus station—it can readily be inferred that A 'has' it, in particular). Plurals usually denote multiplicities of various kinds, which are semantically different from singular things or substances, but the structure of a sentence does not generally depend on the singular/plural distinction (other than grammatical subject/verb number agreement). "The district has small buildings" and "The district has one small building" use the same verb to express both situations.

When partitioning the sentence, nothing should be left out, so as to enable the next stage of the analysis to pick up from where the current stage leaves off.

Second-stage Analysis

Having established the overall structure of the sentence, it becomes easier to analyze the complex constituents from the first stage.

Analyzing A = 'Seattle's central business district': "Seattle's central business district, bounded by Yesler Way, Route 5, Stewart Street, and the waterfront" produces:

> (A 'is a' 'district') &
> (A 'belongs to' 'Seattle') &
> (A 'is central') &
> (A 'is for' 'business') &
> (A 'is bounded by' C)
> where
> C = 'Yesler Way, Route 5, Stewart Street, the waterfront': "Yesler Way, Route 5, Stewart Street, and the waterfront".

Here, the concept labeled A is the same concept previously labeled A. The first atomic sentence specifies what kind of thing A is. Every constituent is a kind of something. The predicate 'belongs to' is inferred from the possessive construction "Seattle's central business district". However, other possibilities might include 'is connected with', 'is possessed by', or perhaps 'is located in', depending on the interpretation. (In this case after expanding "Seattle's central business district" to "the central business district of Seattle", the 3rd sense of "of" is chosen from the Concise Oxford Dictionary.)

The notion that the district is "central" is rendered as simple attributive predication: 'is central'. An attribution, in general, is shorthand for a more complicated semantic structure, in this case that the district is probably located in the downtown area of Seattle. Such dissections are more reasonably part of an in-depth later analysis. Attribution is distinguished from classification as in (A 'is a' 'district'), which classifies the kind of thing A is. Also, 'business' is interpreted as an activity; and C represents a multiplicity of features whose arrangement, it is said, bounds the district. The sentences are logically joined by conjunction (&).

Similarly, analyzing B = 'A's smaller buildings': "among its (the district's) sleek glass monoliths smaller buildings with gargoyles and other lively early-twentieth-century decorations that give it (the district) an interesting texture" produces:

(B 'are' 'buildings') &
(B 'belong to' A) &
(B 'are among' D) &
(B 'are smaller than' D) &
(B 'have' F) &
(F 'give' [to] A E)
where
D = 'A's sleek glass monoliths': "its (the district's) sleek glass monoliths",
E = 'an interesting texture': "an interesting texture",
F = 'B's gargoyles and other lively twentieth-century decorations': "gargoyles and other lively early-twentieth-century decorations".

The first clause identifies what kind of thing B is, and the second spells out an inference that the buildings belong to the district (along with the monoliths). The other clauses talk about the smaller buildings being among the monoliths; about the smaller buildings having decorations (the predicate 'have' is a replacement for the "with ..." construction); and about how the decorations give the district an interesting texture. The atomic sentence organized around 'give' is of the form (<subject> <predicate> <complement> <object>).

Third-stage Analysis

Continuing recursively with the results from the second stage—

Analyzing C = 'Yesler Way, Route 5, Stewart Street, the waterfront': "Yesler Way, Route 5, Stewart Street, and the waterfront" produces:

(C 'are' 'boundaries') &
(C 'has members' 'Yesler Way, Route 5, Stewart Street') &
(C 'has member' G)
where
G = 'Seattle's waterfront': "the waterfront".

The list is interpreted as a simple unstructured collection of boundaries, three of which are atomic (grouped together for convenience) and one is a complex description. The sentence does not provide specifics of how the members are arranged (ie, related to each other), although a more thorough subsequent analysis might surmise that the arrangement forms a quadrilateral.

Analyzing D = 'A's sleek glass monoliths': "its (the district's) sleek glass monoliths" produces:

(D 'are' 'monoliths') &

(D 'belong to' A) &
(D 'are sleek') &
(D 'are glass').

Analyzing E = 'an interesting texture': "an interesting texture" produces:

(E 'is a' 'texture') &
(E 'is interesting').

Analyzing F = 'B's gargoyles and other lively twentieth-century decorations': "gargoyles and other lively early-twentieth-century decorations" produces:

(F 'are' 'decorations') &
(F 'include' 'B's gargoyles') &
(F 'are lively') &
(F 'date from' 'early-twentieth-century').

Fourth-stage Analysis

The only remaining constant is G.

Analyzing G = 'Seattle's waterfront': "the waterfront" produces:

(G 'is a' 'waterfront') &
(G 'belongs to' 'Seattle').

Here, the phrase "the waterfront" implicitly refers to Seattle's waterfront.

In summary, the extracted complex concepts are

A = 'Seattle's central business district'
B = 'A's smaller buildings'
C = 'Yesler Way, Route 5, Stewart Street, the waterfront'
D = 'A's sleek glass monoliths'
E = 'an interesting texture'
F = 'B's gargoyles and other lively twentieth-century decorations'
G = 'Seattle's waterfront'

And the atomic sentences are

(A 'is a' 'district') &
(A 'has' B) &
(A 'is bounded by' C) &

(A 'is for' 'business') &
(A 'is central') &
(A 'belongs to' 'Seattle') &
(B 'are' 'buildings') &
(B 'are among' D) &
(B 'are smaller than' D) &
(B 'have' F) &
(C 'are' 'boundaries') &
(C 'has members' 'Yesler Way, Route 5, Stewart Street') &
(C 'has member' G) &
(D 'are' 'monoliths') &
(D 'belong to' A) &
(D 'are glass') &
(D 'are sleek') &
(E 'is a' 'texture') &
(E 'is interesting') &
(F 'are' 'decorations') &
(F 'date from' 'early-twentieth-century') &
(F 'are lively') &
(F 'give' [to] A E) &
(F 'include' 'B's gargoyles') &
(G 'is a' 'waterfront') &
(G 'belongs to' 'Seattle').

To complete the preliminary analysis, it remains only to identify the background classification type for each concept (ie, what kind of thing it is). The types of the complex concepts have already been identified during the analysis: (A 'is a' 'district'), (B 'are' 'buildings'), (C 'are' 'boundaries'), etc. The types for the remaining literals are listed below:

('business' 'is an' 'activity') &
('Seattle' 'is a' 'city') &
('Yesler Way, Route 5, Stewart Street' 'are' 'roads') &
('early-twentieth-century' 'is a' 'period') &
('B's gargoyles' 'are' 'gargoyles') &
('has' 'is a' 'having') & —ie, (A has B) is an instance of, or a kind of, having situation
('is bounded by' 'is a' 'being bounded by') &
('is for' 'is a' 'being for') &
('is central' 'is an' 'attributing') & —ie, is an instance of attributing centrality
('belongs to' 'is a' 'belonging to') &
('are among' 'is a' 'being among') &
('are smaller than' 'is a' 'being smaller than') &
('have' 'is a' 'having') &
('belong to' 'is a' 'belonging to') &

('are glass' 'is an' 'attributing') &
('are sleek' 'is an' 'attributing') &
('is interesting' 'is an' 'attributing') &
('date from' 'is a' 'dating from') &
('are lively' 'is an' 'attributing') &
('give' 'is a' 'giving').

The predicates 'has members', 'has member', and 'include' are left out of the list, because they are primitive and predefined. All three express the same basic superordinate/subordinate (ie, general/specific, class/subclass) relationship between a concept and its subtypes or instances, and are equivalent to the 'is a' (or 'are' for plurals) predicate.

The following two figures sum up the preliminary analysis. They are graphics showing how the various new concepts relate to each other and to a preexisting background of concepts. The foreground serves essentially as a scratchpad for building an interpretation. The horizontal numbered arrows show the arguments for each relation extracted through the four stages of analysis, and the heavy vertical arrows show how all new concepts tie into background knowledge specified through the type classifications.

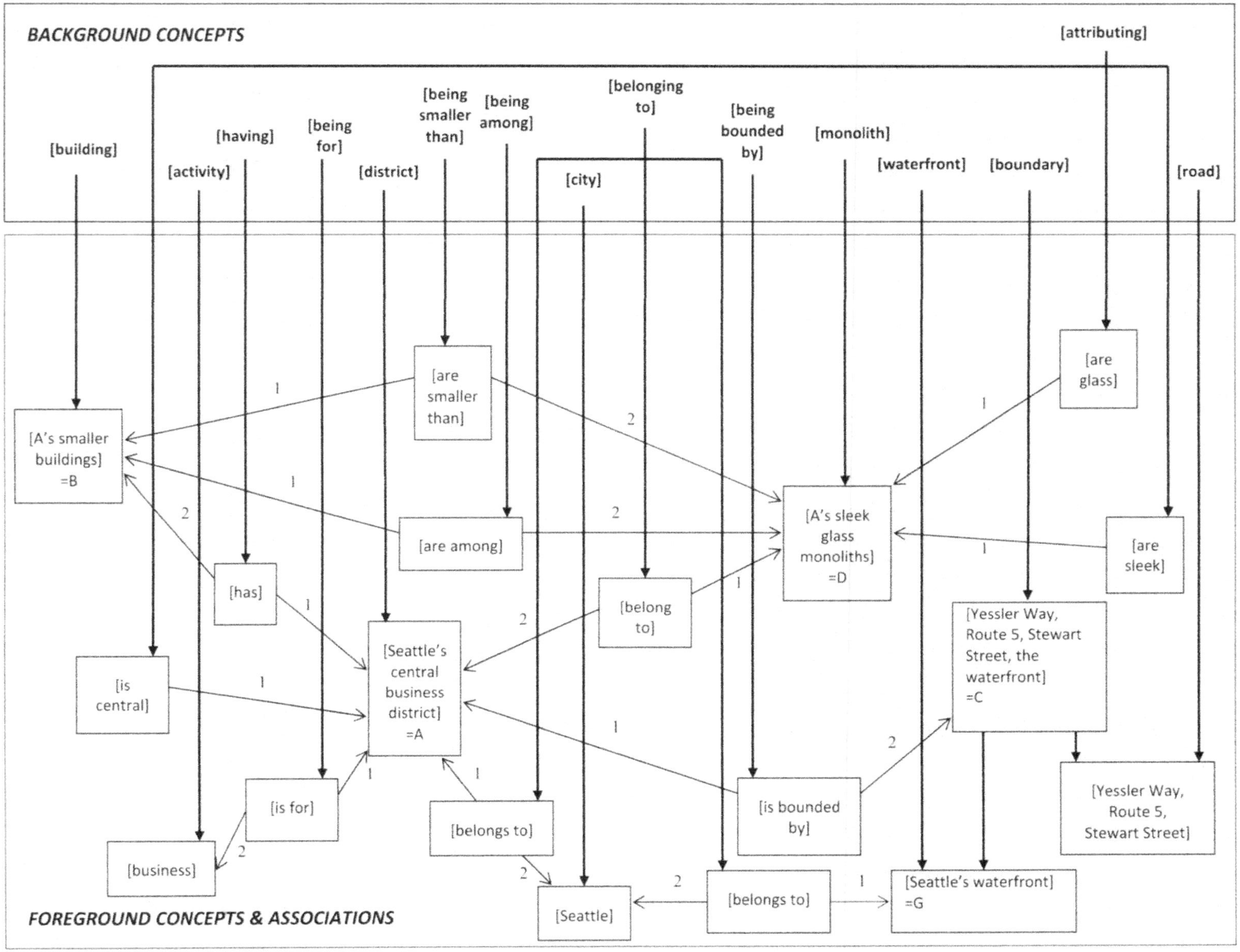

Concept Graph for "Seattle" Sentence (part 1)

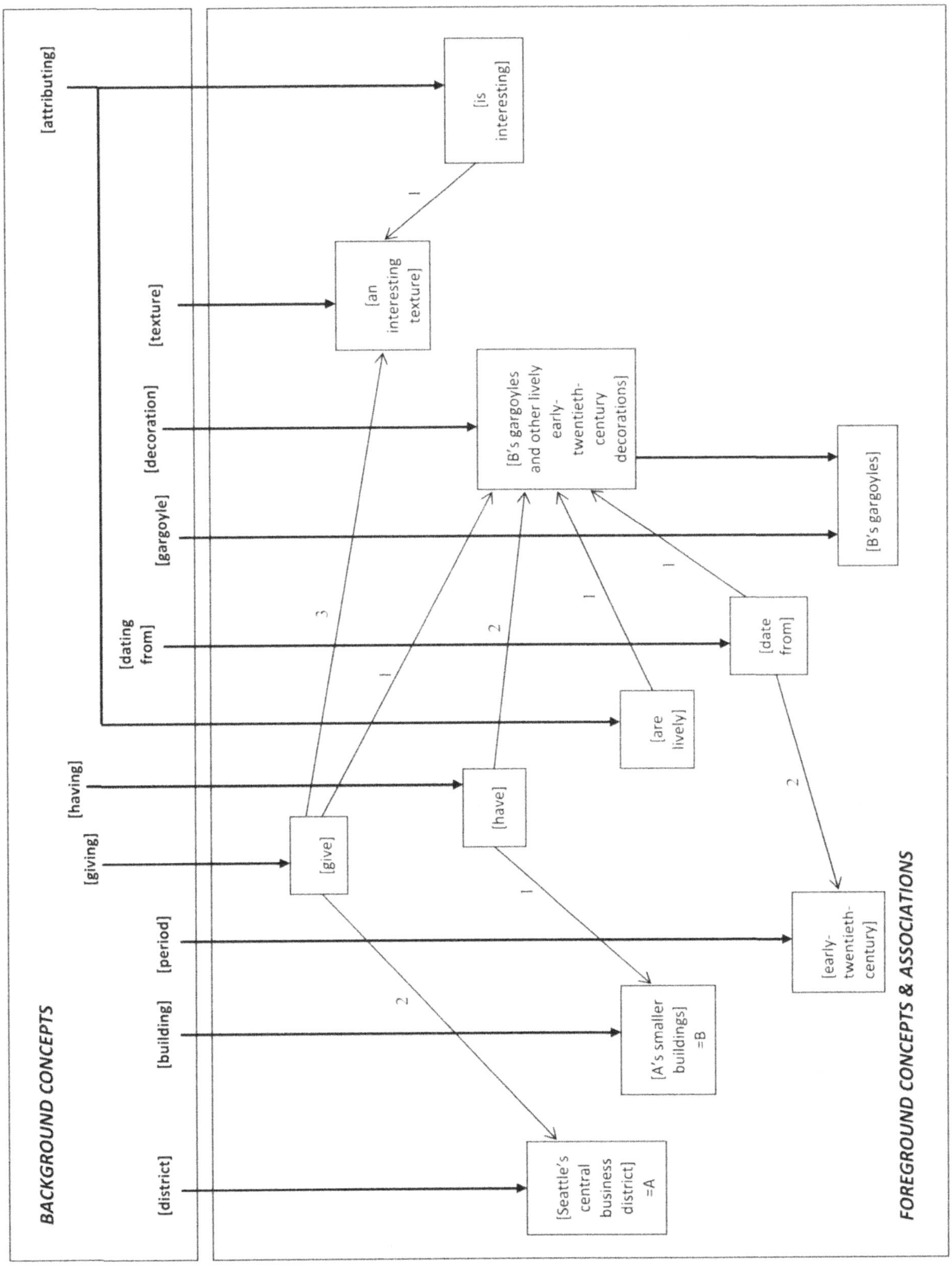

Concept Graph for "Seattle" Sentence (part 2)

As a second, more abbreviated, example of propositional sentence analysis, consider:

"Analyzing the meaning of a sentence is troublesome for a variety of reasons."

- A: "Analyzing the meaning of a sentence is troublesome" =>
  "The analysis of the meaning of a sentence is troublesome"
    - C: "the analysis of the meaning of a sentence"
        - E: "a sentence"
        - F: "the analysis of the meaning of E"
            - G: "the meaning of E"
            - H: "the analysis of G"
    - D: "C is troublesome"
- B: "There are a variety of reasons for A" =>
  "The reasons for A are varied"
    - I: "the reasons for A"
    - J: "I are varied"

Terminal (Leaf) Sentences:

G is the meaning of E.
H is the analysis of G.
C is troublesome.
C is H.
A is "C are troublesome"
I are the reasons for A.
I are varied.

Identification Sentences:

E is a sentence.
G is a meaning.
H is an analysis.
D is being troublesome.
I are reasons.

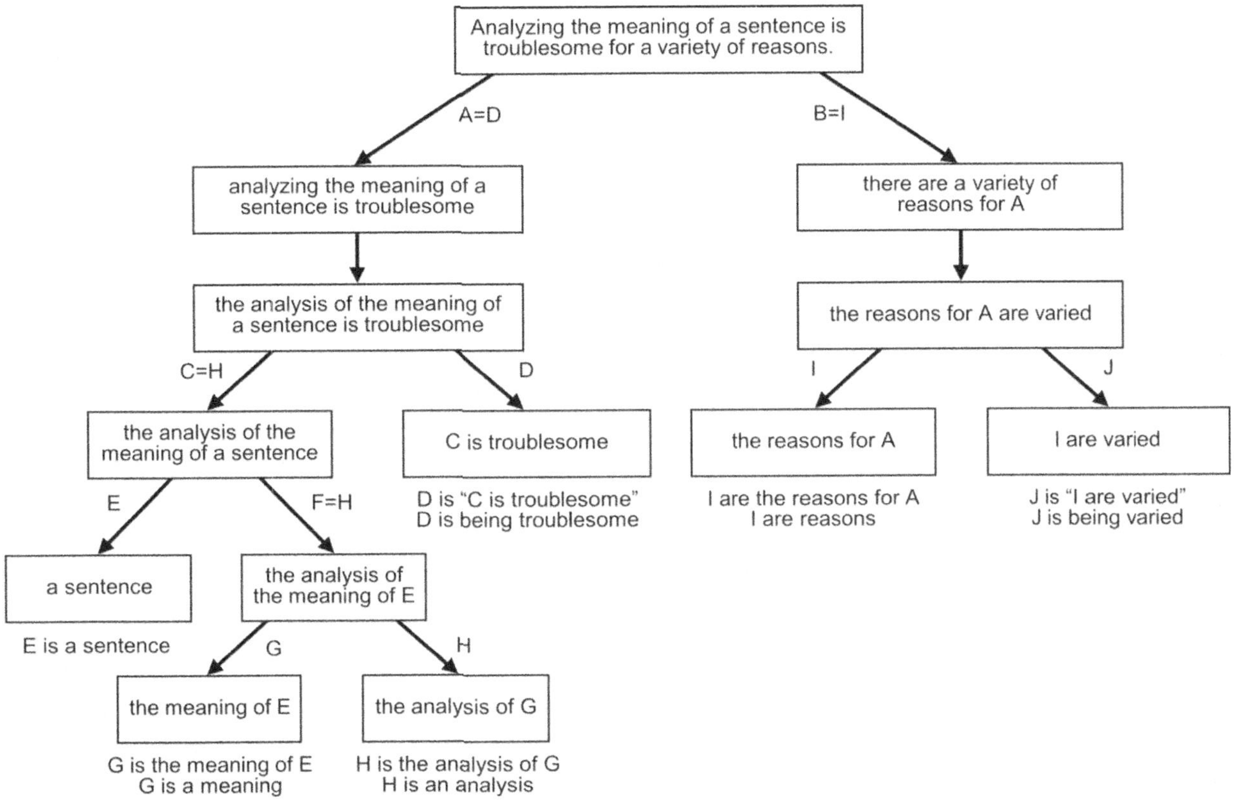

Figure 55: Propositional Decomposition for "troublesome" Sentence

6. CONCEPT NET ORGANIZATION PRINCIPLES

Hierarchical Organization

Most of the concepts in a concept network exhibit some form of hierarchical layering. Lower-level concepts are subkinds (or subclasses) of higher-level, more inclusive, concepts. Organization according to the kind → subkind relation is only one of many potential organizing relations for concepts, but it seems to capture one important way we systematize knowledge about the world. One justification for such a hierarchical structuring is representational efficiency. Attaching a characteristic to a concept means that all concepts below it can be characterized the same way by default. [dog] then can automatically inherit all of the characteristics associated with [mammal] (eg, normally having fur) because it is a kind of mammal. This is the mechanism of conceptual inheritance. The default characterization does not have to be repeated for each applicable concept in the concept net, because it is inherited from above. A lower-level characterization, however, can still override upper-level default values to handle inevitable exceptions.

The philosophy behind the hierarchical organization of knowledge about the world seems to derive from a latent psychological belief, the belief that natural kinds of objects have an inherent distinctive nature. An object's true nature is inscrutable, it is generally assumed, but it can be approximated by identity characteristics. The ideal hierarchy then expresses a skeletal taxonomy of natural kinds, organized in terms of identity characteristics, with new characteristics added at each lower level, expressing greater and greater specificity. Furthermore, an object's nature is normally deemed a privileged criterion for classification. Other criteria like function, appearance, relationships, or history are often useful, but are secondary, and not part of the central taxonomic skeleton. For example, a concept like [statue] is not part of the core skeleton, but the concept of a [material object] functioning as, or having the form of, a statue is. The major consequence of organizing concepts by a single identifying criterion is that each core kind then has only one superordinate kind, and the hierarchy is a branching tree shape rather than a tangled lattice. Allowance however is made for nonskeletal concepts (like common social constructs) to divide their allegiance among several superordinates, depending on their mix of inherited characteristics.

Three Tiers of Knowledge

It is well known that folk conceptual taxonomies tend to be rather shallow, extending naturally to only a few levels. Cruise offers a discussion of lexical taxonomies in (Cruse, 2000) and numerous examples in (Cruse, 1986):

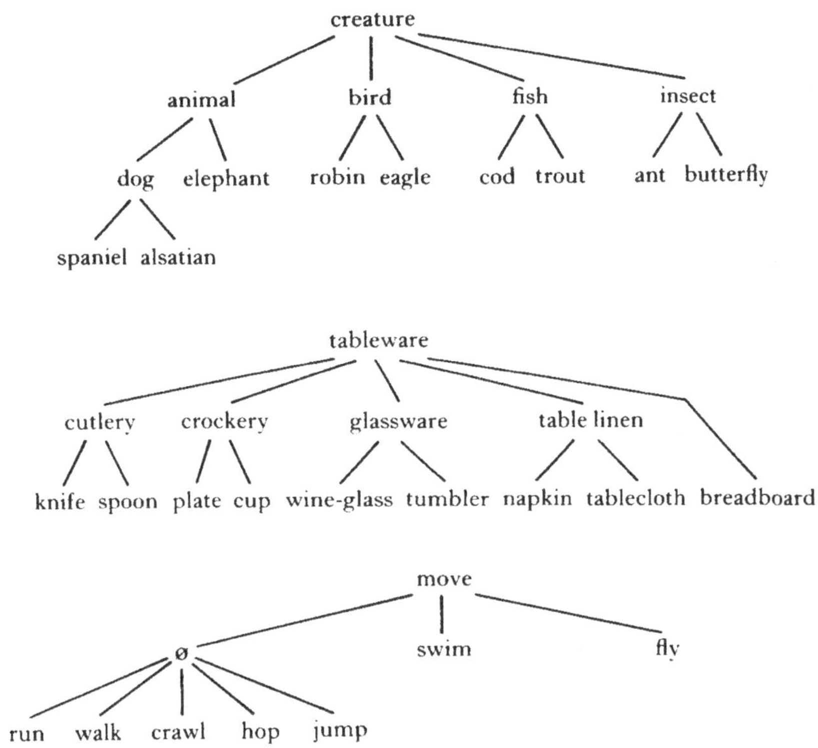

Figure 56: Typical Lexical Hierarchies from (Cruse, 1986)

One possible explanation for the shallowness of human conceptual hierarchies is that the representational efficiency provided by a hierarchy significantly diminishes as one moves down the hierarchy to the lower-level concepts. The potential number of feature variations of highly specific categories (eg, of spoon) that might become relevant in some context (eg, removing a drippy teabag from a cup using a spoon with small drain holes in the bottom) is so large and so combinatorial as to obviate the advantage of a static hierarchical organization. More general concepts on the other hand are typically distinguished by relatively few characteristics—eg, tableware simply covers implements often used at meals set on a table. Concepts like those illustrated above are called "domain concepts", because they address a limited domain of knowledge about some topic of interest. Some other common domains of human interest would include kinship concepts, occupational and recreational concepts, and familiar artifacts, among many, many others. Domain concepts typically are near the bottom of a concept hierarchy, and encapsulate knowledge used to talk directly about particular things in the world. Some examples from the domain of medicine might include [pancreatic cancer], [cancer patient], [fatigue], [insulin], etc.

Another major grouping of concept types resides above the domain concepts in any comprehensive concept hierarchy. These are the more intangible "theoretical concepts", sometimes associated with so-called framework theories (Wellman & Gelman, 1992). Theoretical concepts usually incorporate knowledge which is relatively independent of the lower level perceptual and functional domain knowledge necessary to interact with things. Theoretical knowledge is oftentimes scientific in nature, has achieved a wider scope of acceptance than domain knowledge, often deals with hidden or nonperceptual characteristics, and can usefully inform domain knowledge—as when the domain concept of [whale] was shifted from [fish] (a domain concept) to [mammal] (originally a theoretical concept which was subsequently merged into the domain hierarchy for animals). An example of some theoretical concepts from a medical domain would probably include [organism], [disease], [symptom], [vector], etc.

However, many theoretical concepts develop as extensions of common terms too, and are used to group domain subkinds together for more global objective reasons. Suppose the concept of [fruit1] as a domain concept generally covers the fleshy, juicy, sweet products of certain plants and (according to Rosch's studies) prototypically includes apples, peaches, strawberries, and the like for most people (Rosch & Mervis, 1975). The primary purpose of this folk domain classification is likely based on a combination of dietary and perceptual similarity considerations, and probably varies somewhat by region. The concept of [fruit2] as a theoretical concept, however, may become associated with the ripened ovary of a flowering plant. In general, [fruit2] is a broader concept than [fruit1] because it also includes nuts, beans, grains, and even milkweed. But it is also simpler, in the sense of being less *ad hoc,* and is motivated by its place in a global taxonomy. Fewer characteristics are required to characterize the more theoretical as opposed to the perceptually/functionally-oriented domain concept, and [fruit2] is a useful concept in ways that [fruit1] is not.[195] It might, for example, lead to the discovery of alternative methods of pollination. A further extension of the concept to [fruit3] might permit classification in terms of genetic (DNA) properties, and be used to develop new kinds of fruit or perhaps insect protection strategies. In general, there is no reason why a variety of related concepts, such as those for fruit, could not coexist at any level of the hierarchy.

The third major grouping of concepts is found in the uppermost echelons of the concept hierarchy. The top-level, most general "epistemological concepts" in a concept net can always be distinguished from each other by just one characteristic. For example, the criterial characteristic that distinguishes the concept [particular] from [[universal]] is simply that a universal can have instances, while particulars cannot—eg, |Fido| is a particular, while [dog] is a universal. What distinguishes a [collective] from an [individual] is that only a collective can have members. Additional characteristics may accompany the primary characteristic, but they are not used for knowledge organization. As a group, epistemological concepts relate to the organization of knowledge itself. Such concepts include other generic concepts like [entity], [[kind]], [[relation]], etc, used to classify any possible concept, including epistemological concepts themselves.

An epistemological concept tends to be analytic (or nominal in kind), because it has no content over and above the characteristic that defines it. In contrast, so-called natural kind concepts (covering mostly theoretical and some domain concepts in the hierarchy) are inherently indefinable, because they

---

195 Note, however, that mixing domain and theoretical concepts in the same hierarchy will usually make the hierarchy "tangled" with cross-references. An [apple] then is both a kind of [fruit1] and [fruit2], and can be viewed from either perspective.

possess an indeterminate number of potential characteristics. They have meaning content which is always expandable, but not bounded. More can be discovered about them, but they can never be completely characterized.

Thus, there is an increasing gradation of combinatorial characteristic density moving from the most general categories like [concept] and [thing] to the more specific like [money], [penny], [1943 zinc-coated steel penny] on down to |the 1943 zinc-coated steel penny in my collection|.[196] The concept hierarchy most efficiently represents the upper echelon concepts, what I have called epistemological concepts, as opposed to lower-level theoretical and domain concepts. But there seems to be no sharp boundary dividing the various tiers, and no disadvantage to extending the epistemological hierarchy as far as possible into the theoretical level, and the theoretical level into the domain level.

The Decomposition of Epistemological and Theoretical Concepts

Besides inheritance efficiency, there is another, more important, justification for a concept hierarchy, grounded in how we analyze complex phenomena, in general. A useful way of analyzing an initially ill-defined and difficult concept is to decompose it into exactly two disjunctive subconcepts according to *complexity* considerations. One subconcept to the left (say) covers "simple" instances of the concept, and the other to the right covers all remaining "complex" instances. What counts as a simple instance will normally depend on the content of the parent concept. Between them, the two subconcepts ideally exhaust all potential instances of the parent concept.[197] The objective is to split off the more simply characterized and prototypical cases from all the others. Then, from those remaining complex cases, the simpler complex prototypical cases are split off at the next level, and so on. A similar splitting may also be applied to the original simple cases. The successive splitting off of simple instances from the more complex instances at each level of generality produces a two-way branching tree of subconcepts.

This analytical method of concept decomposition has a well-known analogue in the process of concept differentiation in human development. Early childhood researchers—eg, (Carey, 2009; Mandler, 2004)—envision an innate maturational process of conceptual development in the infant beginning with a more-or-less amorphous conceptual substrate (with regard to conceptual content). The earliest differentiations of the substrate, probably triggered jointly by biology and exposure, are thought to produce a core appreciation of such elementary distinctions as spatiotemporal vs abstract, physical vs nonphysical (ie, unreal), material vs immaterial (eg, heat), object vs substance, organic vs artifact, etc. The criteria infants use for making such distinctions at such an early age are controversial, but these well

---

[196] If the particular concepts appearing at the bottom of a concept network are regarded as a distinct level, then there would be perhaps four tiers of knowledge total. However, particular concepts are qualitatively different from the abstract universal concepts making up the upper three tiers.

[197] It might be thought that a middle-ground category between "simple" and "complex" is needed to cover instances of intermediate complexity. However, given the way simple concepts are specified above, concepts of intermediate complexity can be thrown in with all the other complex (ie, nonsimple) concepts. Intermediates can then be split off subsequently, if the complex concepts are decomposed further.

documented distinctions seem to be part of our evolutionary heritage. The development of a sophisticated adult concept hierarchy (or multiple hierarchies), further informed by scientific concepts, may stem from a continuing process of differentiation, along with integration and reorganization, throughout our lifetime.

Applied to the top-level concept [concept] in a concept net, binary decomposition implies that the subkind [particular] covers all instances of "simple" concepts, and [[universal]] covers all other "complex" concepts. Each of these subkinds can then be analyzed in the same way, recursively, thereby generating the top-levels of the concept hierarchy. The splitting process, furthermore, is required to be compositional, such that a complex subkind can have simple subkinds as *constituents*, in a nesting fashion. For example, universals can have particulars as constituents, which in this case are called *instances* (of the universal); sets can have elements as constituents, objects can have properties, events can have participants, relations can have arguments, and so on.[198] Enforcing this decompositional principle for organizing concepts helps select which concepts, out of all potential concepts, appear in the upper levels comprising the epistemological hierarchy.

The decomposition procedure has potential for analyzing certain theoretical concepts, as well. For example, we might analyze the parent concept of fruit (specifically the concept [fruit2] discussed above) into a left concept labeled [simple fruit], plus a right concept labeled [complex fruit], where the simple fruits are regarded as theoretical constituents of complex fruits. This actually turns out to be the case in the botanical classification of fruits, depending on whether the fruit develops from a single (cherry) or multiple (raspberry) ovaries. As a more abstract example, we might analyze (whole) sentences in terms of elementary simple sentences (left) containing one clause, like "Kate loves Colin", plus complex sentences (right) containing multiple clauses, like "Kate loves Colin or Kate loves Darin" as constituents. In all cases, the complex instances of the parent concept can be thought of as being built up out of the simpler instances using a composition principle which is unique to the content of the parent concept.

The figure below shows the schema for decomposition using the concept of an arbitrary generic kind of [concept]. It is accordingly analyzed into simple and complex subkinds. The simple subkind covers elementary instances of [concept]. The complex subkind covers the remaining instances. Instances of the complex subkind, moreover, are thought of as structured, such that they have constituents which are simple instances of the parent concept. The generic relation [1[has constituent]2] in the diagram shows the destructuring of complex instances by pointing to the parent concept. The constituent of a complex instance, therefore, can be either a simple instance, or other complex instances (just like a sentence may have other sentences as constituents, as well as simple constituents—eg, "Elias believes Kate loves Colin or Darin"). Note also that the decomposition may be continued in a straightforward recursive manner for the complex concept. In other words, the complex concept itself is often usefully decomposed further into simple and complex notions, but always covering more complex instances than the initial split-off of the simplest instances. These complex-complex concept instances themselves can

[198] Note that constituency, as used here, is a much broader concept than parthood, as used in mereological studies—parts are only one kind of constituent, usually subject to an extra principle of connectivity, which gives the whole some degree of internal integrity.

have constituents, as before. In fact, the complex-complex instances may also have the simplest instances as constituents (as well as simple-complex instances), as can be checked by the way the constituency relation is defined. The general rule is that any arbitrarily complex instance can have constituents that appear anywhere to its left or below in a concept hierarchy.[199]

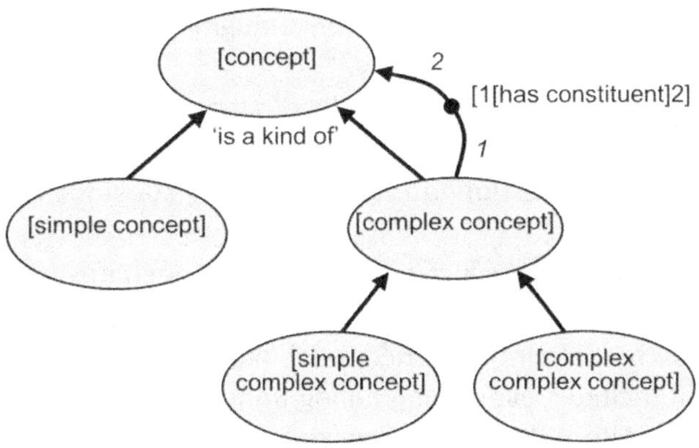

Figure 57: Decomposition of Concepts According to Constituency

The precise constituency relation associated with the decomposition of different concepts will vary, depending on how simple instances are combined to produce more complex instances. However, the constituency relation will always be irreflexive (no instance can ever have itself as a constituent), asymmetric (if one complex instance has another simpler instance as a constituent, then the complex instance cannot also be a constituent of the simpler instance), and intransitive (if a simpler instance is a constituent of a complex instance, and the complex instance is a constituent of a still more complex instance, then it may or may not be true that the simpler instance is a constituent of the most complex instance).

As a consequence of the way decomposition is defined, the most general concepts, with the greatest coverage, are at the top of the hierarchy. As subconcepts progressively split-off from these, the lower-level concepts become more specific, and the total number of their constraining constituent (and other ancillary) relations, inherited from above, increases. The most complex concepts therefore are the most specific—a somewhat counterintuitive result, since it is the familiar every day, highly specific, concepts that seem simple to us. By this measure, [dog] is a more complex concept than [animal], because it inherits all of the characteristics associated with [animal] and adds more of its own. But for the same reason, [animal] is also a more general concept than [dog].

---

199 The way the constituency relation is diagrammed suggests that "backwards" left to right constituents are possible, in addition to right to left. However, this is obviously incoherent since a simple object cannot contain a more complex object (otherwise it would no longer be simple). Left to right constituency relations are excluded by the constituency rule.

The following items summarize the decomposition principles:

- Decomposition can, in theory, be applied to any universal (ie, nonparticular) target concept.
- Some instances of the target concept must be thought of as being "simpler" than all the other "complex" instances in a compositional way.
- Some of the simple instances must be thought of as being constituents of some of the complex instances.
- The target concept is then partitioned into exactly two subkinds covering the simple instances and the complex instances.
- The subkinds must be mutually exclusive and collectively exhaustive of the target concept.
- In the absence of pertinent labels for subconcepts, an appropriate generic label at least for the complex subkind may be the simple concept prefixed by "non" (eg, [[nonparticular]] instead of [[universal]]), since it includes everything falling under the parent concept which is not included in the simple concept. Alternately, an appropriate generic label for the complex subkind may be the parent concept prefixed by "meta" (eg, [[metauniversal]]), since it includes all compositional complexities falling under the parent concept.
- Decomposition may again be applied recursively to each of the subkinds, if useful.

Examples of Concept Decomposition

The foregoing discussion presents the gist of the decomposition idea, rather than a worked out theory. It is still rather weakly defined, and its generalized significance when applied to a broad spectrum of knowledge is unverified. In this context, the following diagrams sketch some sample applications of concept decomposition to several domains of knowledge—specifically, set theory, predicate logic, chemistry, number systems, botany, Newtonian kinematics, vector quantity, and number theory. The examples provide one perspective on the different ways we see complex entities as being "built" out of simpler entities according to rules of composition.

Set Theory

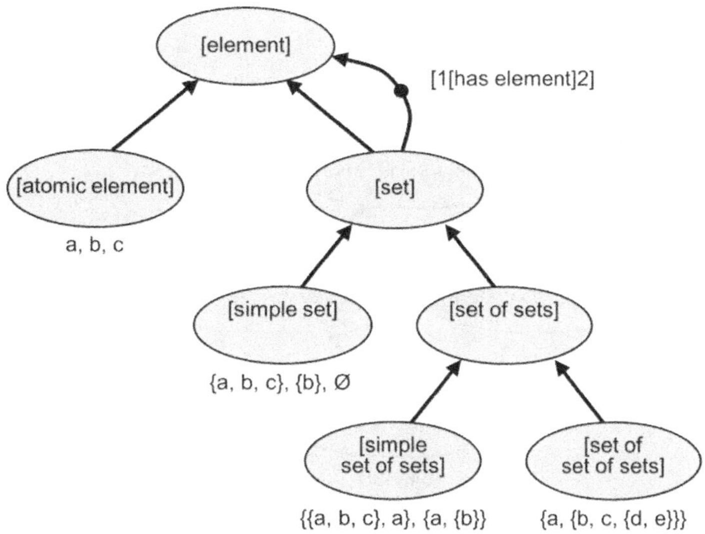

- A set can have elements as members, where an element is either an atomic element or a set.
- A set is any element except an atomic element; a set of sets is any element except for an atomic element or simple set (or any set except for a simple set).
- A simple set can only have atomic elements as members; a simple set of sets can have only simple sets and atomic elements as members; etc.

Predicate Calculus

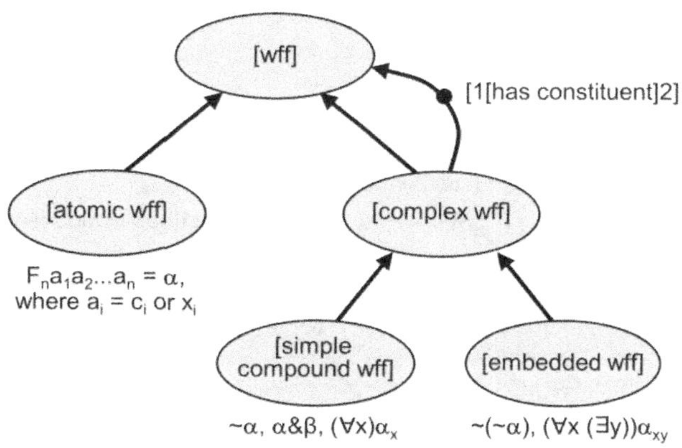

- α is any well-formed formula (wff), consisting of a predicate and its arguments.
- Complex wffs have simpler wffs as constituents.
- F is a basic predicate, taking any number of arguments.

- The arguments to a predicate can be either constants or variables.

Number Systems

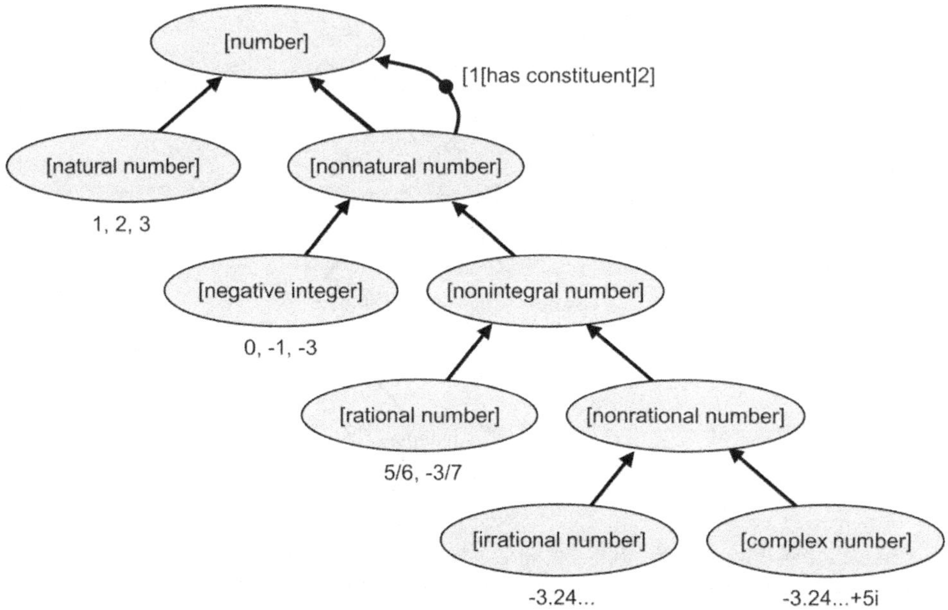

- A negative integer is a structure composed of a natural number plus a negation operator.
- A rational number is a structure composed of two natural numbers or negative integers, plus a quotient operator.
- How, precisely, is an irrational number composed of a rational part (along with an irrational part)? Perhaps, digit by digit. May also be expressible as Dedekind cuts.
- A complex number is an ordered pair of numbers.

Chemical Substance

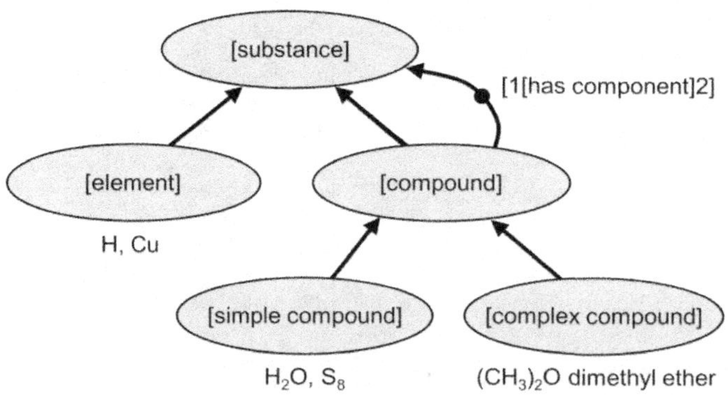

- Chemical substances are either elements or compounds, where complex compounds are made up of either simple compounds or elements.
- Substance structure is physically determined by covalent and ionic bonding constraints.

Botanical Fruit

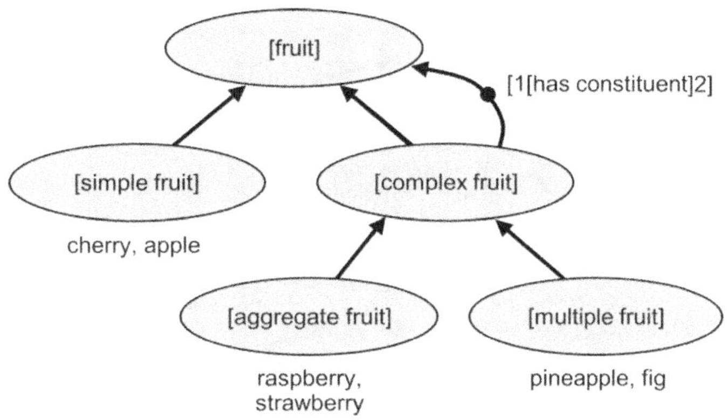

- The distinction between simple, aggregate and multiple fruit is normally based on two attributes: the number of ovaries and the number of flowers. Simple fruits have a single ovary, leaving all complex fruits with multiple ovaries. Aggregate fruits additionally develop from a single flower, while multiples develop from multiple flowers. The constituency relation between [complex fruit] and [fruit], therefore, may be somewhat different from the relation between [multiple fruit] and [complex fruit]. Both are subsumed in the diagram under a single constituency relation.
- Simple fruits might be further subdivided into fleshy and dry fruits, depending on how the ovary wall of a simple fruit expresses itself. An apple or orange is a good example of a fleshy fruit, and dry-walled fruits include examples like legumes and nuts.

Quantity

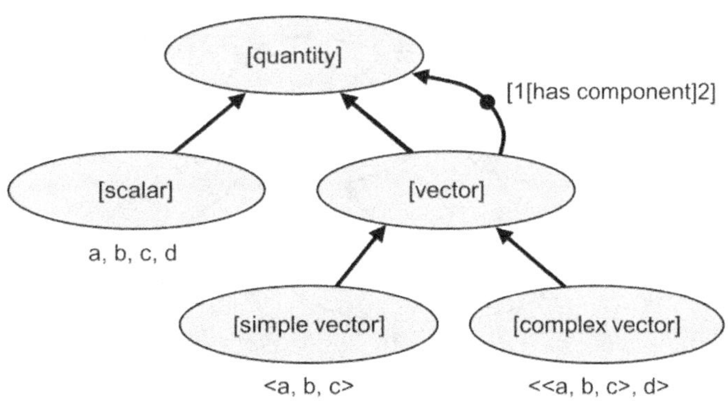

- A quantity is either a simple scalar or a vector.

- A vector is a list which is made up of either scalar or other vector components.

Kinematics

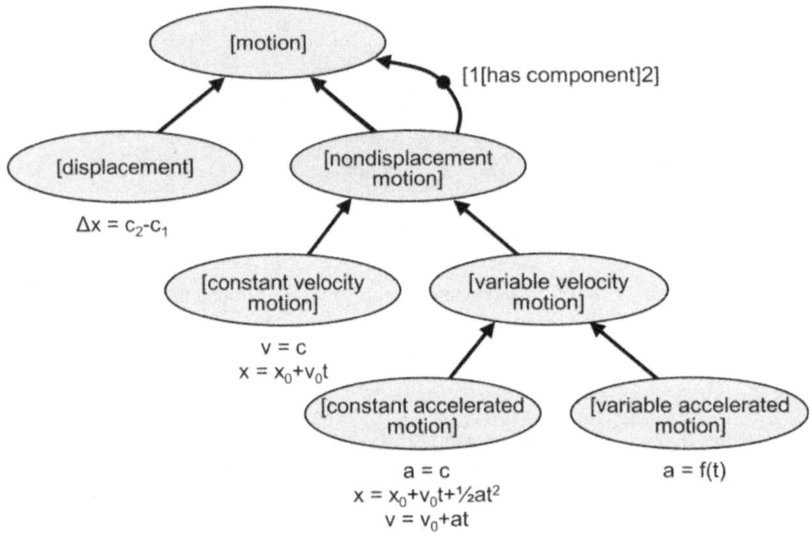

- Displacement is instantaneous from position to position; includes stationary (no motion).
- Constant velocity motion is made up of infinitesimal displacements; constant acceleration motion made up of infinitesimal velocity changes; etc.

Number Theory

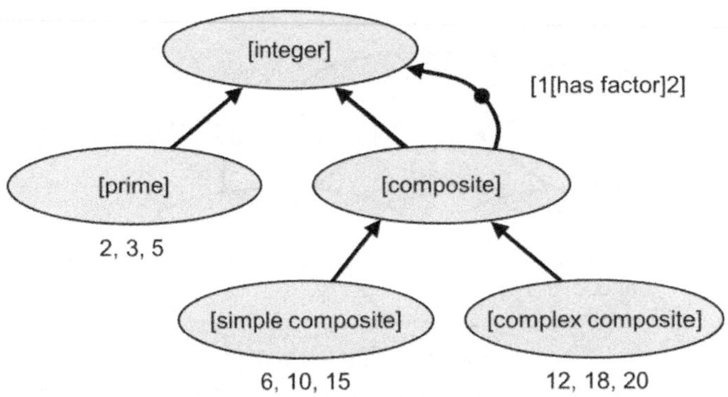

- A prime has only itself and 1 as a factor.
- A simple composite has only two primes as factors.

- A complex composite has three or more primes as factors.

Representing Conceptual Change

The term "conceptual change" covers a broad territory of research including the historical evolution of scientific concepts (Kuhn, 2000; Thagard, 2012; Nersessian, 2008; Andersen, Barker, & Chen, 2006), developmental changes in the way children and adults come to understand core concepts (Carey, 2009; Spelke & Kinzler, 2007), and the acquisition of new concepts in educational settings (Chi, 2008). This section is aimed specifically at how conceptual change can occur in a concept net, as a result of any learning experience that requires changes to existing knowledge representations. Any change generally involves the incorporation of new knowledge in a way that maintains concept net internal consistency. The following discussion sketches a variety of shifts in concept representation due to learning, but does not attempt to discuss nuances of the attendant concept learning mechanisms. Three different kinds of conceptual change are examined ranging from the relatively simple to the relatively complex: 1) learning that some birds do not fly, 2) learning that dolphins are mammals, and 3) learning that water is a chemical substance.

Birds

Toddlers seem to enjoy learning what things are called.[200] Assuming they are aware of some of the salient characteristics of what they are pointing at, the underlying purpose of the activity seems to involve the classification of exemplar individuals. When you tell a child the category name for a particular individual ("That's a bird!"), it enables the child to link the just formulated concept of the particular bird (eg, |bird#1|) with a concept of the category (eg, [bird]) via instantiation ("|bird#1| 'is an instance of' [bird]"). By further drawing attention to the fact that it is flying ("Look, the bird is flying!"), the child can assign a characteristic to the bird (ie, ||bird#1||is flying||). By repeating this scenario multiple times, the child, it is proposed, builds up a set of exemplar individual concepts (|bird#1|, |bird#2|, etc), all of which are characterized by flying. Subsequently, the capacity to develop prototype concepts can lead to a consolidation of common instance characteristics (eg, that birds fly) on the concept [bird] for the sake of efficiency (eg, [[bird][is flying]]), as shown in the following figure, part a. The redundant characteristics on the exemplar individual birds can then be eliminated, since they are recoverable by default inheritance from [bird]. Finally, given further observations that common birds sometimes walk and sometimes fly, the flying characteristic may transform into a modal behavioral characteristic (ie, [[bird][can fly]]), instead of a necessary characteristic.

---

[200] My son's favorite phrase at this stage was "eyes dat" (ie, what's that). This little language game would sometimes go on for what seemed like hours.

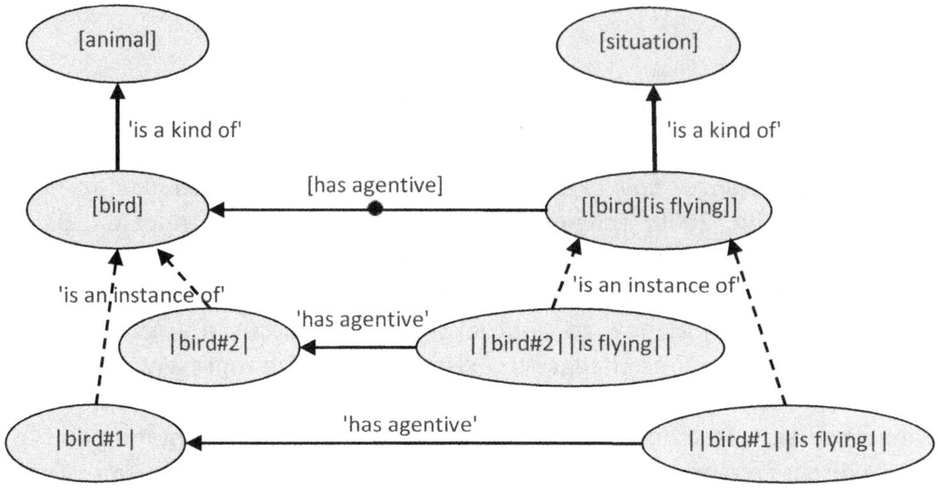

Figure 58: Representing Some Birds Don't Fly (part a)

If the conceptual representations established in the foregoing scenario are roughly correct, then the basic elements are in place for adjustments reflecting the fact that some nonprototypical birds cannot fly. If the child is visiting a zoo, and is shown a bird (eg, an ostrich) and told it cannot fly, then rational consistency requires some modification on the flight characteristic of [bird]. In cases like these, it is proposed that the characteristic undergoes a differentiation into two (or more) property values, where the values indicate the range of possibilities that instances of the concept can take on. Here, the characteristic might become something like [[bird][can fly][|true|,|false|], since there are only two complementary possibilities. But in the more common case, there may be several explicit values (eg, for color), or a scale of values (eg, for weight), indicating possible values of instances. Furthermore, the prototypical representation can include useful distributional (statistical) information regarding how common each value is expected to be. In the current example, the child might assign a rough probability of 0.9 to |true| (ie, can fly), and 0.1 to |false| (ie, cannot fly), leading to the prototypical characteristic [[bird][can fly][|true|(.9),|false|(.1)], as shown in part b of the figure below. (See Chapter 3: *Prototype Effects* for additional discussion of prototype structure.) What this intuitively means to the child is that birds typically can fly. The full prototypical structure for a concept characteristic thus indicates what is the most typical value for the characteristic, but also what other values are reasonable to expect on nontypical occasions.

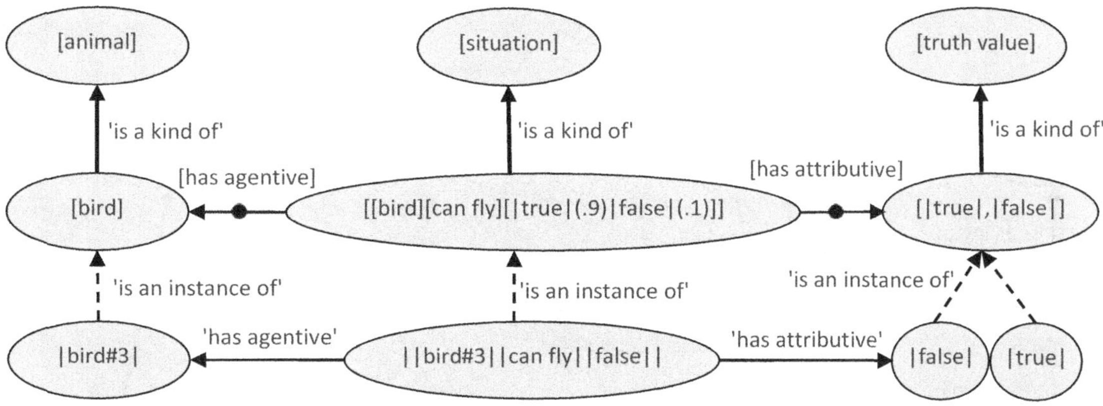

Figure 59: Representing Some Birds Don't Fly (part b)

This next example illustrates a deeper conceptual change involving the reclassification of a prototypical concept, based on learning some additional theoretical information. Dolphins possess many fish-like characteristics, and so the initial untutored conception likely represents the prototypical [dolphin] as a [fish] (ie, "[dolphin] 'is a kind of' [fish]"). In this way, instances of [dolphin] (like |Flipper|) will inherit fish-like characteristics by default, such as feeding in the water. But [dolphin] will then also require extra nondefault characteristics (like breathes air) as exceptions that override fish-like characteristics. This initial prototypical organization of knowledge about dolphins is indicated in the following figure, part a.

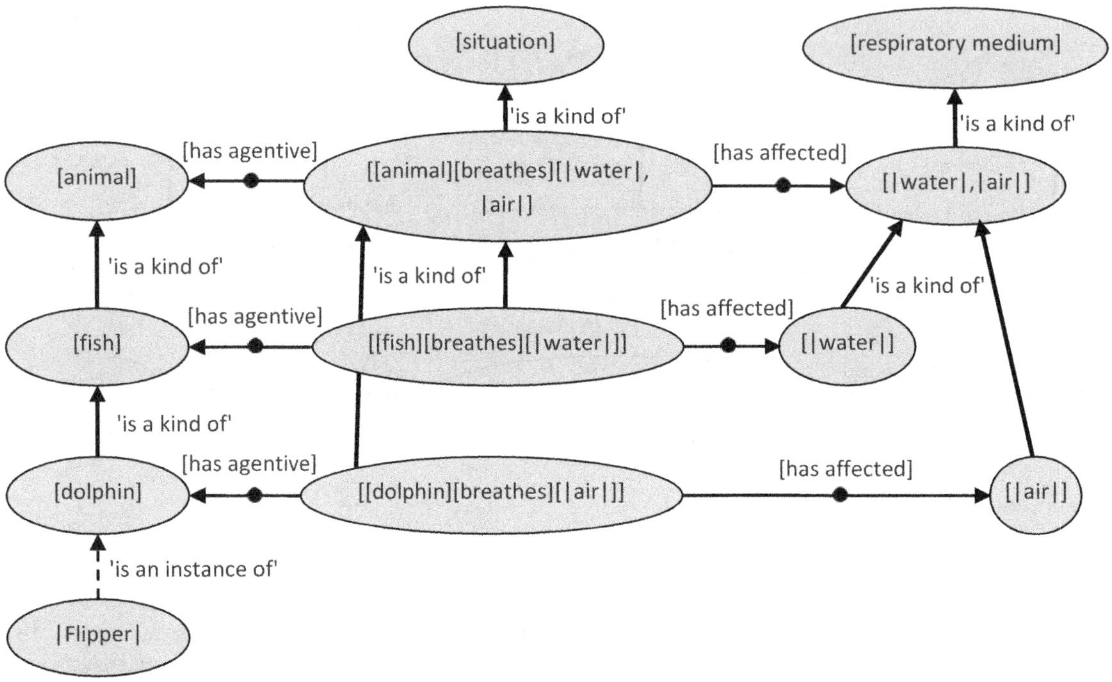

Figure 60: Representing Dolphins Are Mammals (part a)

However, biological species are more systematically classified in terms of lineage. And this kind of theoretical/historical knowledge places [dolphin] in the category of [mammal], rather than [fish] (ie, "[dolphin] 'is a kind of' [mammal]"). The classification, originally based on prototypical knowledge, becomes now based on theoretical knowledge, centered around newly assigned default [dolphin] characteristics like having mammary glands and a neocortex, inherited from [mammal]. But a lot of the default prototypical knowledge about dolphins, originally inherited from [fish] (like feeding in the water) necessarily remains attached to the concept [dolphin]. Maintaining overall consistency in the knowledge hierarchy, therefore, requires updating/verifying that [mammal] has a characteristic sanctioning mammals to feed in the water (in this case inherited from [animal], along with other confirmed habitats—ie, [[animal][feeds in/on][|land|,|water|,|air|]]). The revised representations are shown in the following figure, part b.

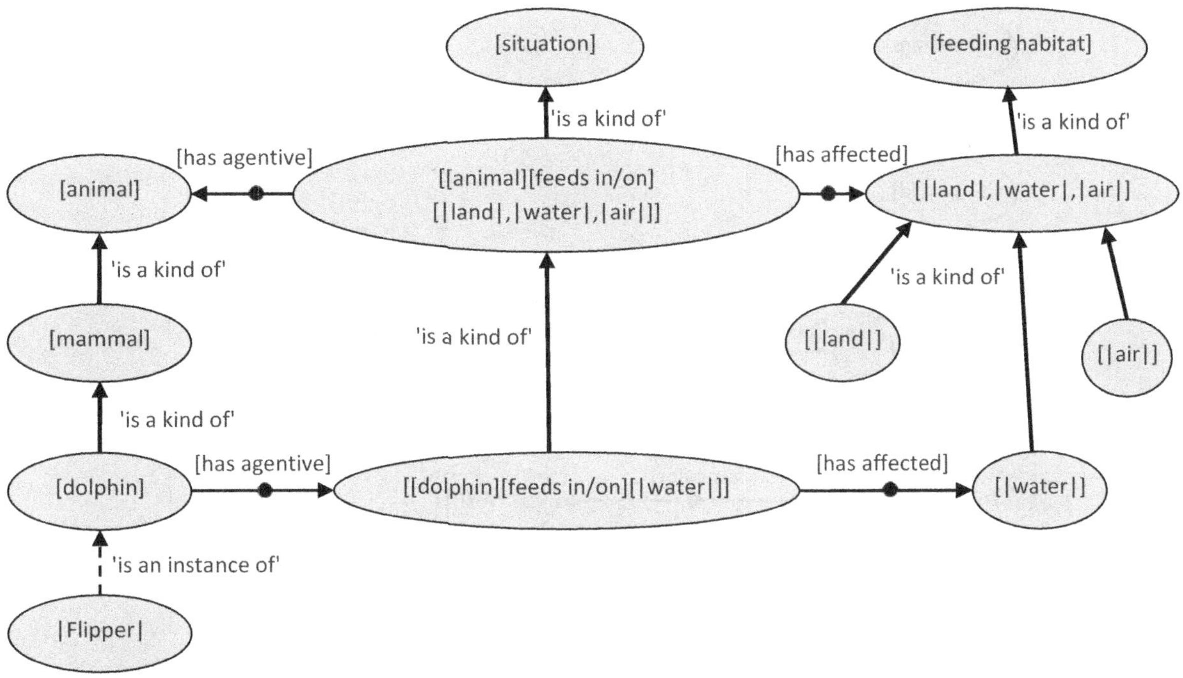

Figure 61: Representing Dolphins Are Mammals (part b)

Water

This last example reveals the rippling effect of conceptual changes brought about as new knowledge becomes available. Presumably we all acquire in childhood a largely prototypical concept of [water] as a colorless, tasteless, odorless liquid found in seas, lakes, and rivers. The prototypical structure flags the foregoing characteristics as typical, but with allowance for a variety of other less typical characteristics, as well, that encompass muddy water, tap water, and puddle water, among others.[201] But the classification of [water] as a kind of [liquid] imposes a normal restriction in use to the freely flowing, shapeless, but constant amount of cohesive stuff we are all intuitively familiar with as a liquid. Water can readily change into ice or steam in the right conditions, but [ice] and [steam] are different concepts exhibiting markedly different prototypical characteristics from [water]. The following figure (part a) outlines the proposed initial prototypical [water] representations. (The figure includes the shorthand dot notation for mode/situation nodes to simplify the diagrams.)

[201] As a reminder from Chapter 2: *The Three Standard Approaches to Concept Representation*, a concept like "prototypical water" is probably better conceived of as "common water", since it may include nontypical, as well as typical characteristics, along with their approximate statistical distributions. The term "prototypical concept", however, is the standard research idiom.

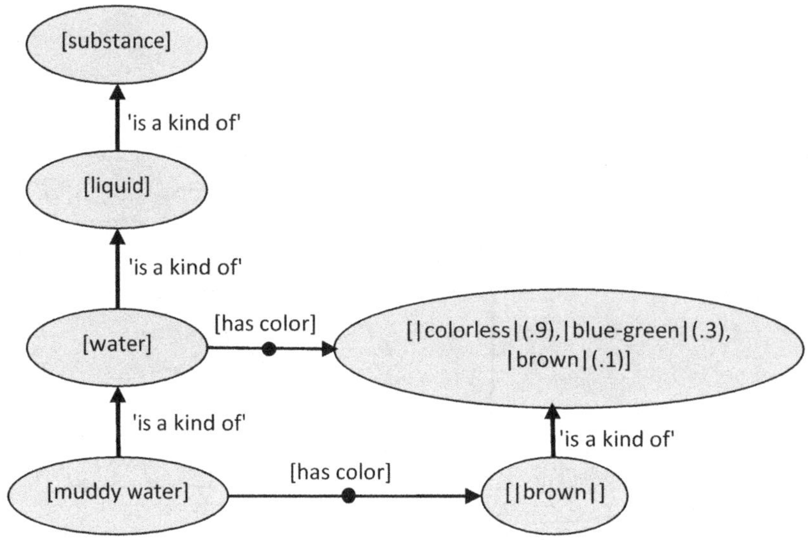

Figure 62: Representing Water Is A Chemical Compound (part a)

Once the prototypical concept of [water] is in place, the conceptual stage is set to incorporate more fundamental theoretical knowledge, an adjustment which requires more than a simple reclassification of the concept [water]. The chemical analysis of water reveals it is a compound substance; ie, a substance which is made up of more than one element. The name we generally give to this compound is [H_2O]. Chemically speaking then, [H_2O] is a kind of [compound substance], and [compound substance] is a kind of [substance]. Note that both [water] and [H_2O] are substances in this understanding, but that prototypical [water], as a [liquid], is classified differently than theoretical [H_2O] as a [compound substance]. The two are therefore different concepts, because not all liquids are compounds and not all compounds are liquids.

This conclusion, however, seems to be at odds with the mainstream philosophical presumption that water just *is* H_2O. In other words, that common water is identical to H_2O, or equivalently, that common water and H_2O are the same substance. This view seems mistaken, as it fails to take into account the conflicting characterizations of liquids and compounds. The theoretical substance [H_2O] can exist as a solid, liquid, or gas; while prototypical [water] is typically understood as being only liquid, as recounted in more detail below. (Conflicts do not occur in the case of [dolphin] above, since there, prototypical and theoretical characteristics are independent.) In this case the dilemma is resolved by creating a new concept [H_2O] alongside prototypical [water], so the two do not mix conflicting characteristics (of liquids & compounds) into one concept. The relationship between prototypical water and theoretical water then is one of composition, rather than identity, where prototypical [water] is composed of [H_2O] (perhaps along with other impurities such as calcium and magnesium, or even particulate matter, such as in muddy water).

As mentioned above, coming to grasp the concept of theoretical water requires conceptual change over and above simply creating a new [H_2O] concept, classifying it as a compound substance, and linking it to prototypical [water] via composition. [H_2O], being a physical substance, can exist in any one of the three states (or phases) of matter, namely solid, liquid, or gas. Therefore, phase must be a property of [H_2O], taking any one of the three *property values* solid, liquid, or gas on different occasions. But in the case of prototypical water, [liquid] is a *kind* (since prototypical water is a kind of liquid), not a property value. This inconsistency on the meaning of liquid (as a kind vs a property value) would seem to require some sort of conceptual resolution, because they are different means of representation. If you come to realize that phase is a property of all substances, then it can apply to both [H_2O] and to prototypical [water] (having liquid phase instead of being a kind of liquid). In this understanding prototypical water is just another kind of substance along with ice and steam, which are characterized by different phases. Given this realization, you can then eliminate [liquid] as a kind (making it a property value instead). With these conceptual adjustments, the concepts of common water and theoretical water can peacefully coexist, separate but connected, as shown in the following figure (part b). It is not necessary to marry the odd couple, which understandably would not work out very well.

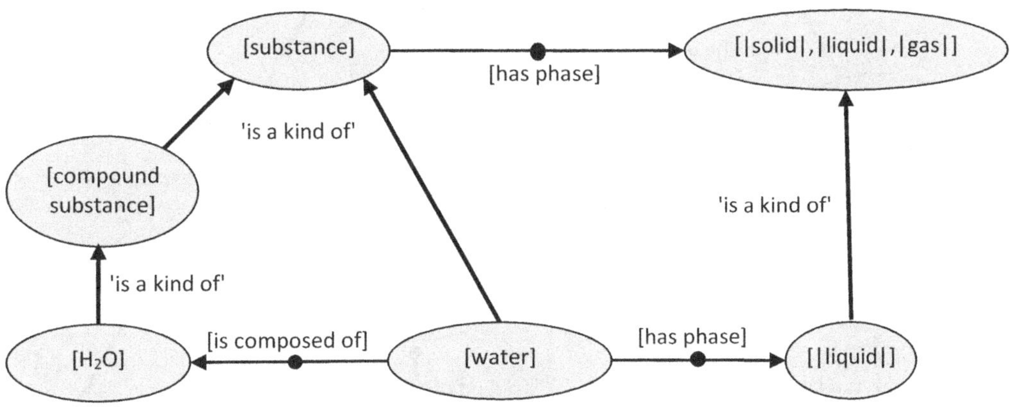

Figure 63: Representing Water Is A Chemical Compound (part b)

There is yet a further conceptual change required to fully comprehend water and H_2O, involving the concept of substance. The intuitive prototypical [substance] concept we initially acquire seems based on a property of infinite divisibility. However, we later learn about theoretical chemical substances having atomic structure. Atoms and molecules are microstructural components of all chemical substances. Prototypical [water] still fits within the rubric of prototypical [substance], but [H_2O] (theoretical water as a compound substance) isn't really like this, instead requiring characterization in terms of H_2O molecules. H_2O molecules are one kind of microstructure, where microstructures, and chemical substances with finite divisibility, constitute theoretical knowledge, analogous to prototypical substances exhibiting infinite divisibility. (Note that there is nothing in the intuitive concept of prototypical water that suggests it has a microstructure.) With this change, prototypical [water] is still composed of [H_2O], but [H_2O], because it is a [compound substance], now has H_2O molecules as components. Any sample of [H_2O] also consists of elemental substances like [hydrogen] and [oxygen], in addition to having compound components (namely, H_2O molecules), but this lower-level of microstructure is normally only important in understanding chemical reactions. Finally, a [H_2O molecule] is just one kind of [molecule], which in turn is presumably classified as a kind of generic [structured material] that includes microstructures and chemical substances, as well. A sketch of these new theoretical concepts merged with the original prototypical concepts is outlined in the following diagram. With these sorts of conceptual change in mind, one gradually comes to appreciate the usefulness, difficulties, and consequences of maintaining a rationally consistent store of knowledge in the face of new information.

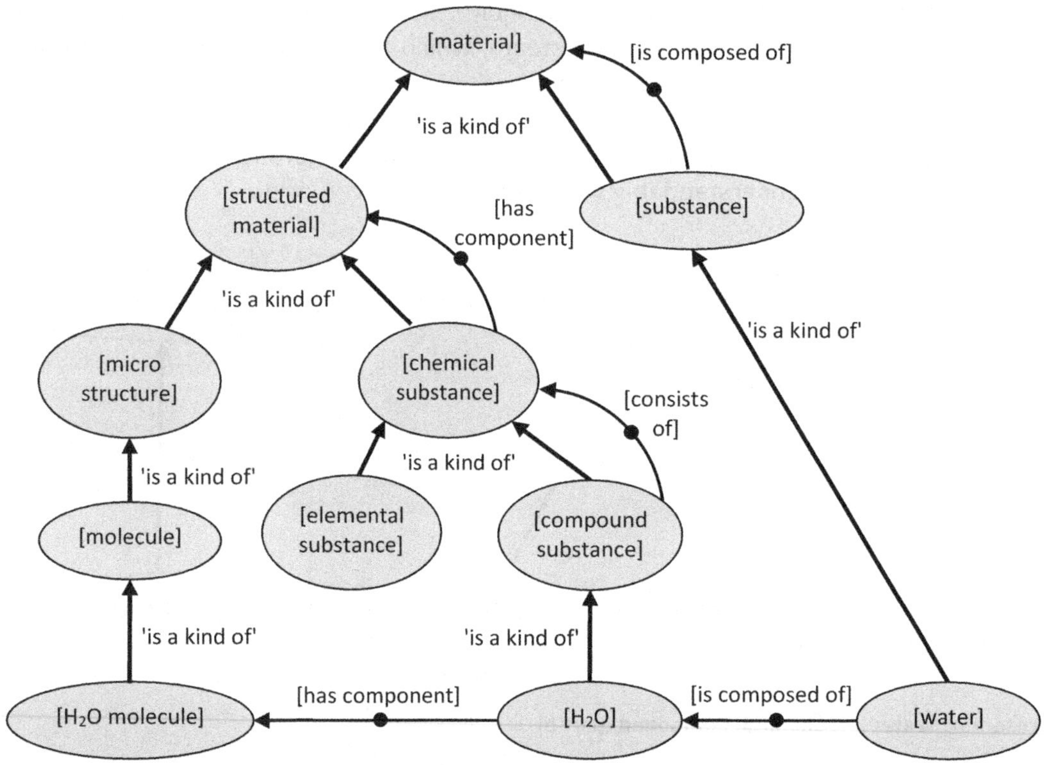

Figure 64: Representing Water Is A Chemical Compound (part c)

The Organization of Chemistry Concepts

How can the working concepts in a particular domain of knowledge be organized using the decomposition principles discussed in prior sections? In spite of the fact that knowledge can always be organized in a variety of ways for different purposes, do the decomposition principles actually help clarify and facilitate the representation of knowledge in some generically useful way? This section applies the principles to fundamental concepts in chemistry.

The field of chemistry offers a promising context for exploring conceptual decomposition, because its basic concepts are well understood and relatively stable, and because the organization of matter in the universe seems to follow a natural hierarchy—ie, atomic particles make up atoms, which make up molecules, which make up substances, which make up materials, etc. If conceptual hierarchical decomposition is a credible approach to knowledge representation in general, it should at least be credible when applied to the natural hierarchy of chemical concepts. More specifically, can it present a reasonable organization of the various forms of matter discovered and cataloged by chemical analysis over the last 200 years. Of course, a "reasonable organization" does not necessarily mean an "objective organization" completely independent of our thought processes. It remains an open question whether the natural kind hierarchy in a field such as chemistry exists independently of us in some objective sense, or whether we structure our understanding the way we do because of the way our conceptual processes work. Like most conceptualizations, it is probably a little of both, but the successes of chemical models attests to their regularity and usefulness.

The basic chemistry concept outline below summarizes a chemical hierarchy produced by conceptual decomposition. The information is taken from the first chapter of Linus Pauling's well-known introductory chemistry text: *General Chemistry* (Pauling, 1970). The text is somewhat dated, and some of Pauling's terms are likely out of fashion today, but it seems a good test case for conceptual decomposition. The whole chapter reflects the excellence characteristic of a Nobel author, both in content and organization. However, Pauling does not take great pains to impose an overall schema for understanding the concepts introduced. While the concepts are carefully defined, it takes a fair amount of effort to work out how they all relate to each other. Perhaps this is Pauling's intention as an educator, but more likely, he may not want to unduly bias beginning students by making conjectures that could be controversial.[202]

By and large, Pauling introduces new concepts in a progression from most general to more specific. His first paragraph contains the extremely abstract and undefined concept of [mass-energy], which the subsequent discussions of chemical properties characterize in various ways. More specific concepts like [mineral] and [solution] follow. The conceptual decomposition therefore begins with the root concept of [mass-energy] (which decomposes into [matter] and [radiant energy]), and branches on downward to the more specific concepts, each specified according to its own characteristics. Each concept in the derived hierarchy is decomposed into a simple subconcept and a complex subconcept, as required by the binary decomposition principle. The simple subconcept always appears first in the hierarchy outline schema, and the complex subconcept second. For example, the concept of a [material] decomposes into [mineral] and [mixture], where mixtures are composed of multiple minerals. A [mineral], in turn, is decomposed into [substance] and [solution], with a solution being composed of multiple substances. Properties are listed after a concept—eg, minerals are homogeneous and have a phase. Properties

[202] There is an alternate position that too much organization is pointless, either since it depends on one's point of view and therefore represents opinion, or since it rarely has an impact on actual practice. Uncritical acceptance of this position fails to recognize that our depth of understanding of a concept depends on proper classification in a wider context. Depth of understanding takes a back seat to efficiency of expression in everyday discourse, but is eminently useful for thoughtful discourse.

attached to a concept are automatically inherited by all subconcepts falling under it, so for example, any substance or solution is also homogeneous and has a phase.

Two notable additions to Pauling's concepts are included in the hierarchy for [nonparticulate matter] and [nonatomic matter]. These new concepts are required to cover, respectively, any kind of matter above the level of an elementary particle (which is almost all matter relevant to general chemistry), and any kind of matter above the atomic level. The significance of these added concepts is debatable in the context of an introductory text. However, the concept of [nonparticulate matter] may be relevant to the study of plasmas, which are chaotic collections of particles (ions and electrons) not organized into atoms. Similarly, the concept of [nonatomic matter] (including unusual nonmolecular grouping of atoms) may assist in the classification of certain extreme states of matter like the Bose-Einstein condensate, strange matter, or super fluids. In any event, the new concepts are on the fringe of Pauling's core topics for the first chapter, but the addition illustrates how a formal hierarchy can pinpoint potential gaps in knowledge. Also note the decomposition for [radiant energy]. Pauling does not discuss radiant energy until later in the book, but its subconcepts are added to the hierarchy to give the concept a minimal measure of meaning.

In looking at the total hierarchy generated from Pauling's discussion, it would seem that decomposition offers a fairly good model for his concepts. The organization makes the relationships between the concepts explicit, allowing them to be easily checked for soundness, consistency and completeness. For example, every [compound] should have a melting point. Virtually all of the concepts he has introduced fit somewhere in the hierarchy, and only two new concepts had to be added for completeness. Moreover, all of the properties associated with the various concepts have clear encodable descriptions, and the assignment of properties to concepts is consistent throughout with no concepts left dangling due to uncertainties about which properties it has. One comes away from the analysis with an appreciation for Pauling's implicit attention to thoroughness and ability to express abstract knowledge cleanly.[203]

There is one topic, however, that the hierarchy does not address, which is crucial to all of chemistry. This is the topic of chemical reactions. Pauling gives a number of examples of chemical reactions toward the end of the chapter, and emphasizes their importance to chemistry. Reactions are not included in the hierarchy, mainly because this is a preliminary analysis, and because they might be included in at least two different ways.

The first approach would make them properties of the substance concepts already specified in the hierarchy. This is consistent with the way Pauling seems to think about reactions. For example, the reaction $Fe + O_2 \rightarrow FeO_2$ which describes the rusting of iron, might be included as a property of iron. In this approach the concept of iron [Fe] would be added under the concept [elementary substance], and would be given the characteristic "reacts with [O_2] to produce [FeO_2]". It's a straightforward

[203] Many brilliant authors and teachers seem to lack the ability to translate highly abstract thought into a lucid linguistic framework. Thought and language may well constitute distinct mental abilities, but they both interface with concept representations.

representation of a characteristic, but obviously favors the iron over oxygen and FeO_2, which for completeness should probably have their own corresponding characteristics.

Another more global approach is to represent the reaction as a separate process, rather than as a characteristic of substances as in the first approach. Then there would be another hierarchy with [chemical reaction] as the root concept, and a decomposition according to the various kinds of chemical reactions (guided as usual by the decomposition principle). The rusting reaction would probably fall under a concept something like [oxidizing reaction], and be grouped with other similar reactions characteristic of oxidation. The substance concepts [Fe], [O_2], and [FeO_2] then would appear as participants in the reaction.

In a follow-on study, it would be worthwhile to work out a conceptual decomposition for all of the characteristics appearing in the mass-energy hierarchy (eg, [substance's crystalline form]), since characteristics are concepts too. Pauling makes a start in this direction by singling out one class of characteristics as intrinsic (or "specific properties", in chemical lingo). Forms of [mass-energy], [chemical reaction] processes, [chemical property] and possibly a few other major conceptual categories would all figure prominently in a comprehensive organization of chemical knowledge.

BASIC CHEMISTRY CONCEPTS

[mass-energy] (every phenomenon in the universe is mass-energy)
    has mass [mass-energy's mass]
    has energy [mass-energy's energy]
        is equal to [[matter-energy's mass]*|the speed of light squared|]
    has speed [mass-energy's speed]
  [matter] has speed [matter's speed] is less than |the speed of light|
    [atomic particle] (ie, electron, proton, neutron)
    [nonparticulate matter]
      [atom] (eg, copper atom)
      [nonatomic matter]
        [molecule] (eg, water molecule)
        [material] (ie, a configuration or collection of molecules or atoms or particles)
          has temperature [material's temperature]
          has homogeneity [material's homogeneity]
          [mineral] (occurs naturally as a product of inorganic processes)
            has homogeneity [mineral's homogeneity] has homogeneity value |homogeneous|
            has proportional composition [mineral's proportional composition]
            has phase [mineral's phase]
            [substance] has proportional composition [substance's proportional composition]
              has variability |constant|
              has solubility [substance's solubility]
              has density [substance's density]
              has melting point [substance's melting point]
              has electrical conductivity [substance's electrical conductivity]
              has thermal conductivity [substance's thermal conductivity]
              has malleability [substance's malleability]
              has hardness [substance's hardness]
              has crystalline form [substance's crystalline form]
              has taste [substance's taste]
              has odor [substance's odor]
              [elementary substance] (eg, mercury)
              [compound] (eg, water; limestone)
            [solution] (aka mixture)
              has proportional composition [solution's proportional composition]
                has variability |variable|
              [gaseous solution] (eg, air)
              [liquid solution] (eg, sugar water—variable concentration)
              [crystalline solution]
                [alloy] (eg, bronze)
          [mixture] (aka aggregate; eg, granite)
            has homogeneity [mixture's homogeneity] has homogeneity value |heterogeneous|
  [radiant energy] has speed [radiant energy's speed] is equal to |the speed of light|
    [quantum] (eg, photon)
    [quanta of energy] (eg, light, composed of many photons)

The Current Top-Level Epistemological Concepts

This section maps out territory in the upper (epistemological) layers of the concept hierarchy introduced previously. The following definitions, diagrams and notes are intended to provide a more-or-less reasoned justification of the current developmental status of the concept hierarchy, although it will always be a work-in-progress. Note that many of the concept names in the hierarchy have rather abstruse and abstract labels, which nevertheless attempt to provide modest intuitions about their content. However, the defining features of each concept resides in its relationships to other concepts, not in its name.

Basically, the epistemological levels of the concept hierarchy define a two-way branching taxonomy, moving from general to more specific concept categories at each lower level. Each subconcept then 'is a kind of' its parent concept (and all its ancestor concepts) on up to the root of the tree. Since each parent category subdivides into exactly two child categories, the recursive divisions produce a downward branching binary tree rooted at the most general concept which is labeled [concept]. The subdivisions are all based on some aspect of a constituency relationship (different at different levels), such that instances of the right subconcept (appearing second in the outline at any given level) can have constituents of the parent kind while those on the left (appearing first in the outline at any given level) cannot. The constituency relations associated with the right-hand concepts (eg, "has instance" for [[universal]] or "has member" for [collective]) indicate the primary selectional restriction on the concept that defines its epistemological function (eg, that a [[universal]] can have instances, or that a [collective] can have members). Other important, but nondefining, selectional restrictions are in parenthesis. Having only one primary characteristic (ie, the constituency relationship) is what distinguishes epistemological concepts from other lower-level theoretical or domain concepts, the latter typically being distinguished by a cluster of correlated characteristics.

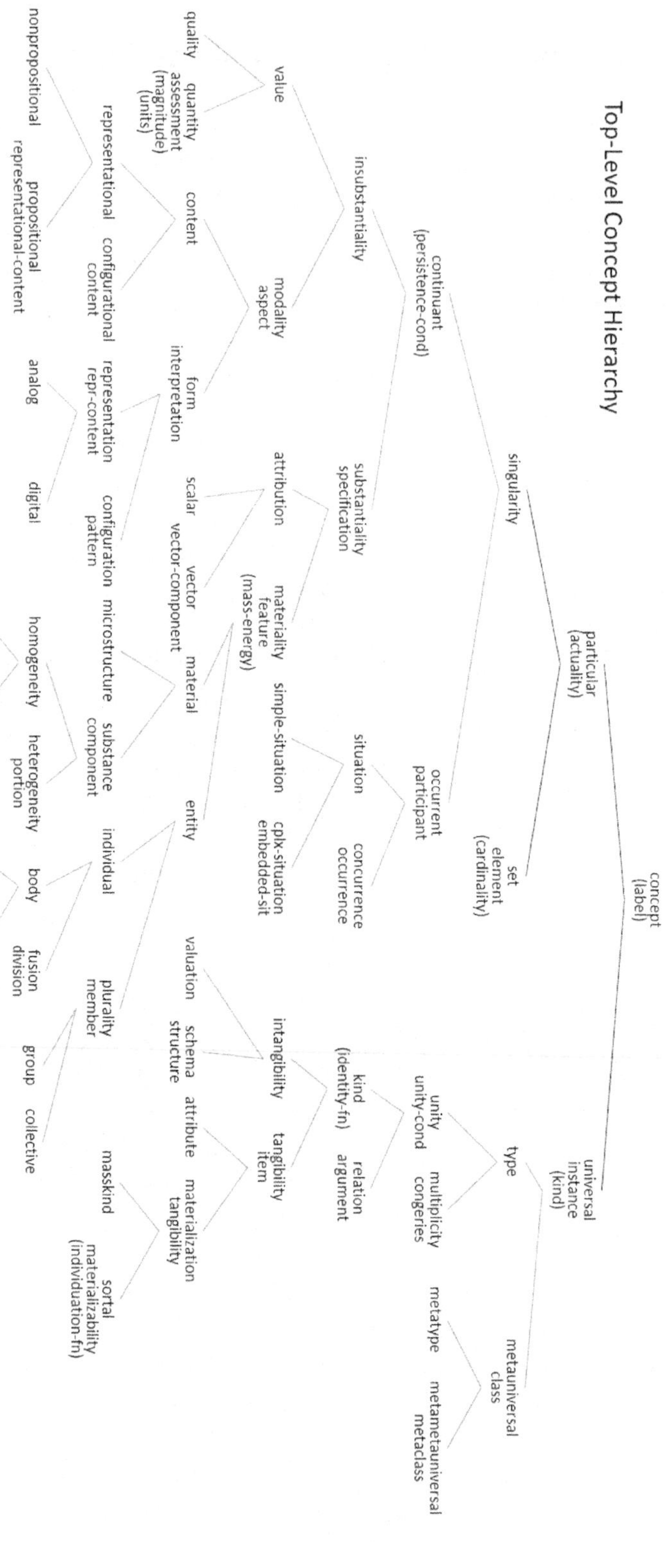
Top-Level Concept Hierarchy
concept (label)
particular (actuality)
universal instance (kind)
singularity
set element (cardinality)
type
metauniversal class
continuant (persistence-cond)
occurrent participant
unity unity-cond
multiplicity congeries
metatype
metametauniversal metaclass
insubstantiality
substantiality specification
situation
concurrence occurrence
kind (identity-fn)
relation argument
value
modality aspect
attribution
materiality feature (mass-energy)
simple-situation
cplx-situation embedded-sit
intangibility
tangibility item
quality
quantity assessment (magnitude) (units)
content
form interpretation
scalar
vector vector-component
material
entity
valuation
schema structure
attribute
materialization tangibility
representational
configurational content
representation repr-content
configuration pattern
microstructure
substance component
individual
plurality member
masskind
sortal materializability (individuation-fn)
nonpropositional
propositional representational-content
analog
digital
homogeneity
heterogeneity portion
body
fusion division
group
collective
element
compound consists-of
part
object partition
fragment
unit segment

TOP-LEVEL CONCEPT HIERARCHY (OUTLINE FORM)

- [concept]
  - [particular]
    - [singularity]
      - [continuant]
        - [insubstantiality]
          - [value]
            - [quality] eg: |wise #1|, |heavy #6|
            - [quantity]
          - [modality]
            - [content]
              - [representational]
                - [propositional] eg: 'Fido is a dog', 'There are 2 dogs in the room'
                - [nonpropositional] eg: |a perception of redness|
              - [configurational] eg: |the direction this arrow is pointing|
            - [form]
              - [representation]
                - [analog] eg: |the symbol π|
                - [digital] eg: |the word "hello"|
              - [configuration] eg: |this triangle|, |this hole|
        - [substantiality]
          - [attribution]
            - [scalar] eg: |Lynne's weight|
            - [vector] eg: |the velocity of that jet|
          - [materiality]
            - [material]
              - [microstructure] eg: |this atom of iron|, |this molecule of water|
              - [substance]
                - [homogeneity]
                  - [element] eg: |the copper in this pipe|, |the gold in that mountain|
                  - [compound] eg: |the water in this glass|, |the salt in the ocean|
                - [heterogeneity] eg: |the granite in this rock|, |that chocolate chip cookie dough|
            - [entity]
              - [individual]
                - [body]
                  - [part]
                    - [fragment] eg: |the tread on this tire|, |the corona of the Sun|
                    - [unit] eg: |the screw in this picture frame|
                  - [object] eg: |this TV|, |Carol|
                - [fusion] eg: |Rosie's head & neck|
              - [plurality]
                - [group] eg: |the LA Philharmonic|, |that flock of seagulls|
                - [collective] eg: ||Moe|,|Larry|,|Curly||
      - [occurrent]
        - [situation]
          - [simple-situation] eg: |Elvis lives|, |Jen loves Max|
          - [complex-situation] eg: |a bomb exploded in Pakistan|, |Zac believes Eli loves Drea|
        - [concurrence] eg: |the Renaissance|, |WWII|
    - [set] eg: {these apples}, {|Larry|, |Moe|, |Curly|}

TOP-LEVEL CONCEPT HIERARCHY (OUTLINE FORM-continued)

```
[[universal]]
    [[type]]
        [[unity]]
            [[kind]]
                [[intangibility]]
                    [[valuation]]           eg: [134 degF]
                    [[schema]]              eg: [surface]
                [[tangibility]]
                    [[attribute]]           eg: [weight]
                    [[materialization]]
                        [[masskind]]        eg: [gold], [water]
                        [[sortal]]          eg: [dog], [table]
            [[relation]]                    eg: [1[loves]2], [1[is red]]
        [[multiplicity]]                    eg: [three rabbits]
    [[[metauniversal]]]
        [[[metatype]]]                      eg: [[species]], [[sortal]]
        [[[[metametauniversal]]]]           eg: [[[taxon]]]
```

Explanations for the Top-Level Epistemological Concepts

A theoretically straightforward way to work out the intended meaning of the top-level concepts in a concept net is to think about instances of the complex right-hand concept as being made up of simpler instances falling under the left-hand concept.[204] A complex concept then is just composed of simple concepts, where the kind of composition (or more precisely, constituency) will vary depending on the concept. But the devil is very much in the details, contributing to many subtleties; and informal language often fails in accurately expressing the nuances. The following descriptions nevertheless attempt to gloss the intuitive meaning associated with these relatively nebulous concepts, most expressing a high degree of generality (for particulars) or abstraction (for universals).

[concept] – The concept whose instances encompass every concept, including itself. In other words, every concept in a concept net is, directly or indirectly, an instance of [concept]. [concept] is the root of the concept hierarchy, and therefore is the most general (or least specific) of all concepts, such that every universal concept (including [concept] itself) is a kind of [concept]. The concepts in a concept net are usually given perspicuous labels (ie, names) to facilitate thinking about them, but their operational meaning depends on their interrelationships.

[particular] vs [[universal]] – Any concept is either a particular or a universal. Universals encompass all those concepts that can have instances (eg, [person], [animal], [1[loves]2], [concept], [[species]], [[universal]]). Universals also can have subkinds, which furnishes the basis for organizing all universal concepts into a kind hierarchy. The concept [[universal]] is a subkind of itself, since only universals (and

[204] This expedient is not entirely accurate because the right-hand side concept includes instances of itself in addition to the left-hand side, and because the right-hand side may include no simple instances in the null case, but it helps capture the gist of the distinction between left and right concepts.

not particulars) can have universals as subkinds. Particulars (eg, |Andreas|, |Andreas' gecko|, |Elias loves Andreas|) cannot have instances, because they represent specifics, not abstractions. Particulars are always unique—ie, denote one thing—even if they denote specific things like sets or events possessing other things as constituents. The fundamental characteristic of a particular is its existential value (eg, whether it is physical like Abraham Lincoln, or fictional like Sherlock Holmes).

[singularity] vs [set] – A set is a multiplicity of arbitrary things considered together only for their own sake (ie, without any binding constraints in the context of use). Each thing in the set then is an element of it. A mathematical set is prototypical, where the set is *defined* in terms of its elements. If an element is added or removed, it becomes a different set. A singularity (as a nonset) is anything conceptualized as a single thing, whether or not it contains multiple parts. |Drea| is a singular concept, but {|Drea|,|Eli|} or {those people} is a set (unless some interconnection is implied). Cardinality is a primary characteristic of a set.

[continuant] vs [occurrent] – A singularity is either a continuant or an occurrent. A continuant is a particular whose instances have more-or-less stable characteristics that enable it to be identified and tracked as the same individual, for example between appearances. Concrete enduring objects are prototypical continuants, and are often depicted in language using noun phrases. A continuant is fully present at any point in its lifetime, although its characteristics may change from time to time. A primary characteristic of a continuant is its so-called persistence condition, which specifies the terms under which the continuant continues or ceases to exist. An occurrent is typically a particular state, condition, event or process. Occurrents (like WWII), unlike continuants, do not reappear once they have happened; although, as particular phenomena, they can have a time and place of occurrence and be associated with mass-energy. They generally unfold over time, although the duration of a punctual occurrent may be vanishingly brief. An occurrent is typically described by a proposition, containing a verbal action and other participants in the occurrence.

[insubstantiality] vs [substantiality] – An insubstantiality expresses a way some given substantiality can be. An ocean wave, as a persisting thing, is an insubstantiality, because, while the water composing the wave is substantial, the wave itself, as a particular configuration of the water, is not. Both however are conceptualized as distinct phenomena with locations. Similarly, a knot expresses a delimited insubstantiality (ie, a disturbance) in a rope; the donut hole reflects a certain absence of the dough in a donut. Values, like |the redness of this apple| or |34 kg #6| (ie, a particular instantiation of [34 kg] attributed to some particular object), are also insubstantialities, in that they are used to describe substantialities.

[attribution] vs [materiality] – A materiality is any kind of substance or object, but an attribution is generally regarded as some specific characterization of that substance or object, often thought of as a feature. A materiality can normally be characterized in terms of its mass-energy. Concepts like |the mass of the Sun|, |my cat's personality|, and |the density of gold| are attributions ascribed to various materialities. Attributions are often informally labeled as "properties" or "tropes", but these words come with connotations far beyond those intended here for attributions.

[scalar] vs [vector] – An attribution is either a scalar, like temperature, possessing a single value on a simple scale; or a vector, like velocity, with component values described by multiple scales, namely speed and direction. A vector may contain other vectors or scalars as components, but a scalar cannot.

[value] vs [modality] – Values and modalities express the way something is (in distinction from its parts or what it is made of). As continuants, they represent persisting regularities in objects, rather than dynamic occurrents in which objects participate, and so technically do not describe object states, as the terms are used here. A value associates a qualitative or quantitative assessment with the object, while a modality characterizes it in terms of some arrangement, configuration, or pattern exhibited by the object.

[quality] vs [quantity] – A quality is a kind of value, like hot or red, which characterizes an object on a qualitative (categorical) scale, like (cold, cool, lukewarm, warm, hot) or (red, yellow, green, blue). Qualities are a simple kind of value assessment, commonly used for everyday object descriptions. A quantity expresses a valuation on a quantitative (measured) scale, like 100 degC or 50 lb, where the value typically possesses a magnitude and units. Quantitative values provide a more precise way of characterizing objects than categorical assessments.

[content] vs [form] – A modality can be either content (eg, a piece of information) or a form (eg, a pattern of some kind, where the immaterial pattern is distinct from the underlying host substrate). Some kinds of forms may be given an interpretation, which is their content (or meaning). Content is often expressed in terms of propositions (eg, for digital representational forms), and equations (eg, for analog representational forms), but not all content is explicitly expressible.

[representation] vs [configuration] – A configuration, as a kind form, is simply the arrangement of parts of some materiality, where the configuration specifies their relationships. The pattern of a brick wall is a configuration, as is the arrangement of matter to form a crack in the wall. A representation is an intentional form, which can be assigned a meaning during interpretation. It is created to fulfill a communicative purpose. The meaning is expressed as the content of the representation.

[analog] vs [digital] – Representations come in either analog or digital forms, depending on whether the elements of the representation are continuous or discrete variables. The typical analog thermometer is the material substrate for a continuous representation of temperature, according to the height of the mercury. A digital clock features a digital representation of the current time.

[representational] vs [configurational] – Representational content is content derived from representations; ie, content which was created for some purpose by an information processing producer (either personal or subpersonal). Configurational content is a wider concept applying to any content extracted from naturally occurring (nonintentional) arrangements, patterns, or configurations of matter/energy. Both types of content can be meaningful to varying degrees to an information consumer.

[nonpropositional] vs [propositional] – Nonpropositional and propositional content are intentional kinds of representational content associated with nonpropositional and propositional representations (ie, propositions), respectively. Propositional content gives rise to linguistic meaning resulting from the interpretation of propositions, while nonpropositional content arises from the interpretation of other nonlinguistic representations such as diagrams, images, and models. Nonpropositional content is commonly produced via perceptual processes.

[material] vs [entity] – Materials like 'the gold in this ring', and entities like 'this ring' are all forms of a persisting materiality. Entities (ie, most objects and groups of objects) are usually composed of materials or other entities. Entities are characterized by their form (eg, shape), configuration (eg, relative layout of parts), and orientation (eg, with respect to a reference frame).

[microstructure] vs [substance] – A substance is a material thought of as being either continuously divisible (such that any portion of it is still an instance of that same substance), or finitely divisible, (made up of elementary microstructures, namely, atoms or molecules). Substances are normally characterized by amount (eg, weight or volume for continuous substances, or moles for finitely divisible substances). Since substances encompass various kinds of matter, many physical characteristics (such as density) are associated with them, rather than directly with objects. For example, weight is technically a derivative property of an object, in virtue of the object being composed of a certain amount of substance. Substances can also have components, which are either microstructures or other substances, as in a mixture.

[homogeneity] vs [heterogeneity] – Homogeneous substances are conceptualized as substances with a consistent composition (except perhaps for trace impurities); while heterogeneous substances, like granite (containing mica, quartz, and feldspar) are viewed as heterogeneous aggregates, and contain portions of other homogeneous (or other heterogeneous) substances. Context is often required to disambiguate whether some substance is being viewed as homogeneous or heterogeneous for the purposes at hand.

[element] vs [compound] – A homogeneous substance can be viewed either as an element or a compound, reflecting the lowest-level of the homogeneous/heterogeneous distinction. For example, a homogeneous substance like salt (at the molecular level) may be viewed as a compound consisting of sodium and chlorine, where sodium and chlorine are elemental substances.

[individual] vs [plurality] – An entity is either an individual or a plurality. Most everyday things we encounter are conceptualized as individual objects, not containing other individuals as members. But some things are pluralities, defined as an entity which can have members. A plurality is a holistic singular thing conceptualized as an assemblage of individuals or other pluralities, where the members are connected together in some way according to some principle. Collective singular nouns like "flock" do not denote a set, because its members are connected together into a functioning unit. Normally, such a plurality will maintain its identity in spite of changes to its members, unlike a set.

[group] vs [collective] – Groups and collectives are kinds of pluralities. While a collective is a simple enumeration of items brought together for a purpose, a group is a coherent plurality whose members can change and still remain the same group.

[body] vs [fusion] – An individual is either a body or a fusion (ie, an aggregate of individuals). A fusion is inherently viewed as a sum of individuals, tightly bound together, that make up a larger individual. It constitutes a mereological sum. The United States could be classified as a fusion, since it is typically conceptualized as an aggregate of states bound together under common customs and laws. The individuals making up a fusion are usually not interchangeable. A body is normally regarded as an integral unit of some kind, rather than as an aggregate of individuals.

[part] vs [object] – Most everyday articles are whole objects usually understood as having integral parts. Parts are not considered as whole objects in and of themselves, but rather as contributing (often functionally, spatially, or temporally) to the conceptualization of the whole object. A part is conceptually dependent on the object of which it is a part. For example, a doorknob in most contexts is considered a part of a door, although in specialized contexts it could be modeled as a whole object. The parts of an object are normally interchangeable.

[fragment] vs [unit] – Fragments and units are different conceptualizations of parts. While units are parts readily individuated from their governing objects (like a drawer in a chest of drawers), fragments are arbitrary subdivisions (like the top of a page in a book). The left side of a brick, or the tread on a tire are similarly fragments of their respective object bodies.

[situation] vs [concurrence] – An occurrent is either a situation or a concurrence. A situation is standardly represented by a proposition, while a concurrence is represented by a multifaceted collection of propositions representing a complex occurrent such as WWII or the Renaissance. In a concurrence the constituent occurrents are typically expressed in terms of coordinate and subordinate clauses joined by operators like "and", "or", "because", etc. Whole embedded situational attitudes (eg, A believes that B) fall under [situation] rather than [concurrence].

[simple-situation] vs [complex-situation] – Simple and complex situations are situations standardly denoted by individual whole propositions in language (as opposed to more expansive concurrences, which group numerous multiple situations together in a description). A simple situation corresponds to an atomic proposition, consisting of a single predicate and any number of arguments. A complex situation may contain quantifiers and embedded propositions, describing the so-called propositional attitudes like believing, hoping, wanting, etc.

[[type]] vs [[[metauniversal]]] – A universal is either a type or a metauniversal. [[type]] defines the first level of abstraction above particulars, typically corresponding to common nouns in language. All instances of [[type]] are labeled with single-pair brackets (eg, [ghost] or [value]). [[[metauniversal]]] is the concept whose instances encompass all universals at or above the second level of abstraction—ie, those concepts labeled with double-pair or more brackets like [[species]], [[type]] and [[[metauniversal]]] itself.

[[unity]] vs [[multiplicity]] – Universal types are subdivided into unities or multiplicities. A unity represents a category which expresses a conceptually unified class of things, while a multiplicity typically expresses a disjunction of classes. Instances of a [[unity]] are kinds of [singularity] (eg, [dog]), while instances of a [[multiplicity]] are set categories (eg, [three dogs] representing the category whose instances are sets of three dogs). A unity also seems to confer a criterion of unity for its instances, so a class like [dog] would incorporate criteria for distinguishing dogs from other kinds of things.

[[kind]] vs [[relation]] – A unity is either a kind or a relation. [[kind]] is the universal concept whose instances reflect typical common noun categories of things (eg, [student], [particular], [dog]) and do not take arguments to complete their meaning. Many [[kind]] concepts are also understood as carrying an identity function, which goes beyond counting, allowing one to say when two appearances of some kind are the same individual. A relation is the concept whose instances can have arguments and a relationship—eg, [1[loves]2] with two arguments, [1[is ticklish]] with one, [is raining] with 0, etc. The meaning of a relation is wrapped up with the meaning of its relationship and the selectional restrictions on its arguments. Kind concepts, expressing variables over instance concepts, are the prototypical arguments of relations.

[[intangibility]] vs [[tangibility]] – Tangibilities and intangibilities are kinds, where a [tangible] kind of thing is manifested in terms of mass-energy (eg, a [cat]), while an [intangible] kind of thing is not (eg, a [surface]). Instances of [[tangibility]] and [[intangibility]] are categories of [material] and [immaterial] things, respectively.

[[valuation]] vs [[schema]] – A valuation is a kind of intangibility associated with objects characterizing them in terms of attribute values. [137 lb] is an example of a valuation. A schema similarly characterizes objects, but in terms of their modality. [wallpaper pattern] is a schema for characterizing the modal pattern in various kinds of wallpaper.

[[attribute]] vs [[materialization]] – An attribute, like [weight], is a kind of tangibility used to characterize something. A materialization, like [person] or [gold], is anything which can be characterized by attributes.

[[masskind]] vs [[sortal]] – A materialization can be either a masskind of divisible substance like [gold], or a kind of sortal like [dog]. Sortals are concepts which carry a criterion of individuation, which permits distinguishing one individual thing from another, and therefore support counting. Bulk masskind cannot be so individuated or counted. Note however that a piece or portion of a masskind is a sortal because it can be individuated.

[[[metatype]]] vs [[[[metametauniversal]]]] – A metauniversal is either a metatype or a metametauniversal. Metatypes encompass all concepts at the second level of abstraction, while metametauniversals include concepts at all higher levels. The concept [[species]] is a metatype, whose instances are sortals like [tiger]. Normal discourse refers to these very high level kinds of abstraction only rarely.

The Organization of Situation (ie, Event) Concepts

The variety of situations we can express linguistically is extraordinary. If we can explicitly think about something, we can usually represent it in a proposition expressing that situation. The sweeping scope of autobiographical episodes we recall as detailed events in our everyday lives attests to the importance our psyches attach to representing situations. At the same time, it is difficult to accurately examine the content of individual situations, since they are subject to the same ambiguity, vagueness, complexity, and multilevel generalities as all other concepts. The attempt to bring some kind of semantic order to the panoply of situational meaning has a long history (ranging from interpretation based on temporal characteristics, to case roles, event completion/culmination parameters, various logical forms, syntactic verb alternations, and probably many others), but the goal of a structured, comprehensive and workable approach, potentially supporting automated natural language understanding, remains elusive.

The approach to situational meaning and representation proposed here adheres to the hierarchical decomposition method developed previously for other concepts. That is, the progressive binary decomposition into various kinds of situation is based on the idea of constituency. But a constituent of a simple one-predicate situation goes beyond identifying the participants in the situation, which generally applies to any occurrent expressed as a proposition. It rather applies to constituents of the particular action indicated by a verb. In this more restricted sense, the constituents of a simple situation are other simple situations, which are customarily implied by the action. For example, some implicational constituents of the situation "Kathy baked a cake" would include at least the normal implications "Kathy is a person", "Kathy is female", "a cake is food", "baking is an accomplishment" (presuppositions), and "the cake is now fully baked" (the final state of the cake). These are constituents of the *meaning* of the situation, whereas participants are constituents of the situation itself. Also note that the constituent implications, even though they are themselves situations, are not the same as subsituations (ie, subevents or component stages/phases of the baking activity, such as mixing the batter, preheating the oven, cooking the batter, etc). Any situation, like any concept, may be so subdivided, but this is not the meaning of *constituency*, as explained in Chapter 5: *Primary Relations of the Material Object*. The constituents of a simple situation include the background beliefs required to understand the propositional expression of the situation, as well as the inferences that should hold true if the simple situation is accepted as a fact.[205]

In the previous example, "bake" was interpreted as an *accomplishment*. This is an intuitive way of classifying the meaning of the situation, based on feature analysis, consistent with one approach discussed in Chapter 5: *An Alternate View of Situation Types*. The principal characteristic of an accomplishment is its completion. However a constituency analysis, in an attempt to ferret out the nature of the underlying process, would classify "bake" instead as a type of *progressive change*, where the constituent state of the affected object is slowly changing over a time interval. The final state of the cake as baked is just one state, following a succession of other states, any of which may be singled out as a constituent of the whole baking process (eg, the cake is half-baked). The core meaning of "bake" then is tied directly to the action itself, rather than to the culmination or conclusive state of the action

[205] For propositions, these presuppositions and implications are sometimes called "truth conditions".

as in the alternate view. Classifying the verb as expressing a situation of progressive change thus represents its default core meaning, which can be altered by context. The default meaning of "bake" is best expressed syntactically by the progressive aspect, as in "Kathy is baking a cake", indicating a progressive change kind of action sans culmination.[206] But in "Kathy baked a cake" (a different context) the default meaning is supplemented to include a culminating state.[207] Modulation of the core meaning can cover repeated bakings as well, as in "Kathy had to bake the same cake twice", in which there are two bakings but probably only one culminating state.

Focusing only on the core actions associated with verbs, the situations they express can be organized into a dichotomous concept hierarchy just like all other epistemological concepts, as shown below. Note that some verbs may possess more than one polyseme, and in this circumstance, each polyseme will fall under a different category. The two major kinds of situation then work out as *states* and *changes*, such that changes can have states (and other changes) as implicational constituents, depending on the context of verb use. Likewise, each kind of situation can have, as implicational constituents, any situation to the left or below itself in the hierarchy.

[206] However, the core meaning of other verbs may be better expressed through a different tense or aspect. For instance, the normal interpretation of "the light is flashing" probably involves repeated flashes. In this case the core meaning is more clearly expressed using the past tense "the light flashed", where the default is more likely a single flash.

[207] This is not to say that culmination is not part of the expected meaning of "bake", but it is not part of its *core* meaning.

Situation Concept Hierarchy

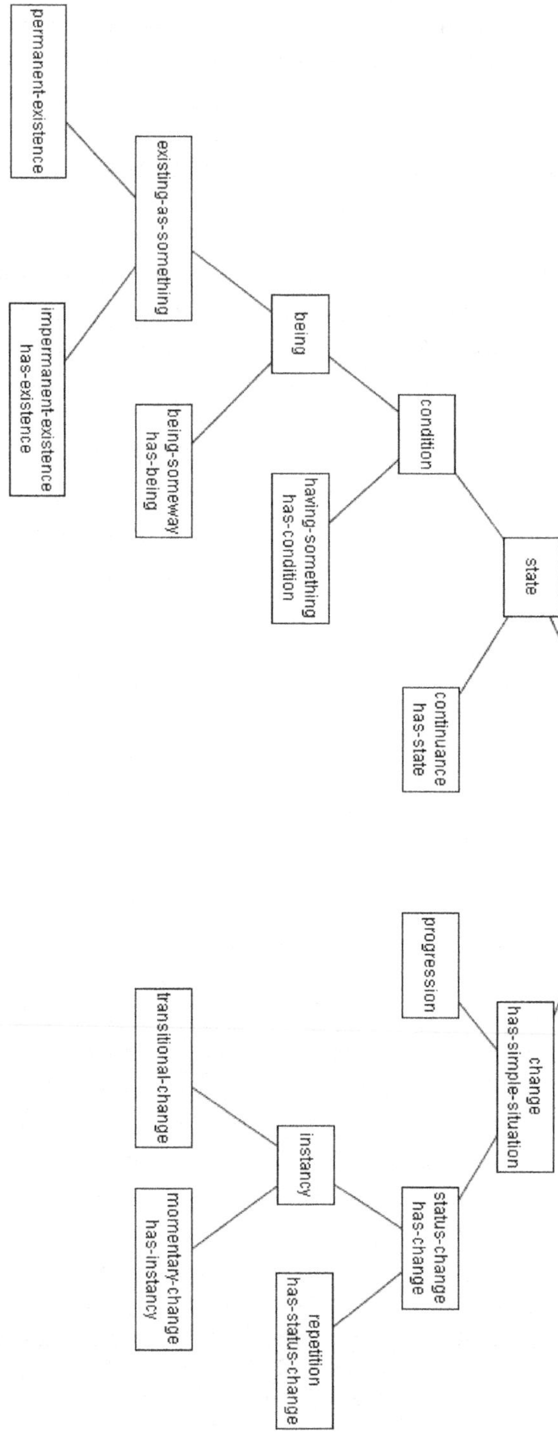

SITUATION CONCEPT HIERARCHY (OUTLINE FORM)

- [simple-situation]
  - [state]
    - [condition]
      - [being]
        - [existing-as-something]
          - [permanent-existence] eg: is a mammal, is a disease, is a time
          - [impermanent-existence] eg: is a student, is a bachelor
        - [being-someway] eg: is tall, is sick, is charming, is greater than 3
      - [having-something] eg: has a dog, has measles, has a tail, has charm
    - [continuance] eg: lives in LA, rain, boil, hunt, talk, hurt, see
  - [change]
    - [progression] eg: improve, grow, eat, ripen
    - [status-change]
      - [instancy]
        - [transitional-change] eg: turn off a light, catch a ball, arrive, die, begin
        - [momentary-change] eg: flash, shoot a gun, hit
      - [repetition] eg: beat a rug, repeat, iterate

Explanations for the various situation categories may help to clarify the intended semantics:

[simple-situation] – A simple situation is any instance of a relation representing a particular state, event, process, act, etc; commonly expressed in language by a simple single predicate along with some number of participants (eg, "Paul likes birds"). It contrasts with a [complex-situation], which is expressed by a complex predicate that can take sentential arguments (eg, "Mimi believes Paul likes birds"). Situations, in general, are said to "occur" or "hold" in time and at some place (usually expressed via tense, aspect, or spatial & temporal arguments), although there are also timeless and placeless situations (see below). In addition to verbal predicates, nouns can often express situations as well (eg, "the destruction of Rome").

[state] vs [change] – A state is a simple situation in which nothing is effectively changing in a global sense. For example, to say "Phoebe is smiling" is to report a constant state of smiling, for whatever period of time is relevant. Similarly, to say "the pot is boiling" means the water is in a constant state of boiling. Although great agitation of the water (and therefore local, random, or cyclic change) is implied, there is no effective global advancement. On the other hand, a change does indicate some global movement or alteration in its constituent states. A strawberry ripening is a progressive change, as are most events which have or can culminate (eg, Bob's dying prematurely).

[condition] vs [continuance] – A condition is a static state in which no overt change is implied (eg, "Zachary is sick" says nothing inherently about whether the sickness is coming on, going away, or just is, even though we might normally conclude that he will eventually get better). A continuance is a dynamic state which implies an underlying process of some sort, but the process does not result in an effective alteration or global change in status. Examples of a continuance include verbal expressions like "boil", "run", and "sit", where the action is dynamic, but there is no necessary implication of ultimate change due to the process itself. For a verb like "run", the running will often take the runner someplace else, but this is not a requirement, since the starting point and destination may be the same. If a horizontal constant line in time were to depict a static condition (———), a horizontal wavy line in time might depict a continuance (~~~~).

[being] vs [having-something] – The distinction between being and having is relatively straightforward in English, since the two verb constructions (be, roughly in the sense of existing; have, roughly in the sense of possessing) are quite dissimilar. Confusing the use of "be" and "have" is probably rare: "John is an artist" is meaningful, but "John has (an) artist" is anomalous in a default context. Having something can variously cover many situations of possession, relationship, parts, containment, experiences, and some qualities.

[existing-as-something] vs [being-someway] – The distinction between existing as something and being someway is likewise usually marked by verb construction, where "is a" signals existing as, and "is" signals being someway. In "Marcalan is a university graduate", he is classified as a member of a category; but in "Marcalan is responsible", he is being characterized, where the characterization can be thought of as a mode of existence.

[permanent-existence] vs [impermanent-existence] – The final dichotomy of states denoting existing-as-something, distinguishes between situations which are essentially timeless (and placeless), and those which meaningfully hold for a period of time. To say "Lynne is a woman" is to make a timeless assertion, in spite of the fact that she is mortal, while asserting "Lynne is a nurse" would normally be interpreted as a temporal state-of-affairs, since she was not always a nurse. Permanent (ie, invariant) states tend to carry ontological significance; they can serve as a stable framework for knowledge organization. Impermanent states usually carry functional significance, important for the myriad functional classifications we use to describe how the world works.

[progression] vs [status-change] – A progression is a process that is "going somewhere" over a period of time; it is gradually heading for some final result or potentially a state of some kind which is different from the starting state. The verb "ripen" provides a good example, since the process is normally expected (barring interruption) to culminate in something being in a ripe state. "Spin" however expresses a kind of motion which is not headed for a culmination or global change in state, and therefore expresses a simple dynamic state (ie, a continuance). An intuitive diagram of a progression might remind one of an increasing (/) or decreasing (\ ) function. A situation which expresses a status change reflects a marked shift in state, and is typically of short duration, although multiple status changes can take place over an extended period. Generic events like "drop", "put", "give", are prototypical examples of a change in status.

[instancy] vs [repetition] – An instancy is a single status change event of short duration, while a repetition implies multiple occurrences of an event (but still expressed by a single verb). "Flash" is a single instancy, since there is but a brief status change (from off to on, to off again). Verbs indicating a repetitive situation are probably rare, but "beat" (as in "beat a rug") seems to express a repetitive hitting action, where "hit" is an instancy. Using an instancy verb with the progressive aspect (eg, "the light is flashing" or "Drea is hitting Eli") may transform it into expressing a repetitive situation. In this case, the progressive form represents a distinct polyseme of the underlying verb, since the repetitive action has instant actions as constituents.

[transitional-change] vs [momentary-change] – Transitional and momentary changes are kinds of instancies, normally regarded as instantaneous changes of state. A transitional change expresses a sudden shift from an initial state to a final state (eg, "receive"), while for a momentary change the normal implication is that the sudden shift occurs sometime between the initial and final states (eg, "flash"). Thus a momentary change can incorporate two transitional changes as constituents, one up and one down, before returning to the ground state. If an intuitive visual aid for a transitional change looked like a step function (_|¯), then a momentary change might look like a pulse (_|_).

APPENDIX: COGNITIVE BEGINNINGS - PERCEPTION

A. SENSATION

Our ability to think abstractly is grounded in concrete encounters with the world. Sensory information becomes transformed into meaningful perceptual information, and perceptual information, in turn, becomes transformed into explicit conceptual knowledge. Concepts inherit their character from the antecedent sensory and perceptual processes which shape their development.

This appendix traces the origins of conceptual content, and puts forward some basic hypotheses about sensation and perception that are consistent with the view of cognition developed in the main text. Concentrated research over the last 100 years has already explored much of the sensation/perception landscape, yet an integrated, step-by-step information processing account about what sensation and perception are actually doing is lacking. For example, two landmark textbook summaries by Palmer (for vision) and Coren (for generalized sensation and perception) outline a number of processing stages, but provide few details about how information progresses from one stage to the next (Palmer, 1999; Coren, Ward, & Enns, 2004). It still seems premature to propose a testable global theory, but attempting to explain the flow of information spawns unexpected insights, and exposes some constraints on how all the pieces might fit together.

Functional Organization of Sensory/Perceptual Systems

In the absence of major defects, we all seem endowed with more-or-less the same sensory and perceptual equipment, organized in more-or-less the same way according to genetic codes. Individual differences in design and expression do arise repeatedly, but within established norms these variations typically have only local impact on the structure and function of the nervous system.

Our common-sense perceptual experience of objects and events in the world is also more-or-less the same, under a relatively wide range of circumstances, and in the absence of interfering stressors or major developmental differences. Eyewitness accounts of the same events are notoriously contradictory, but this is mostly due to lack of attention. We can usually agree about what we perceive if we are paying attention to the same thing, and are not overloaded. Perceptual experiences are different from conceptual ideas, about which there is often no "fact of the matter" to reliably distinguish between alternative viewpoints.

Our sensory and perceptual systems are organized into six large-scale modalities: vision, hearing, touch, taste, smell, and pain. These modalities are part of the somatosensory nervous system, and monitor the body's external environment. Some pain receptors technically monitor internal conditions (eg, in the joints), but can be thought of as monitoring external sources or causes of pain. Somatosensory information is normally available to consciousness.

In contrast, the autonomic nervous system monitors the body's internal environment—eg, pH levels in the blood and basal body temperature. The information it provides remains largely unconscious. Autonomic functions are mainly reactive, and do not involve the higher level representations of information characteristic of somatosensory functions. The autonomic system is not discussed further.

The somatosensory modalities tend to process information in parallel functional channels. A functional channel is an information processing pathway, generally extending from peripheral neural circuits, to numerous intermediate waystations, and on into the central nervous system. Basically, a channel performs an identifiable sensory/perceptual function.

For vision, there is good evidence for four distinct functional channels: color, form, depth, and motion (Livingstone & Hubel, 1988), summarized in the figure below from (Palmer, 1999). The labels are generic—eg, generic color includes complex texture analysis, and generic depth includes inclined surface analysis.

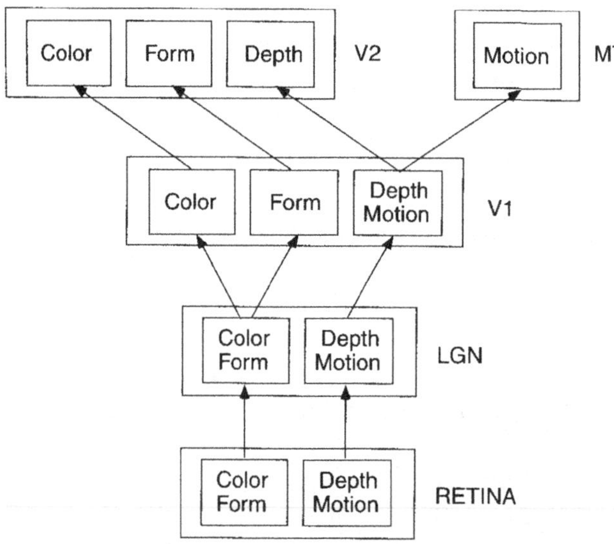

For hearing, the two main auditory channels involve the analysis of sound quality (pitch, loudness, timbre, etc) and direction. There are also high-level auditory channels dedicated to speech processing and music. A medical syndrome called amusia adversely affects the ability to recognize music, but not speech or other sounds.

For tactile information, there are four channels for analyzing vibration, flutter, buzz-like sensations, and pressure (Johnson, 2001). Temperature is apparently distributed over multiple pathways. For taste, six different taste qualities have been identified: sweet, sour, salty, bitter, umami (meaty flavor), and fatty. The details of how these sensations are processed are still uncertain, but a perceived taste involves a mix of these qualities. There is some evidence that sweet, bitter and possibly umami are processed together (Montmayeur & Matsunami, 2002). For smell, there is evidence of two major channels

(Bartoshuk & Beauchamp, 1994): a proximal food smell channel for analyzing food in the mouth, and a distal smell channel for odorants sniffed through the nostrils. The two principle pain channels, corresponding to fast-conducting Aδ fibers and the slow-conducting C fibers, process noxious and dull/burning pain, respectively.

Each functional channel processes a different kind of information. A channel essentially cultivates information related to some separable facet of the world embedded in its accustomed perceptual stimuli. But talk about functional channels is something of a misnomer, since the "functions" are associated mostly with the middle stages of the information flow from receptors toward the frontal lobes. The flow of information tends to diverge and refine as it progresses up a channel, and there is a certain amount of cross-talk between channels. (This divergence of information is in contradistinction to the physiological convergence of peripheral nerve fibers onto the central nervous system; the receptive field for neurons grows progressively larger moving toward the deeper brain areas.) The information analysis in a modality's channels might be described as progressively unscrambling highly entangled facets of a stimulus, eventually attaining qualities that exhibit significant stability, reliability, and invariance under a wide range of background conditions—eg, an object normally keeps its same perceived size even if it moves farther away.

Information Processing Stages

The information flow through a functional channel typically follows nerve fibers that run between tightly bundled nuclei (including the columns of the neocortex). Most processing takes place in the nuclei, suggesting a staged approach to extracting information from the stimulus. For example, most theories of vision roughly divide the processing stages into image processing, surface processing, object processing, and category processing. Similar stages are proposed for the other senses.

A staged approach to information processing offers evolution a practical way to perform complex computations. Instead of developing one, highly integrated, but very complex faculty, evolution incrementally cobbles together smaller processing modules that have proven worth. But the modules must effectively communicate with each other in order to work together. The adaptive modules generally seem to develop a strategy of producing *meaningful* units of information that other modules can use and combine together into other meaningful units.[208] This functional principle of meaning *compositionality* operates throughout the nervous system, and is likely responsible for both its organization and structure. The examples of composition below illustrate the staged production of meaningful units in the visual system.

- Representations of individual *points* in the visual field combine together to produce a representation of a 2D environmental *surface* oriented in 3D space

[208] Meaningful units of information are "meaningful" only to consumer modules that can use the information in some way—see Chapter 2: *Conceptual Meaning*.

- Representations of environmental surfaces combine together to produce a representation of a whole *figure* as viewed from one perspective
- Representations of the same figure from different perspectives combine together to produce a representation of a *situated form* in the environment
- Representations of the same situated form on different occasions combine together to produce the representation of a persisting *perceptual object*
- Representations of similar persisting perceptual objects combine together to produce the representation of a similarity-based *category* of perceptual objects

From an operational point-of-view, the final stages of perception seem to involve constructing an effective here-and-now model of the environment. We then interact with our internal model of reality (which is tied to intentional actions), rather than respond to sensory stimulation more-or-less directly, as most other animals do.

In order to create an effectual model, we principally build mental representations of objects, events, and the relations between them. The sensory and perceptual processes construct complex multidimensional representations step-by-step out of simpler unidimensional representations, beginning with sensory information, as discussed below.

Sensory Processing

Sensory receptors of various kinds are scattered throughout the body. Multiple receptors (of multiple kinds) in a localized area are typically connected to a neural (ganglionic) circuit. A local circuit processes information from receptors in its neighborhood (aka its receptive field).

Receptors respond to changes in energy impinging on them. Different kinds of receptors respond to different kinds of energy. Photoreceptors respond to electromagnetic energy from visible light, thermoreceptors to thermal energy from molecular agitation, mechanoreceptors to mechanical energy from local forces, and chemoreceptors to chemical energy from contact with substances. We don't have receptors for detecting other forms of energy like electrical energy, magnetic energy, or nuclear energy. Some other animals, however, have evolved this capability—eg, sharks and platypus (electrical), pigeons and sea turtles (magnetic).

That sensory receptors are our primary, if not only, means of contact with the environment calls into question one of our most cherished perceptual assumptions: that the environment is made up of objects and their interactions. Since receptors respond only to fluctuations in energy, objects must be a useful interpretation, constructed by perception to explain and exploit those fluctuations. The basic existence of the environmental "flux" is not an issue, but the way we represent it in terms of objects is. The flux more accurately corresponds to continuous fluctuating fields of energy in various forms, rather than to discrete bounded objects (at least at our relative size in the scheme of things). Why evolution chose to build our perceptual experience around objects and not fields is open to debate. However, there is considerable evidence for spatial and temporal frequency processing in psychophysical perceptual channels, so perception may use both kinds of representation, depending on utility vs relative design complexity.

When receptors are activated, they trigger sensory processing chains of neurons extending on into the central nervous system. A sensory processing chain acts to summarize a *quality* present in the stimulus at a particular location in the sensory field. A single stimulus typically elicits different qualities via multiple chains, which together act to characterize the stimulus in the vicinity of that location. For example, photoreceptor chains extract information about the degree of lightness, brightness and frequency of received light; thermoreceptor chains gauge hotness and coldness; chemoreceptors are the source of qualities associated with pain, sweetness and pungency; and mechanoreceptor chains signal qualities like pressure and vibration on the skin, loudness of sounds, and muscle tension.

Some sensory processing chains form the basis for subsequent conscious *perceptual* sensations like brightness and coldness. In spite of the fact that we can have perceptual sensations of brightness, coldness, etc, the sensory processes that derive these qualities in the first place are unconscious. It is possible to focus attention on the qualities after they become part of a perceptual experience (ie, a sensation), but this involves deconstructing the later perceptual experience which is built up from the component qualities.

The amount of a quality registered by a sensory processing chain depends on the activation level of the receptors, which in turn depends on the intensity of the stimulus (above a reference background level and above a minimal threshold). Qualities define a base level of neural activity (ie, a reference point corresponding to an average firing rate), with possible activation levels below and above the reference indicating less or more of the quality, respectively—eg, for lightness: darker or lighter than a nominal level. The activation level can then be interpreted as the current value of the quality present in the stimulus.

Qualities are possibly implemented via so-called opponent cells in the nervous system. An opponent cell is a neuron that decreases its firing rate below nominal when stimulated by one kind of receptor, and increases its rate when stimulated by another (opponent) kind. For example, area V1 of the primary visual cortex seems organized into bands (or stripes) of activation, serving to distinguish between concentric circles and radial fan-like shapes (Billock & Tsou, 2010). These shapes are never experienced together simultaneously, and so represent opposites on some internal, subjective shape scale. Similarly, opponent cells have been found in the retina and lateral geniculate nucleus for detecting black/white, red/green, and blue/yellow oppositions (De Valois, Abramov, & Jacobs, 1966).

A black/white cell coding for the quality of lightness responds minimally to completely dark surfaces, maximally to extremely light surfaces, and in an intermediate fashion to various in between shadings. (Actually, opponent cells seem to come in pairs. For the quality of lightness, this means there is one opponent cell coding for black/white (A), and another coding for white/black (B), where the latter's response is the inverse of the former—ie, the excitation levels of each are roughly described by A+B=1. The information conveyed by one is therefore the same as the information conveyed by the other.

Pairing opponent cells in this manner possibly increases the redundancy and reliability of sensory processing, but it would not appear to increase the amount of information conveyed by the cells. Therefore, only one summary quality, in this case black/white, is assumed to matter for subsequent perceptual processing.

In theory, it would seem that a polarized scale, ranging from 0 to 1, could be constructed from any two qualities, x/y, placed in opposition to each other, where the first (x) corresponds to the minimal activation level (0), and the second (y) to maximum activation (1). Intermediate values on the scale reflect tendencies toward one pole or the other, or neither at 0.5.

Psychophysical experiments show that we can reliably classify only about four to eight different values of any one quality present in a stimulus, holding the other qualities constant. Being able to classify a quality value means being able to categorize the value according to prototype or exemplar values encoded in long-term memory.[209] This is different from making a side-by-side discrimination between a pair of quality values, which we can do with uncanny accuracy, and which does not involve long-term memory. For example, holding the volume and other qualities of a tone constant, we can reliably recognize about five distinct levels of pitch (Pollack, 1952):

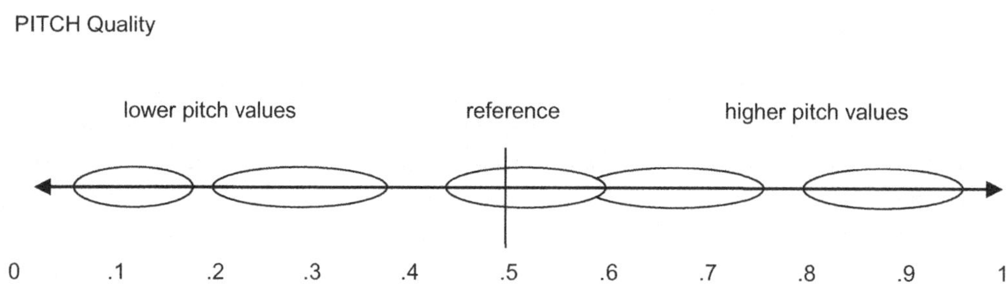

But then, how is it that we have no trouble recognizing thousands of different sounds? One explanation is that we routinely combine the various qualities in any stimulus together (in this case, pitch, volume, temporal and spatial parameters, overtones, background noise level, etc) to achieve reliable recognition. Each sound then is defined in terms of multiple dimensions, greatly increasing the theoretical number of recognizable categories: one dimension ≈ five categories, two dimensions ≈ 5x5=25 categories, three dimensions ≈ 5x5x5=125 categories, etc.

Quality Combination

The combining of sensory qualities into perceptual *attributes* marks the transition from sensation to perception. As an example of how individual qualities might be combined, thereby giving rise to more complex attributes, consider the perception of color. The psychological experience of color is now

[209] See Chapter 3: *Prototype Effects*, for a discussion of conceptual prototype representation.

known to involve three integral components: hue, saturation, and lightness. Colloquially speaking, "color" means hue, but here it means anything having to do with general luminance, including black, white, shades of gray and even texturing derived from luminance.

Assuming the basic qualities of black/white, red/green, and blue/yellow are provided by sensory processes, the problem then is how such qualities become transformed by perceptual processes into a particular sensation of color like red, orange, or pink. Considering the blue/yellow and red/green quality scales together, the pair of quality values associated with a stimulus activating both can be represented as a compound value in a new two-dimensional quality space. (Mathematically, the new value is a vector with component values from each quality scale.) For example, if the compound value were maximal blue/yellow and minimal red/green, they would combine into the new value <yellow, red> = <1,0>, which defines a crux value recognized in most Western societies as orange—a more-or-less equal mix of yellow and red. Filling in the new space with some other compound color values produces the second-order quality shown below, which I will call hue-saturation for reasons explained shortly.

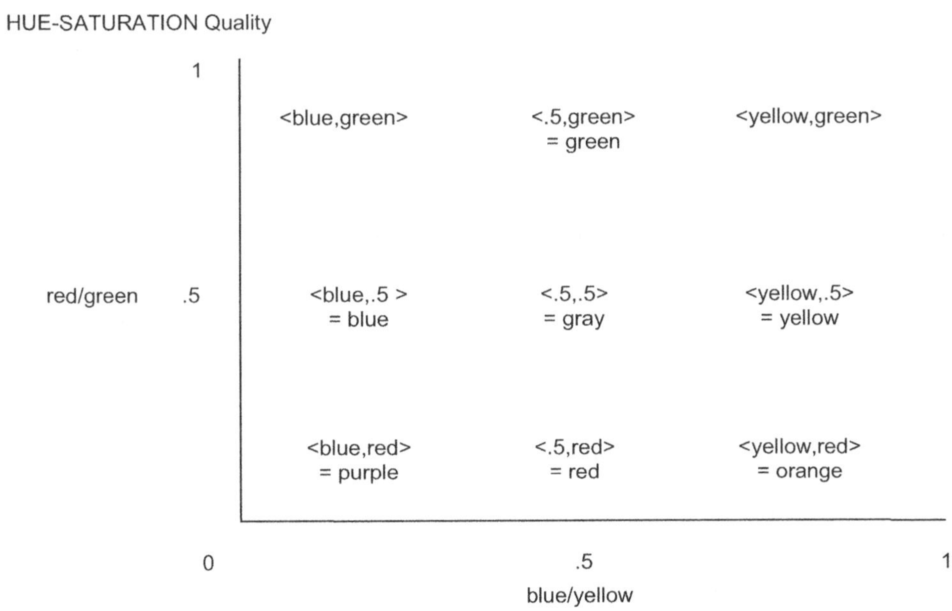

Intermediate quality values like <0.7,0.3> make it possible to represent any value of hue-saturation composed from the two underlying quality values. Filling in all the intermediate values and replacing them with their equivalent subjective colors, produces a map that looks a lot like the standard color circle shown below. The color circle depicts our phenomenal experience of color under medium lighting conditions.

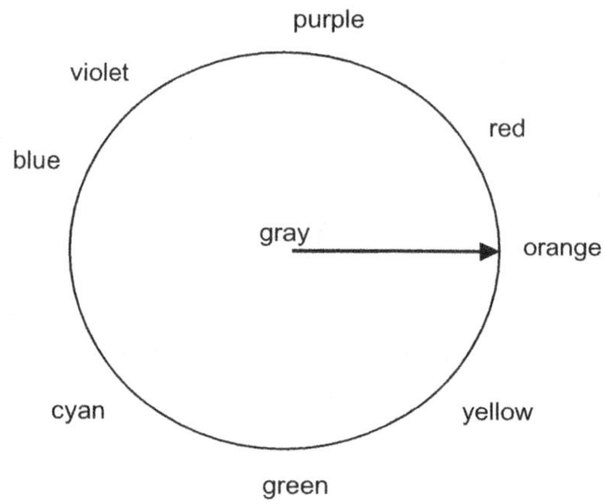

The color hues change gradually moving around the center, just as in the second-order quality space of hue-saturation. The most saturated colors are at the edge of the circle, with saturation decreasing toward gray at the center. To pick up the remaining color sensations like pink and brown, requires adding in information from the lightness (ie, black/white) quality scale to produce a third-order quality containing three components, graphed below. This third-order quality is then equivalent to the attribute of color, which we are able to perceive as a sensation.

To contrast qualities with attributes, note that the previously described hue-saturation quality is not an attribute and does not represent perceived colors, because it excludes the lightness quality. We can perceive the color circle above only because it is drawn for average lightness (0.5 on the lightness scale—midway between pure black and pure white). A full perceptual registration of the color attribute merges all three quality components together as an integrated, separable, representation (which, for color, happens to be a conscious experience of a region in a three dimensional color space), as illustrated below.

Another example of an attribute (of a pure tone) combines pitch and loudness. Values of both are needed to uniquely describe any tone in a two dimensional value space. Similarly, there seems to be another completely separable attribute of color that applies only to luminous sources of light, and incorporates the quality of brightness, instead of lightness (which only applies to light reflected from surfaces).

COLOR Attribute Space

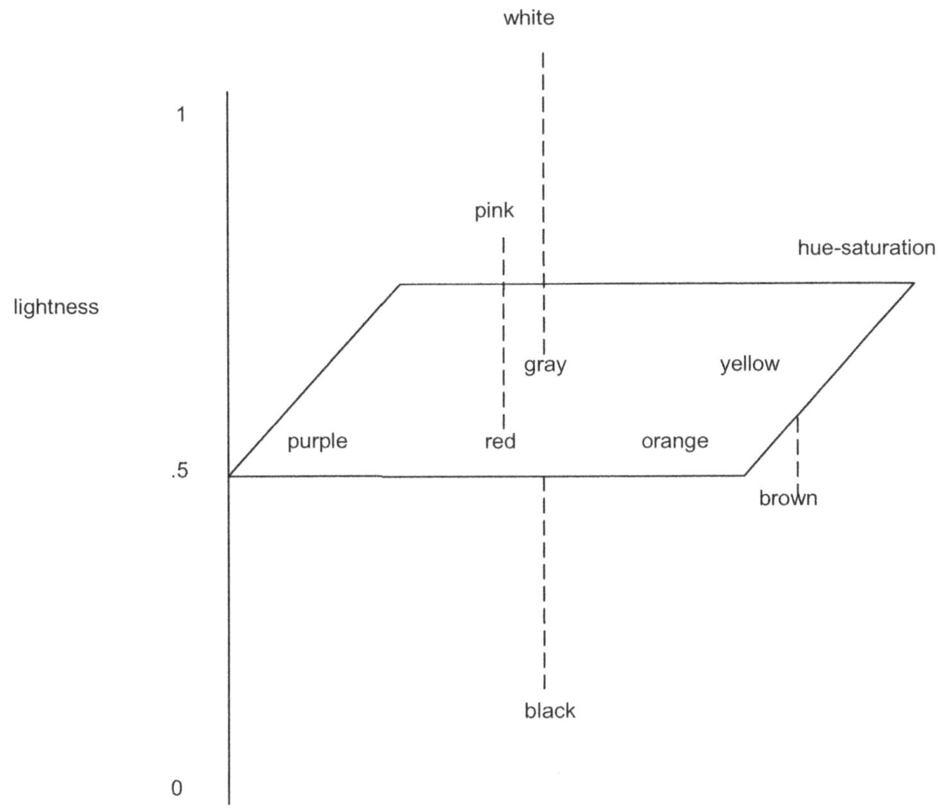

An attribute then is an integral quality space of one (minimal) or more dimensions, where "integral" means you cannot assign a value to one of the qualities without also assigning values to the other qualities. The qualities always go together in a stimulus. The number of integral qualities for color is three, resulting in a 3D *attribute space*. Each point in the attribute space encodes a unique *attribute value*. Some points may become distinguished as *focal values* (like yellow, pink, etc). Nearby points encode attribute values that are more similar to each other than the values of points farther away—eg, brown is more similar to black than to white. It is possible to combine any number of integral qualities together into an attribute space, resulting in higher dimensional representation spaces that defy visualization, but are nevertheless functional from an information processing viewpoint. It is also possible to combine attribute spaces together into even more complex *value spaces* for representing arbitrary (ie, nonintegral) clusters of attributes. Value spaces, it is conjectured, ultimately enable the representation of sophisticated property spaces, like wisdom (perhaps as a combination of experience, knowledge, and judgment attributes?).

Attributes are a more powerful way of describing the world than simple qualities, because they articulate a constraint among certain qualities that appear together over and over again. The processing of sensory and early perceptual information thus can be seen as following a steady progression of steps involving stimulus, intermediate qualities, and culminating in the representation of attribute values, each characterizing a small spot in the whole sensory field.

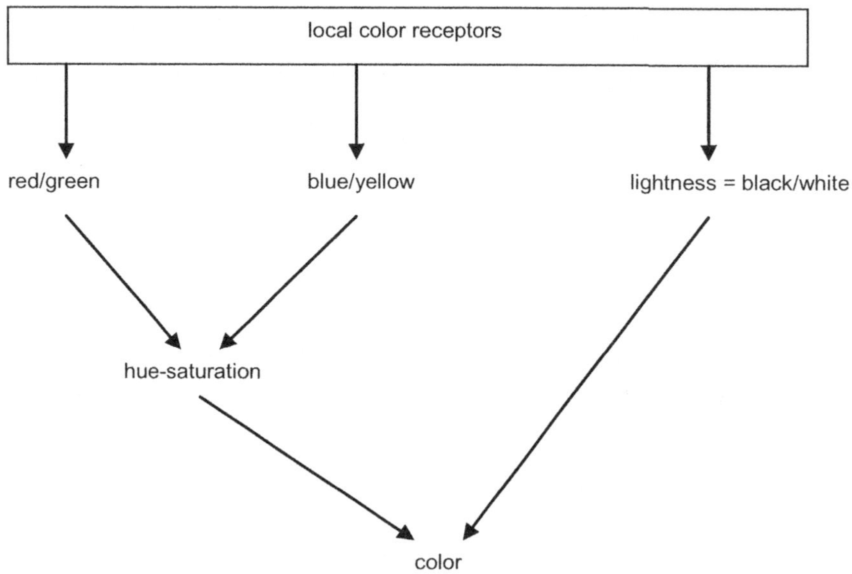

This same kind of progression is assumed to operate throughout the sensory and perceptual systems, eventually leading to representations of perceptual objects (among other things). Postulating attributes brings along a possible (but not necessary) latent assumption—namely, that an attribute can be an attribute of "something", as opposed to a quality which only presumes the detection of environmental conditions at some location. The assumption is not necessary, since the attribute could be free-floating (eg, an unattached percept like simply being cold). However, the something to which an attribute normally attaches turns out to be a small region of a perceptual field.

B. EARLY PERCEPTION

Attribute Maps

The somatosensory system receives information from its sensory fields—namely, the retina (vision), basilar membrane (hearing), skin (touch), tongue surface (taste), olfactory epithelium (smell), and deeper body tissues (pain).

The information provided by each sensory field differs not only in kind, but also in spatial content. The retina and the skin provide 2D information (about points on a surface), the basilar membrane provides 1D information (about points along a line), the body pain receptors provide 3D information (about points inside the body), and the taste buds and olfactory epithelium provide 0D information (about individual receptor points with no particular spatial significance).

The various sensory fields (including those incorporating 0D information) roughly project onto neural *topographic maps* in the central nervous system. The neural maps express the values of attributes corresponding to stimuli appearing in the fields at a particular time. Multiple maps, one for each attribute, are associated with each perceptual field. (Described mathematically, an attribute map is a vector space.) Presumably, attribute maps play a role in most of the sensory modalities, although vision is emphasized here.

A *color* attribute value describes the current color at a point in the visual field, such that each color value registers the three color qualities (hue, saturation, lightness). The current attribute values in a map implicitly segment the map into *regions* with similar values, as illustrated below. The regions are implicit at this stage of processing, because it requires further evaluation to decide which points belong to which regions. In a 2D color map a region might correspond to a surface of an environmental object, while in an auditory 1D frequency map, a (linear) region might correspond to a tone composed of certain frequencies.

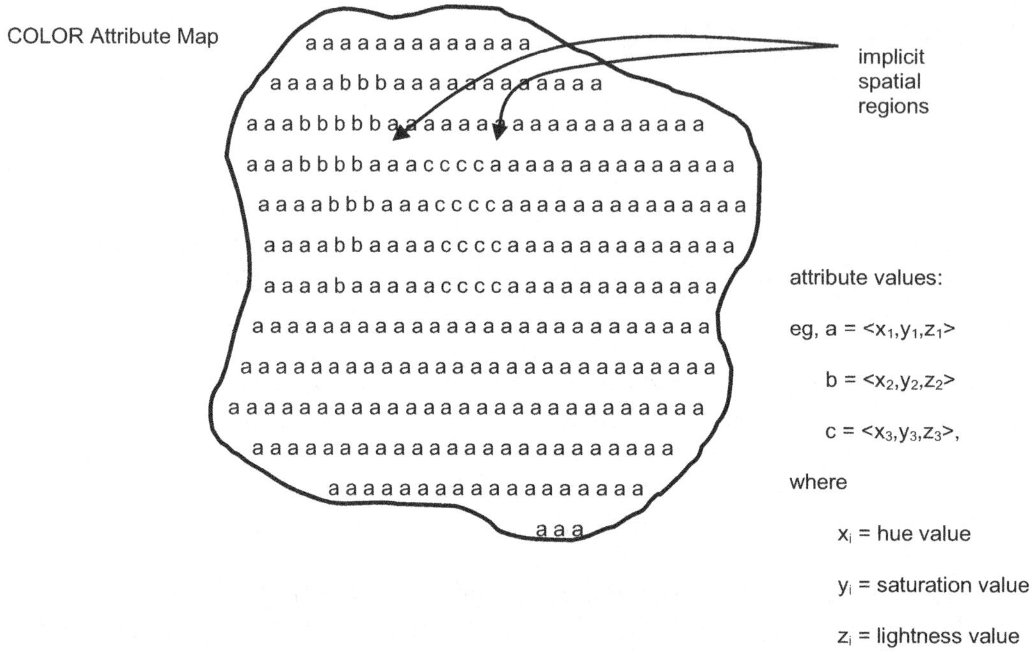

In addition to color, another important source of information about a visual scene comes from *depth* estimates. There is quite a bit of information available in visual scenes with which to accurately determine the distance to points in the scene. Among these are ocular information (eg, lens focus), stereoscopic information (eg, image overlay from both eyes), dynamic motion information, and pictorial information (Palmer, 1999). These various information sources for depth perception combine in order to produce a depth map of the visual field, much like the map described previously for color. Values in the depth map correspond to perceived distances to points in the visual field. By aligning the depth map with the color map, a reading for the relative distance to points in the color map can be obtained. Implicit regions of similar and gradually changing depths roughly correspond to environmental surfaces located at various distances from the observer, just as for color.

Given a depth map it becomes possible to compute an *inclination* attribute map, as well. Relatively small changes in depth from one point to another in the depth map means that the region around those points takes on a local inclination, which can be determined from the rate of change of depth in all directions (with other cues like shading helping out). Gradual changes in inclination characterize an implicit inclination region. The figure below shows estimates of surface inclination provided by one experiment (Koenderink, van Doorn, & Kappers, 1996). Inclination is indicated by gauge circles depicting slant and tilt. Inclination information is particularly useful for recovering environmental surfaces when color attributes are lacking.

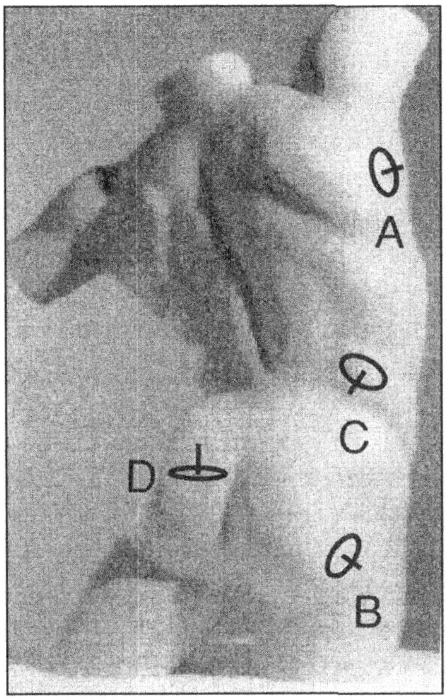

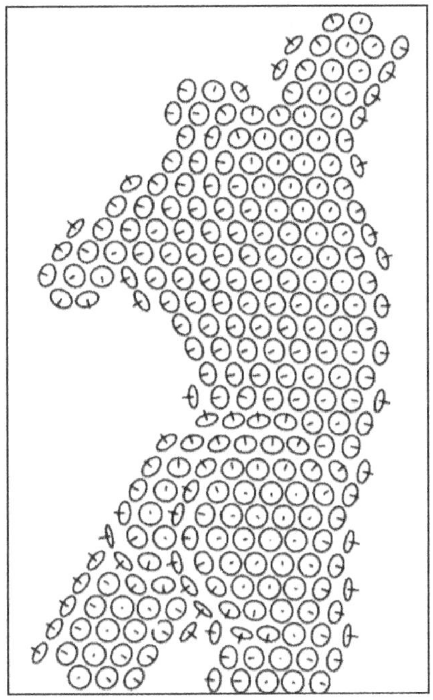

Our perceptual systems are also tuned to accentuate changes in value across a perceptual field, facilitating the detection of boundaries between regions. Most comprehensive theories of vision offered by Marr, Rock, Palmer, and others suggest that edge detection and subsequent perceptual alignment of edges into extended *contours* play a crucial role in early vision. An extension of this thinking maintains that the visual system constructs contour maps along with the other attribute maps, as an aid to segmenting regions (van Tonder & Ejima, 2000). A standard contour map extracted from a scene is shown below. The values in the contour map simply indicate the direction of the contour (0-360°) at each point.

Instantaneous Attribute Value Changes

Attribute values in a map change from point to point, but they also change from moment to moment, due either to shifts in our own stance or to environmental changes. An instantaneous value change is the amount of change in an attribute value at one point from one moment to the next. The amount of change in an attribute's value is a measure of the rate-of-change in that attribute at a point, if the time interval between successive sampling is fixed. (Mathematically, the instantaneous rate-of-change of an attribute value in any map is the first derivative of the attribute's value, approximated by $\Delta c/\Delta t$, where Δc is the amount of change in the attribute's value over the time interval Δt.) For example, if the hue component of color at a point changes smoothly from red to orange (representing, say, a change of 1/8 of the distance around the color circle clockwise, or $+\pi/4$ radians), and this change occurs over 0.1 sec, then the instantaneous rate of hue change is $+5\pi/2$ rad/sec. This rate information describes how fast the color is currently changing and in what direction. Similar component computations can be made for all attribute maps (color, depth, inclination, contour), to derive corresponding *dynamic* attribute maps depicting instantaneous changes (rate and direction of color change, rate and direction—ie, toward or away—of depth change, rotation rate and direction of inclination change, turning rate and direction—ie, clockwise or counterclockwise—of contour change). Together, the two kinds of attribute map provide both a static and a dynamic description of the attributes.

Gauging attribute dynamics requires neural circuits that can compare values over short periods of time, in effect remembering earlier values. But such comparator circuits do not involve organized short or long-term memory representations. They simply delay a value at one time by a fixed interval, before making the comparison with a value at the same position a moment later. The speed and sophistication of dynamic processing is apparent in the way we are able to process phonemes in continuous speech. However, language processing circuits are probably specifically built for that purpose, whereas most dynamic attribute analysis is likely general purpose, covering many kinds of stimuli.

One of the most sensitive and important dynamic evaluations performed by the visual system involves local motion. If a spot slightly shifts its position in the perceptual field, motion detectors in V1 and higher brain areas fire at the location of the spot. Collections of motion detectors in the higher areas like MT and MST may constitute a local motion attribute map, whose values indicate the magnitude and direction of the motion at each point in the field. The motion map is a dynamic attribute map, but it apparently is not derived from an explicit position attribute map akin to the other dynamic maps. Position information is explicit in all static attribute maps, since each point in a map registers a unique position in 2D space. Explicit local motion detectors fire when adjacent attribute values change in sequence by the same amount, which we interpret as motion. An implicit region in a motion map at a particular point in time corresponds to those points of the scene which are moving together (eg, a car driving by). Motion maps previously have been called optic flows (Gibson, 1966) in reference to retinal motion, although here the reference is to a higher level representation. You can get a feel for the content of the motion map by watching a waterfall, or any other consistent motion, for a period of time, and then looking away at a static scene. The static scene will appear to move oppositely to the original motion, illustrating the *waterfall effect.* There is then no factual movement in the scene, but the instantaneous motion map will still register motion until it readapts.

The term "local motion", referring to a point-to-point kind of incremental motion across the receptors in a field, needs to be distinguished from the more global "movement" of an object across the field, interpreted as motion through space. The local motion map registers what is usually called "instantaneous velocity" at a particular point in space and time, including both the magnitude and direction of motion at that point. On the other hand, analysis of the movement of an object through space involves determining an "average velocity" over extended intervals of space and time, where average velocity is the amount of spatial displacement over a time interval. Local motion information becomes useful in later perception for analyzing both kinds of global kinematic object motion—namely, translation and rotation. Analyzing the translation of an object in 3D space involves analyzing local systematic spatial changes in color, contour, and perhaps depth and orientation values over time, as does rotation—eg, a spot on a rotating sphere.

The values in different attribute maps are also spatially and temporally correlated with each other. For example, an implicit region in one map might depict a certain color, while a matching region in a different map depicts a certain motion, and an auditory map depicts a certain sound—eg, a moving red car tooting its horn. Exactly how all the different modalities and maps merge to produce a unified perceptual object remains problematic (the notorious object binding problem), but the process may have its roots in attribute value analysis. Attributes are not so much constructed to characterize an unknown object; it is rather that regions of similar attribute values *determine* fragments of objects, which then are bound together to produce the whole object (from a particular point of view at a particular time). The attributes, in a sense, *are* the object, being the perceptual atoms out of which perceptual objects are made.

Iconic Maps

Merging the various attribute maps for a given modality (eg, vision), produces an integrated, more stable and comprehensive *iconic* map of the perceptual field. The immediate content of iconic maps is what we are currently seeing, hearing, feeling, etc. They are the source of the palpable perceptual experience of "being in the here and now".

For vision, the iconic map partitions the receptive field into small preliminary 2D environmental surfaces based on value similarities in the underlying attribute maps (color, depth, inclination, contour). Motion is apparently not represented in the iconic map, although early motion parameters may assist in the surface partitioning process. The point-by-point attribute similarity evaluations for surface partitioning are nontrivial, but probably reflect an averaging of nearby points. If a point value in an attribute map departs only by a small amount from the local average, then it is part of the local attribute region, otherwise not. The surface building process is therefore probably based on the uniform connectedness (Palmer & Rock, 1994) of neighboring points in the attribute maps. Attribute regions may grow something like volcanic islands until they reach other regions (thereby defining a boundary), eventually completely partitioning the scene.

An iconic map is like an attribute map, except each value in the field of the iconic map is a complex value merging all of the data available from each of the input attribute values at that position. That is, if the value of color at some position is <1, .8, 0> (hue, saturation, lightness), and the value of depth at that same point is <6> (distance), and the value of inclination is <π/4, π/8> (slant, tilt), and the value of contour is <π/3> (direction), then the net iconic value at that point would be < <1, .8, 0>, <6>, <π/4, π/8>, <π/3>, <B> >. The last value component, <B>, is a label indicating that this point belongs to the same surface as other adjacent points with the same label. Labeling reflects the integration of information in all attribute maps aimed at partitioning the scene into environmental surfaces. The explicit iconic partitions typically correspond to fragments of surfaces in the scene.

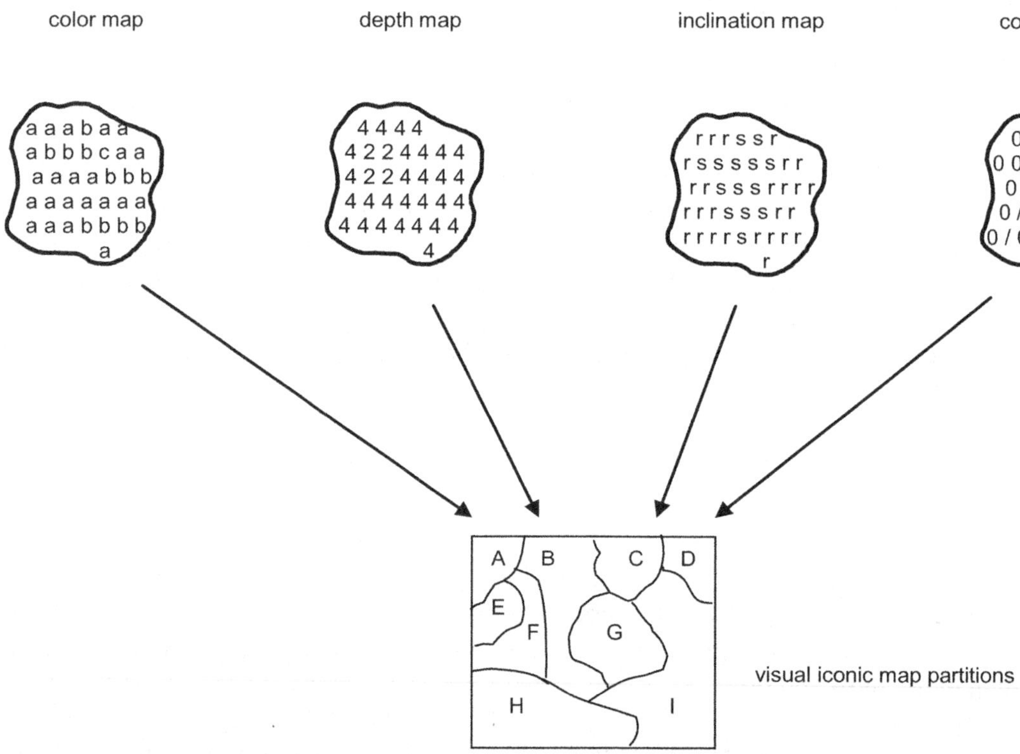

The iconic map accurately reflects the detail carried by the attribute maps, but fuses the information into a higher-level, more persistent map, thereby stabilizing the representation of emergent environmental surfaces. We are normally not consciously aware of this iconic detail, but you can experience it in realtime by closing your eyes and blinking them open for a brief "snapshot". The afterimage fades quickly, but all of the attribute details in the iconic image are available for subsequent processing.

The following photo illustrates the initial partitioning of a scene into surface fragments based on uniform connectedness of attributes. Borders have been drawn around the adjoined points, but do not appear overtly in the iconic map. Note that the black shadows on Lynne's blouse adjoin with the rest of her blue blouse, the small color variations in the lawn merge together, etc. Generally speaking, the smaller the surface, the more distinctive it must be to avoid merging with other surfaces. Wide ranges in attribute values over a surface are also customary, particularly for expansive backgrounds, but the changes must be gradual from point-to-point.

Iconic Memory

In vision, it is well-known that the eyes unconsciously execute rapid saccadic movements around any scene. Saccadic movements are brief fixations at one spot lasting for roughly 300 ms, followed by quick 30 ms ballistic movements to another fixation point (only just noticeable during rapid reading). Most perceptual processes are inhibited during the ballistic movement, so it would appear that the iconic maps (at least for vision) are presenting sequences of snapshots. The snapshots, then, are overwriting the information in the iconic map roughly three times a second.

The most recent snapshots are saved in iconic memory. Experiments show that iconic memory lasts for up to one second, and contains most, if not all, of the information provided by the underlying (attribute) structures. An iconic memory configuration consisting of roughly four sequential maps is one interpretation of these findings, where the information in one map is transferred to the next map in sequence every 1/3 second.

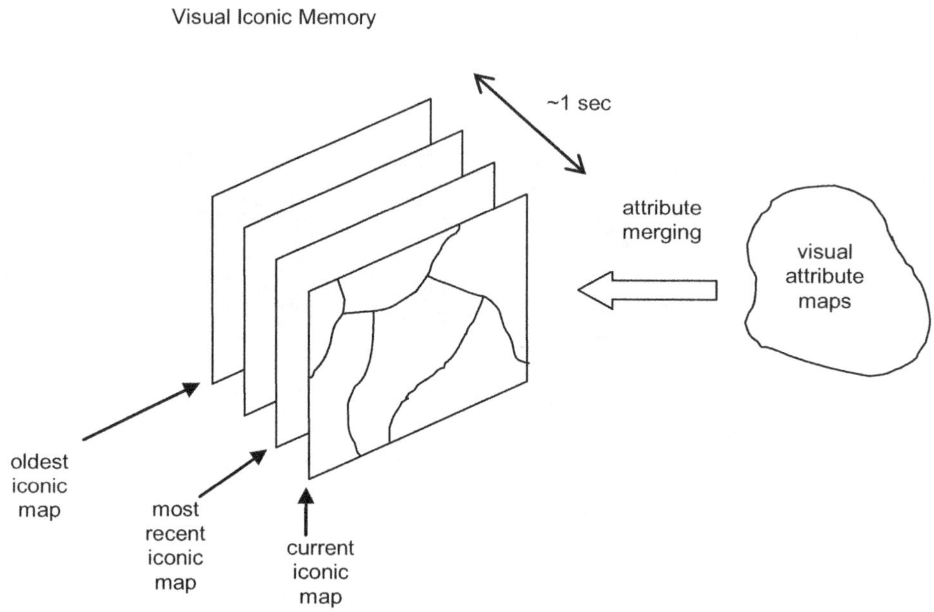

Each saccade establishes a new fixation point in a scene, and registers its information in the current iconic map. This means the positions of the surface fragments in the successive iconic maps are jumping around wildly. But our experience of the environment is stable. The successive iconic snapshots (tied to retinal coordinates) might be combined to produce the stable scene-based spatiotemporal experience. The environmental surface fragments which repeat in successive iconic maps over time but in slightly different places, along with predictable saccade targeting, could provide an alignment reference for integrating the maps. However, experiments argue against this iconic integrative approach (Irwin, 1992). Alternately, the stable experience may arise from later (noniconic-related) processes, which project our attention *into* the scene. The iconic processes reflect a more detached viewpoint—something like viewing a movie where you are aware of seeing the screen, rather than being immersed in the action.

The purpose of iconic memory remains somewhat obscure, but it may have something to do with quickly building up a high-resolution image of a scene. A single iconic map can register the scene for one fixation, but the image resolution in that map falls off rapidly, away from the center point around the macula. Multiple iconic maps would increase high-resolution coverage over more of the scene, enabling more informed processing in subsequent stages. Multiple maps would also allow iconic memory to hold onto very brief exposures, giving a little more time for subsequent analysis.

Figure/Ground Organization

Objects in a scene generally exist in front of backgrounds. This figure/ground assumption is built-in to our basic perceptual experience, and determines how we organize the surface fragments in an iconic map. The pictures below illustrate how adjacent regions encourage perception to regard one as a portion of a foreground surface figure and the other as background, in spite of the fact that in this case it is an illusion (Rock, 1975). We see either a claw-like object or a finger-like object, depending on which side we are perceiving as figure and which as ground.

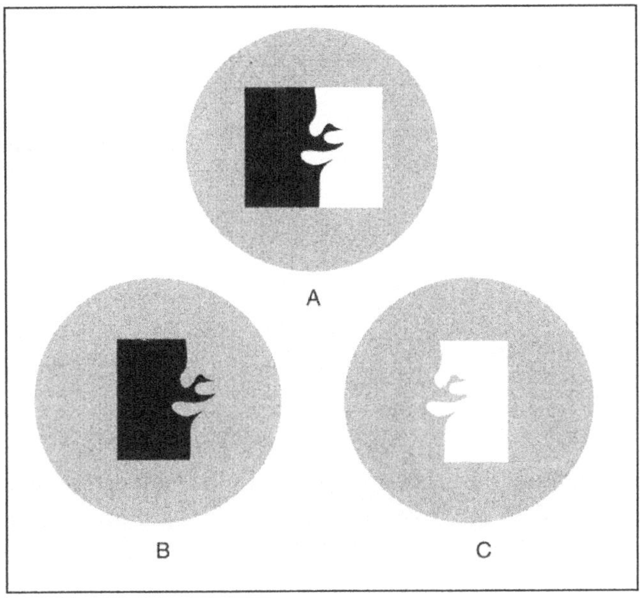

Figure/ground organization operates in other modalities besides vision. For example, there is usually a lot of background clatter while driving, but a new squeak becomes a figural object when it stands out from the ordinary background sounds.

The figure/ground stage of visual processing attempts to connect adjacent surface fragments in an iconic map into whole object surfaces. For example, in the segmented photo of Lynne above, the fragments corresponding to her hair, face and neck, blouse, skirt, left hand, right hand, legs, and shoes belong together because they constitute facets of one figure.

Deciding which adjacent fragments belong together depends on a number of cues, which can be derived by considering how 3D environmental objects should normally appear when projected onto the 2D retina. While all such cues are subject to error, and interact in subtle ways, multiple cues will significantly increase reliability. First, the surface fragments making up a whole figure are generally connected together, which means pair-wise they share at least one boundary. Second, two surface fragments that exhibit a depth discontinuity along a boundary are probably not facets of the same figure, because the nearer fragment is probably occluding the surface of the more distant one. Third, two adjacent fragments that exhibit the same inclination across their common boundary, are likely facets of the same figure. Fourth, if a surface is completely surrounded by another, they are likely not facets of the same figure, since the smaller is probably occluding the larger (although it is possible for a surface to occlude another on the same figure). Fifth, if a fragment has symmetrical or parallel boundaries, then it is likely distinct from adjacent surfaces sharing those boundaries. Sixth, an otherwise isolated, small fragment that potentially could connect to a larger adjacent fragment probably does, since small isolated fragments are unlikely to be whole figures. Seventh, two adjacent surfaces whose boundaries intersect in a double-T (ie, TT) formation are probably different figures, since one is probably occluding the other. Eighth, two adjacent surfaces whose boundaries intersect in an I formation are probably facets of the same figure, one being a continuation of the other. The illustrations below show examples of the last two types of shared boundaries.

TT-Formation (occlusion) I-Formation (continuation)

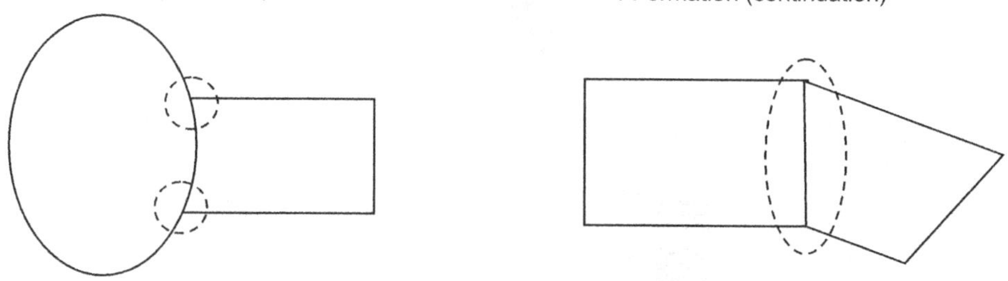

Applying these kinds of surface connection rules to the preliminary fragments in the iconic map leads to the perception of figural and background surfaces in the scene. Some figures and backgrounds are relatively complete at this stage, in that they reflect wholly connected object surfaces (like the figure of Lynne below). Other figures will remain fragmented, since they are occluded by other figures or otherwise indefinite—eg, Lynne occludes the left and right sides of the bench.

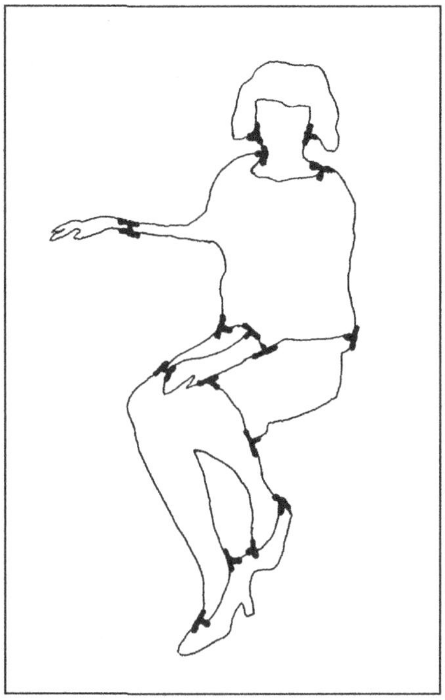

The process of connecting surface fragments is a likely basis for the bistable interpretations of ambiguous figures like the Necker cube. One interpretation is of a cube viewed from above, the other from below.

Two Necker Cube Interpretations

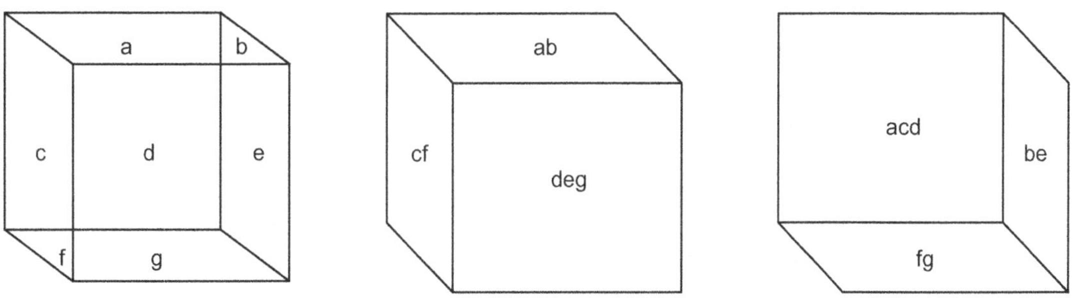

Different ways of grouping the figural fragments accounts for the two interpretations. The first interpretation groups fragments (ab, cf, deg), while the second groups (acd, be, fg). These groupings reflect common pre-established patterns, thereby ruling out unusual groupings like (ac, bdf, eg), unless special circumstances subsequently dictate a novel interpretation. (The Necker cube can also be construed as a wire-frame, as opposed to a solid object. In this case the interpretations would be derived from which wires were occluding which others.)

Only figures and their fragments have definite, meaningful outlines or shapes. The boundaries of background surfaces are generally indefinite, and are typically determined by the accidental arrangement of objects in front of and around them. Some experiments indicate that the shapes of backgrounds may not even be perceptually registered (Rock, 1983). However, a background can be turned into a figure by perceiving it against another more global background. As the organization of a scene evolves, figures come to be indexed at various distances from the observer, where the background for one figure becomes the figure against another background farther away. Complex scenes containing objects located at different distances seem to require numerous passes/saccades to properly extract the important figural information. Motion is also an important attentional cue for picking out figures, but in the absence of motion, figural elements that are closer to the observer usually receive (unconscious) attention first, probably for evolutionary safety reasons.

Figural Surface Grouping

Connected figural fragments frequently correspond to significant portions of a whole object or background surface in the environment, but additional grouping is required to unite those figural and background elements that are disconnected for various reasons. The bench slats and seat, and the grassy area in the photo each possess surface fragments that are close, but not adjacent, to each other. These separated fragments must be grouped together in the right way to produce a meaningful and accurate representation of the figures in the scene.

The classical principles of grouping include proximity (of nearby surfaces), similarity (in color/texture, size, inclination, depth, etc), continuity (smooth continuations of surfaces), and closure (grouping of closed surfaces). More recent grouping principles (Palmer, 1999) include synchrony (of multimodal figures in time), and common region (grouping within a closed area). Another grouping bias, called common fate, likely comes from the motion map (separate from the iconic map), if there is motion in the scene. Elements which move together in the same way (eg, parts of a partially hidden snake in the grass) are an indication they may be connected together.

Proximity grouping of similar nearby surface elements works if the elements are close enough together. However, occlusion of one figure by another can lead to widely separated portions of the occluded figure—eg, the left and right sides of the bench in the photo. Other experiments indicate a process of surface completion (sometimes called relatability theory) operates to connect and fill in contours and similar surfaces across wider expanses, based on simple assumptions about the occluded surfaces (Kellman & Shipley, 1991).

The process of filling in or completion operates over time as well as space. Specific neurons in the temporal lobe appear to fire explicitly when a tracked object temporarily goes out of view (Baker, Keysers, Jellema, Wicker, & Perrett, 2001). The neurons return to baseline levels of activity when the object reappears. In general, spatiotemporal filling in allows perception to maintain an enduring representation of a figure on a trajectory with spatial and temporal gaps.

The final result of grouping is a set of more-or-less complete figures (the surfaces of environmental objects or backgrounds as seen from one viewpoint including filled in occluded surfaces), each described by the spatial detail provided by the static iconic and dynamic attribute maps. But how are the figural surfaces, exhibited in successive iconic time slices, coordinated? In other words, how is a figural surface with one set of attributes at one moment determined to be the "same" figural surface with a different set of attributes at the next moment (corresponding to a single changing environmental object)? At the figural level of perception there are no summary figural qualities (only point-by-point attributes) with which to compare one figure with another to decide if they should be assigned to the same environmental surface. It is (relatively) easy to tell that two spatially separated concurrent figures do not refer to the same environmental surface, but two consecutive shifted figures may or may not refer to the same surface, depending on assumptions about how environmental objects move.

To solve the temporal reference problem, perception appears to assign a simple index (or pointer) to each salient figure in the current scene, up to perhaps 4-5 simultaneous visual indices (Pylyshyn, 2003). Experiments show that people can routinely track up to five independently moving identical spots of light, if the spots' motions are regular, even if they briefly disappear and reappear, as if passing behind an occluding object. An index can evidently jump to a new location when the location of the new figure is predictable from the old, thereby establishing that the figures refer to the same surface. An index therefore does not track by comparing descriptions of successive figures. It simply points to the localized region of the figural surface in the current iconic map, and updates to the location of the next figural surface in the next moment based on built-in constraints about probable object motions.

Our fundamental ability to track figures using a simple index goes some distance toward explaining our advanced capacity for using mental representations to point to perceived "things" in our environment. It is the correspondence between our representations and the environment that makes our representations useful. An index points to a representation that corresponds to a feature of the environment, which we interpret as pointing directly to the environment. Furthermore, indexing may be what gives objects their individuality. Objects are individuated when indices are activated, as when looking at a cloudy sky, some patches may suddenly become indexed. These figures then go on to become individual clouds, and the locus for subsequent analysis that assigns object characteristics of various kinds.

If the foregoing perceptual processing scenario is roughly correct, then it would seem our primal experience of a world filled with objects stems from the construction of perceptual figures. (Use of the word "object" throughout this appendix refers generically to all those things we perceive in the environment as individuable or separable entities—from bats and balls, to lawns and spilled milk, to fog, shadows, and corners.) For each modality (seeing, hearing, etc), a figure evokes impressions of something present. The intuition that figures entail environmental objects is prescientific in the sense that veridical perceptual figures may or may not correspond to actual physical objects. A shadow or a hole can be a figure, even though it does not exist as a physical object. (However, shadows and holes, as appearances, are scientifically *explainable* in terms of physical objects.) Other more ephemeral figures, coupled with emotional states, may give rise to all sorts of figural impressions, including ghosts and flying saucers. One task of advanced cognition under this interpretation is to determine which figures are in fact physical, given that unvarnished perception supports a virtual ontological zoo of immaterial figural objects.

The Situated Form

The perceptual grouping of surface elements together into an indexed figure prepares for the next major step in perceptual processing: creating a characterization of the figure I will call a *situated form*. A situated form amounts to a summary of the iconic and dynamic information contained in a figure, normally experienced as a projection out into the environment. (It also incorporates the index which points back to its underlying figure.) Being a summary, it contains considerably less detail than the figure itself, and admits to some tolerance with respect to viewing conditions. Situated forms are the source of what we "see" (or "hear", "feel", etc) with our mind's eye if we observe a scene, and then close our eyes and imagine the scene. After a few seconds, the resulting imagery comes from the summary information just encoded (assuming we have not seen it before). The image lacks iconic detail, but nevertheless portrays the overall characteristics and configuration of individual figures derived from the scene. The above description of a situated form is broad enough to cover not only contiguous bounded figures, but also more dispersed fluidic figures like puddles, and steam columns, as well as other figural "conditions" like falling snow occurring over large areas of a scene. A situated form is akin to a stripped down, transient version of an active "object file" representation of memory proposed (Kahneman & Treisman, 1984) for perceptually tracking a single environmental object over time.

How perception achieves the situated form of a figure seems almost magical. But it probably begins by compensating for figure variations due to the particular viewing conditions at the time, which have nothing to do with the intrinsic nature of the figure itself. For example, when we shift our gaze, the figures in a scene move across our retina, which is the reference for the iconic map. But we automatically compensate for these kinds of self-movement and perceive the corresponding situated forms as nonmoving in the environment. Similarly, when a shadow falls across a colored figure, we normally perceive the surface as having one intrinsic color, not two. These kinds of compensation are called *perceptual constancies*, and give the situated form a modest degree of observer independence:

- form size independent of distance from observer to figure
- form color/texture independent of lighting conditions
- form shape independent of viewing angle
- form orientation independent of head tilt
- form position independent of eye/head/body position
- form movement independent of eye/head/body movement.

A related mode of perception effectively transforms our perceptual reference frame from one centered on ourselves as observer to a more impartial one centered on the figures. It is the properties, relations, and relative movement of objects in the environment that we end up perceiving most of the time, rather than the incidental properties, relations, and movement (eg, parallax) brought about by our own point-of-view. Moving the reference frame out to the figures in a scene allows perception to develop *intrinsic features* of the situated form, which describe the figures independent of the observer.

There are at least two basic kinds of information extracted from a complete figure: summary features of the individual elements making up the figure (eg for vision, surface element size), and a summary characterization of the figure as a whole (eg, figure size). The processing of each kind of information seems to take place in opposite cerebral hemispheres in parallel, with figural elements analyzed in the left temporal-parietal lobe, and the whole figure analyzed in the right temporal-parietal lobe (Robertson, Lamb, & Knight, 1988). Both processes generate separate situated forms, which become linked via part/whole associations.

The summary information encoded for a visual figural element likely includes a specific set of perceived features. Each feature probably reflects an averaging of the attribute values for the surface element, assuming the variation over the surface is nominal. The average color of a robin's red breast is such a summary for the surface element corresponding to its breast. A surface element exhibiting a wide variation in attribute value is probably too confusing to support a meaningful summary, and such features probably do not generate enough activity to participate in subsequent registrations of the situated form. Experiments show significant interference among multiple values within one feature. For example, it is harder to summarize a mixed two-colored surface element than a one-colored element.

The specific features relevant to a surface element include texture (with respect to background lighting conditions and local edge averages, where texture now includes color as a simple case of texture), size (perhaps with respect to the human body), position (with respect to other figural elements), symmetry (with respect to the number of symmetry axes), orientation (of surface normal with respect to gravitational vertical), movement (speed and direction with respect to background), and most important, shape (with respect to boundaries and surface type—ie, planar, convex, concave, etc). Further surface element features are probably developed from the dynamic "instantaneous" attributes like speed of translation/rotation (eg, when a figural person is turning their head).

The features describing the whole figure are probably the same as for figural elements, except that they now characterize the whole figure rather than an element. We normally characterize a whole zebra as striped, even though some of its body surfaces are not striped or exhibit different variations and degrees of striping. This is a case of the more generalized phenomenon of a salient part standing for the whole. The two distinct characterizations form the basis for subsequent representations of the whole in terms of its parts. But the figural whole is more than the proverbial sum of its parts, since it receives a separate characterization.

Neurologically speaking, a feature is probably a simple association (analogous to a type of directed link) between the situated form and a field of previously learned values. For the texture feature, possible values might be red-&-white-thin-striped, brown-polka-dot, etc; for the size characteristic the values might be spec-sized, pea-sized, fist-sized, body-sized, etc. The types of possible perceptual feature (eg, texture, size) are probably genetically fixed, but the possible values probably depend on experience.

The possible values for a feature can be thought of as a multidimensional *feature space* made up of integral attributes, just like attribute spaces are made up of integral qualities. A set of attributes that typically go together to describe a distal situated form become bound together into a feature space in long-term memory. For example, if the integral attributes have dimensions a, b, c, then the dimension of the feature space will be a+b+c. A characteristic space is simply an aggregate of a number of underlying attribute spaces. A point in a feature space corresponds to a specific value of that feature, which is composed of its integral attribute values, which is composed of its qualities, which is composed of its opponent cell values.

Not all figures (or figural elements) are of equal importance, or warrant full characterization. Which figures are selected seems to depend on how much attention they garner. Figures engender bottom-up attention based on their *salience*—ie, how distinctive they appear with respect to other figures. A strong salience, like a bride dressed in white lace among more modestly dressed people, causes her to automatically "pop out" from the other figures. Salience may also extend to subtle unexpected features like a limp. On the other hand, top-down consciously focused attention can also single out a figure, as when searching for a suitable houseplant in a store display. The selected plant achieves a representational status not afforded to the others. Generally, the number and quality of features generated for a situated form depends on both bottom-up and top-down attention. Koch recounts an episode of Candid Camera where the host engages an unsuspecting passerby in a light conversation (Koch, 2004). Two workmen then walk between them carrying a large door, temporarily blocking their view of each other. One of the workman deftly hands off the door to the host while blocked from view, leaving the workman behind to continue the conversation with the passerby. Half of the passersby filmed didn't notice they were talking to a different person. What features we actually register is usually much less than what we think we register, the illusion being created by our uncanny and sometimes disturbing ability to fill in missing details from expectations and stereotypes.

Characterizing Shape

Shape is perhaps the most important feature of a situated form, since an object's shape is highly correlated with what kind of thing it is, and how it behaves/functions or can be used. A strawberry broadcasts its ripeness by turning red, but it is the strawberry's shape that allows us to classify it as an object of interest.

The shape feature derived from a figure or any of its surface elements consists only of its "front" surface (2D surface oriented in 3D space). The front is the surface that would be seen after all interference from occluding objects is removed. This interpretation of shape is different from the standard 3D model interpretation in many theories of late perception (using geons, generalized cylinders, etc) where the 3D models characterize the whole environmental object, including hidden back surfaces. It is also somewhat different from Marr's proposed 2.5D sketch, which operates in early perception, does not explicitly divide up the object into surface elements, and does not summarize the surface configuration. Here, a shape characterization describes an object as seen from a single vantage.

The usefulness of this intermediate representation of shape information derives from the geometric fact that the shapes of figural surface elements do not usually change much when the viewing angle does not change much. The shape feature, therefore, is a kind of low-level abstraction covering a small range of viewpoints. When the viewing angle does change enough so as to present different object surfaces that previously were hidden, the new shape is captured by a new value of the shape feature. Simple 3D objects with few surfaces translate into a small number of alternative values. The illustration below shows how a cubical solid with six surfaces only requires three basic values to fully register its alternative shapes—and two of these are highly coincidental (rare) views.

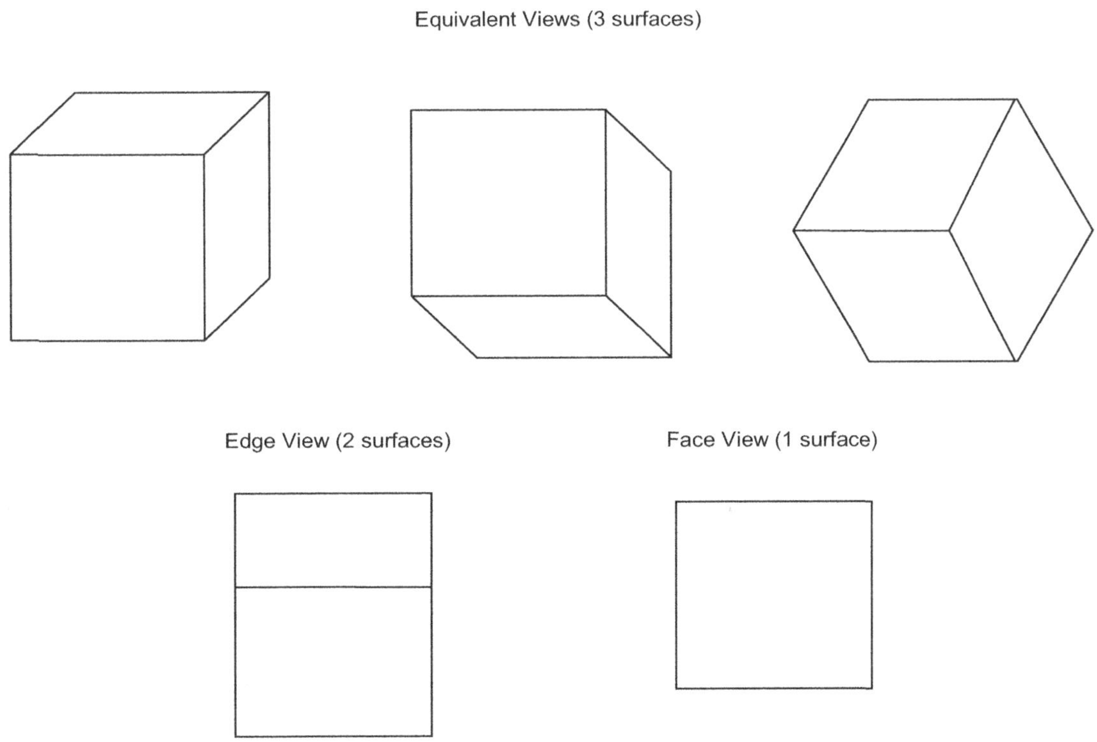

How, then, is the shape of a figure represented? Since the figure is typically composed of figural surface elements, the first question becomes: how is the shape of an individual surface element represented? For a given surface element, there are two basic descriptors that specify its form: the boundary (describing the spatial extent of the surface), and the surface type (describing how the surface curves within the boundary).

The boundary of a surface element normally contains discontinuities, where the boundary changes direction (ie, curvature). In the cube examples above, the boundary discontinuities occur at the corners, with constant boundary direction/curvature between discontinuities. The boundary discontinuities, along with a measure of the amount of curvature between discontinuities, then describe the shape of the boundary, even for relatively complex shapes like those shown below. Dots on the figural boundaries indicate curvature discontinuities. (Mathematically, curvature measures involve either 1st or 2nd order derivatives, depending on the complexity of the curvature. Analogously, perceiving constant velocity movement involves a 1st order derivative, while perceiving constant acceleration, as in an accelerating or breaking car, involves a 2nd order derivative.)

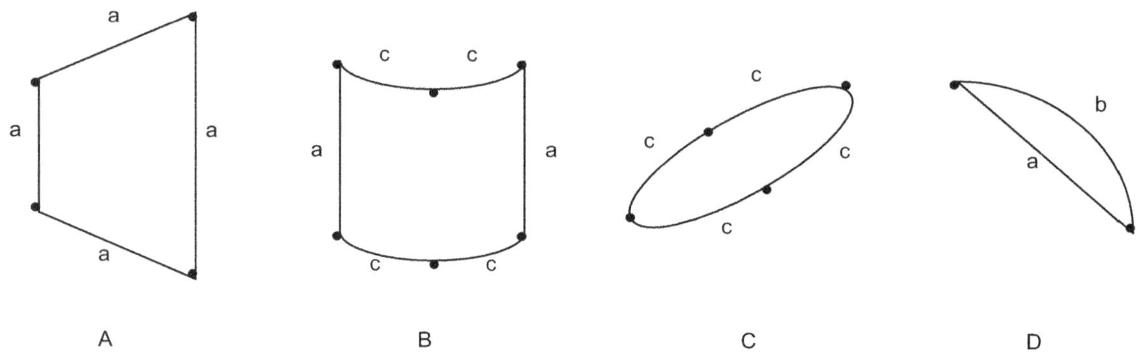

<u>curvature</u>:
a = zero curvature
b = constant curvature
c = increasing/decreasing curvature

As for the type of surface occurring within the boundary, a rudimentary classification is either planar, convex, concave, or both convex and concave. All of the surface elements above admit to any of these type interpretations—eg, surface B is readily interpreted as a piece of paper bent either toward (convex) or away (concave) from the observer, or even a cutout lying flat (planar), or somewhat more difficult to see, both concave and convex at the same time like a potato chip. A more advanced approach to surface classification would be needed to describe element-internal surface features like dents, ridges, and corrugation. However, these kinds of surface features are amenable to analysis by way of surface curvature in the same way that boundary curvature is able to describe the bumps and depressions of boundaries.

Given the layout of the figure surface elements, marked with boundary discontinuities, curvatures, and surface types, it is straightforward to build a graph as an abstract representation of the figure. The graphical representation amounts to an integration of surface element shape (aka parts) and figural shape (aka whole). There is no need for a separate structural description of how the elements are configured in relation to each other, as is common in many theories of shape perception.

The graphs for some common figures are shown below, where A is a convex surface, B is a concave surface, and C is a planar surface. The curvature labels for each boundary connection between nodes are left off. (All surfaces on the person figure are convex—ie, type A—as is the case for most natural solid objects.)

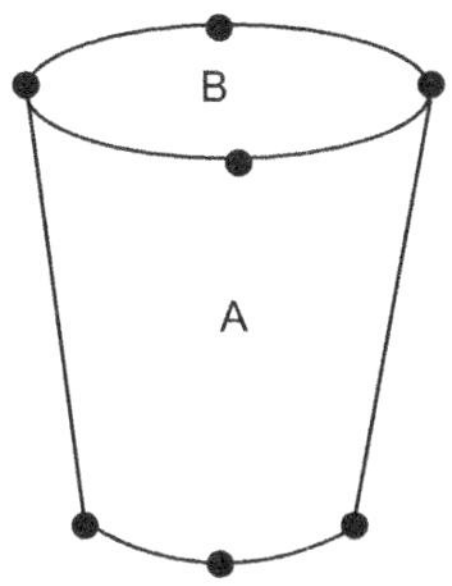

Tumbler

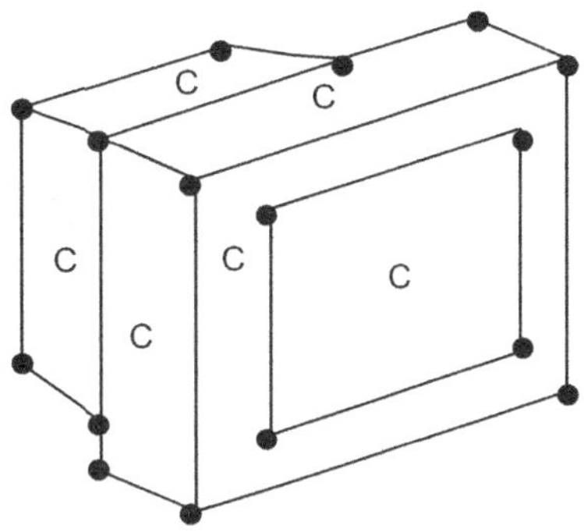

Monitor

Person

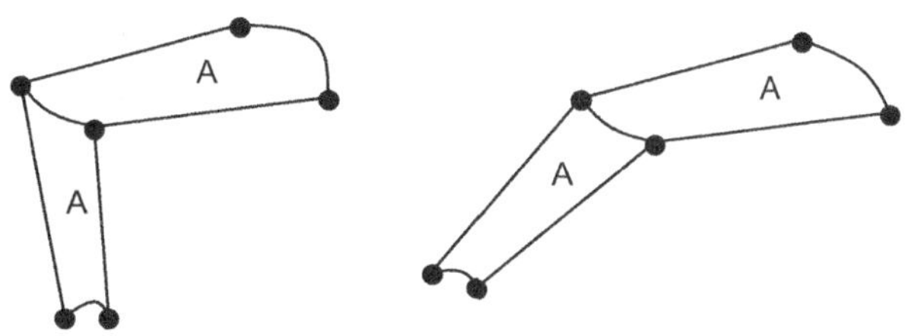

Leg Joint

The nodes in the graph of a figure constrain the general topology of the figure's surfaces, but not their particular location or alignment. (A node has a relative position with respect to other nodes it is connected to, but not an absolute position in space.) Capturing the topology of the figure means the characterization can represent objects with parts that articulate (common for animals) or stretch (due to viewing angle), since they express roughly the same graph. The shape characterization, therefore, corresponds fairly well to environmental objects composed of connected parts fastened to each other in such a way that they can move about somewhat independently—like a marionette. However, an animal like a puffer fish would need two sets of characterizations, since it can have two dramatically different shapes. The graphical approach is even less useful for representing fluid shapes like an octopus, except perhaps from an ideal viewing point. But the shape of an octopus is probably better defined in terms of its 3D body parts (rather than visible 2D body surfaces), which requires the more sophisticated analysis provided by conception.

Elaboration of the Situated Form

As the individual figures in a scene distill into situated forms, those forms are establishing feature associations with other perceptual representational systems—namely, the feature value spaces in long-term memory. But in addition to the intrinsic form characteristics mentioned previously (texture, size, orientation, etc), there are also extrinsic features describing the location of the form relative to other forms in the scene (layout), the connection with a familiar context (setting), and the potential functional meaning of the form with respect to our basic capabilities (affordances). Layout is quite separate, neurologically, from the intrinsic characterization of situated forms—corresponding roughly to the so-called "what" and "where" pathways of the temporal and parietal lobes, respectively. We are usually able to remember the layout of objects in a scene long after we have forgotten particular object details. The setting or "gist" of a scene encodes the current environmental context (eg, backyard, restaurant, business meeting, picnic, etc). The speed with which we can extract the overall gist of a scene is evident in experiments by (Chun & Potter, 1993). Accurate assessments of common settings seem to take only about 0.1 second of exposure. Our direct perception of affordances (Gibson, 1979)—eg, that we can grasp or kick or sit on something without first categorizing it—is possibly tied to associations with motor scripts in the motor cortex. Such scripts might be automatically triggered by the secondary, action-oriented vision system (Goodale & Milner, 1991) that has been linked to conditions like blindsight.

There are also extrinsic features describing the important mereological (ie, part/whole) relationships between a figure and its elements. The independent processing of a global figure and its local elements probably results in separate situated forms for each. The separate forms become linked via an extrinsic mereological feature from part to whole. If a figure has elements, they are specified through this temporary association.

Other extrinsic features of a situated form developed at this stage likely include primitive spatial relationships with other forms in the scene. Exactly what spatial relationships are being encoded at the lowest levels is a matter of active research—eg, (Talmy, 2003; Mandler, 2004)—but they probably underlie impressions of proximity (near/far, at), containment (in/out), relative location (above/below/under, beside, between), contact (on), etc. The spatial primitives appear to express fundamental topological notions of space, independent of metric distinctions like length, area, or volume characteristic of more advanced Euclidean notions (Talmy, 2003).

The primitive mereological and spatial relationships mentioned above are features (ie, simple associations between one or more forms), and do not encode explicit *relations* between forms at this stage. For example, the relational observation that a marble is in a box is not represented as a separate form. Rather, the form representing the marble possesses an "in" association link that points to the situated form representation of the box. This means that one perceptual form is always singled out as a reference or landmark, while the other becomes related to that reference through an association. Thus in the example, the perceptual association is not from the box to the marble (in which case the meaning of the association would be "contains"—ie, the converse of "is in"). The spatial description attaches to the marble (ie, the topic, not the reference object) at the lowest perceptual level. A similar asymmetrical usage is observed in language for prepositions, which also assume a reference object/event to which other objects/events are related: the cow jumped over the moon. The functioning of such spatial linguistic expressions is likely based on underlying perceptual representations.

Concurrently with the emerging visual forms, situated forms are also being created in the other sensory modalities based on their underlying "figures" and "figural elements". Petting my cat, Angus, usually elicits a squawking-like sound accompanied by a formidable tuna stench as he exhales with his mouth wide open. In this case my situated experience includes the feel of fir and body (tactile form), the squawking noise (auditory form), a visual form, a smell (olfactory form), a sequenced movement summary of a head form rearing back with an opening mouth (movement form), and perhaps a setting on the living room sofa.

Once situated forms appear, they are available for triggering stereotyped motor reactions (again, before conscious awareness or classification/identification of the stimulus)—particularly useful in situations requiring immediate action. A large spider-shaped form is likely such a trigger for most people, although a trigger could also be as indistinct as a figural condition (deja vu?), a way situated forms are arranged (feng shui?), a distinctive element (synecdoche?), or an insight about how to use a common object (creative affordance?). Some rare reports of sleepwalking, during which the person dresses themselves or even climbs out a window while in a trance-like state, attest to the presence of a sophisticated sensorimotor loop of stereotyped behavior, considerably above the level of the reflex arc, but independent of consciousness. This sensorimotor loop is operating covertly all the time while we are conscious (eg, maintaining balance, focusing attention, following habits), but we can become aware of it after the fact. Some people can drive long distances completely absorbed in thinking about other business, and suddenly "wake up" without remembering anything about the trip. Such autopilot dissociations, although potentially dangerous in some situations due to the absence of continuous conscious attention, do enable adaptive multitasking.

In summary, a situated form embodies a moderately abstract representation of a figure in terms of intrinsic and extrinsic features, over a brief interval of time. These concise, but comprehensive representations are what is placed in short-term memory, in preparation for further analysis, or possible storage in long-term memory.

Summary of Early Perceptual Representations

The overall picture of early perception that emerges from this study reveals a succession of parallel processing stages, with each sensory modality analyzing information in roughly the same way. Each stage acts to develop a distinctive representation of the information at its level of processing. While the representational content of the information varies greatly between channels and between stages, the processes and intermediate representations employed by different channels are similar because they are all designed to analyze stimulus patterns in a sensory field irrespective of what those patterns ultimately mean.

Each of the main processing stages (four are proposed here) consolidates information from the stage immediately preceding it. The summary representation of that information then serves as input to the next stage. The first stage amounts to a simple point-by-point representation of the sensory field. Each point is characterized by a vector value at any time indicating the magnitude of stimulation present at that point for various attributes. For a visual point in area V1 the value corresponds to a certain level of illumination, dissected mainly into color and depth parameters. The density of point values determines the perceptual resolution of the field. Adjacent points with similar values are summarized in the next stage as regions of constant or smoothly varying stimulation. Regions effectively segment the stimulus field by defining boundaries or edges, making some regions distinct from others. At the next level, regions in the stimulus field are combined into surfaces. The term "surface" has visual connotations, but is used here in an abstract sense to denote any bounded description of activation extending over space and time subject to built-in and attentional constraints. An auditory surface, for example, might characterize one cycle of a fire engine's siren. Finally, individual surfaces are combined to produce a situated form. A form is a physically coherent configuration of surfaces representing an environmental object perceived from one point-of-view.

Each representational level (point, region, surface, form) is perceptually characterized according to characteristic relevant to that level. The potentially relevant features are listed in the table below:

Representation	Feature
Point	Color (hue, saturation, lightness)
	Depth
	Inclination (slant, tilt)
	Instantaneous color change
	Instantaneous depth change
	Instantaneous inclination change
	Instantaneous motion (speed, direction)
Region	Contour (connected edges)
	Average color
	Average depth
	Average inclination
	Average motion
Surface	Boundary (edge position, curvature)
	Size (surface extent)
	Convexity

	Orientation
	Texture
	Surface dynamics
Form	Index
	Projected location
	Projected size
	Projected shape
	Projected color/texture
	Projected motion (translation, rotation)
	Axes of symmetry
	Setting
	Affordance
	Layout

For example, consider the perceptual representations that might be developed while looking at a stationary zebra. A point in the sensory field could have a color value of white or black at a certain depth and inclination depending on the part of the body. A region might correspond to a black or white stripe of points. Separate surfaces might correspond to the pattern of black and white stripes on the zebra's cylindrical body or its legs. And its form constitutes all of its visible surfaces in relation to each other, along with its location, body axes, estimated spatial extent, and overall shape. Each represented unit at each level is assumed to point to multidimensional (probably implicit) neural structures or patterns characterizing the values of its features, as discussed next. Specific points or small regions in feature space represent values of the above features (eg, star-shaped in shape space).

C. LATE PERCEPTION

If early perception encompasses those processing stages leading up to the capacity for stereotypical responses to environmental stimuli, then late perception enables original responses that can take past perceptual experience into account. Late perception continues the progression of representational stages evident in early perception, but the representations are more abstract, longer lasting and convey extended temporal and object kind, as well as spatial, content. The diagram below (Baars & Gage, 2007) maps out some of the principal cortical areas linked with developing these representations.

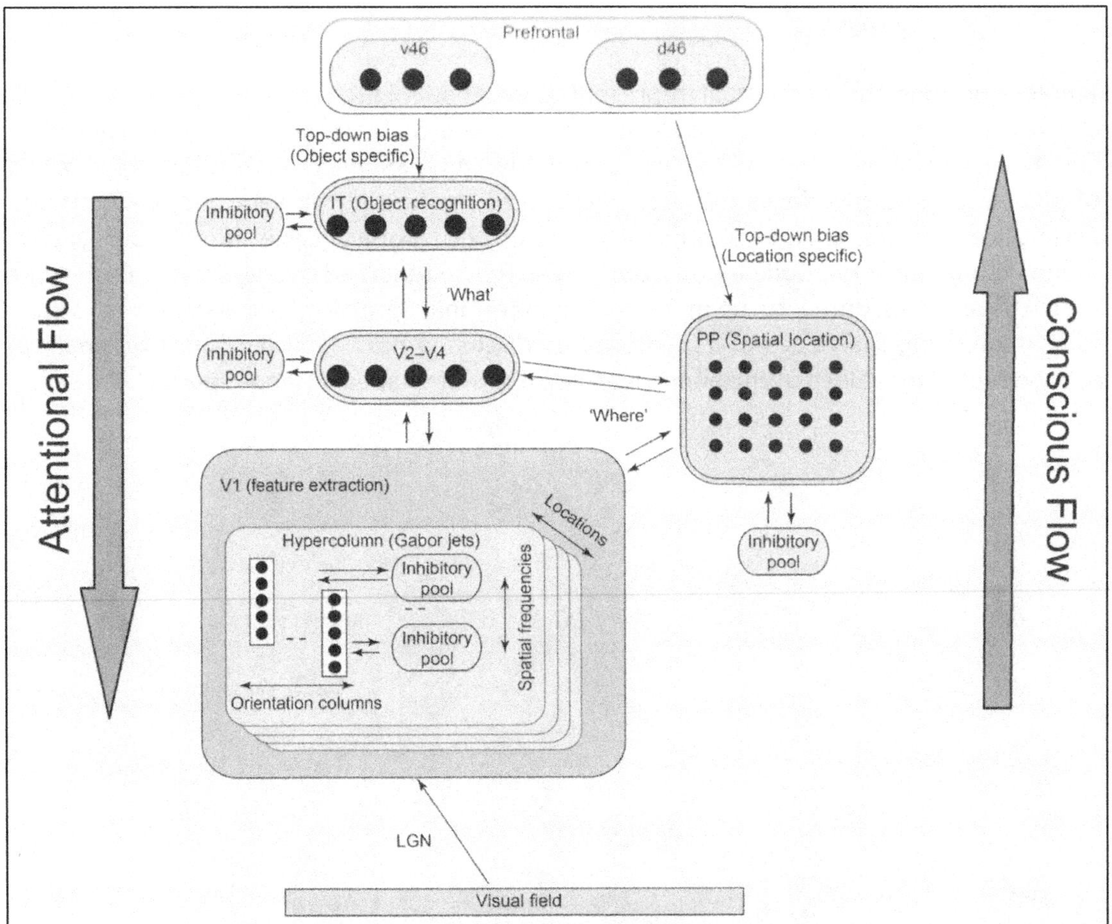

Information generally flows upward from the subcortical visual field and LGN to cortical area V1, where low-level features of the visual field become represented—edges, orientations, etc. (The information flows for the other senses, such as hearing and touch, are thought to roughly follow the same organizational principles as those for vision, with convergence of all modalities occurring in the association area IT.) The low-level features in V1 then serve as input for higher-level representations of color, curvature, junctions, motion, etc—areas V2-V4 and MT (not shown)—and spatial location. The well-known divergence between the "what" and "where" pathways becomes apparent at this level, with the what ventral pathway, specializing in object recognition, extending into the temporal cortex, and the where dorsal pathway, for locating objects, into parietal regions. Information subsequently flows from IT to the frontal cortex and hippocampus for evaluation and integration into short-term and long-term memory.

Information also flows downward from executive functions in the prefrontal cortex, so as to direct attention to specific objects, features, and locations in the lower-level visual representations. Attention acts to either enhance or attenuate those representations depending on our goals, whether conscious or unconscious. The constant rapid interplay between environmental information flows (upward) and attentional flows (downward)—detectable in EEG fluctuations (Freeman, Burke, & Holmes, 2003)—is the stuff of experience and being a situated participant in the world.

The original transient situated forms constructed by early perception (in the primary and secondary areas associated with most perceptual modalities) go on to achieve a measure of durability in short-term memory—probably located in the hippocampus, temporal, and parietal areas (Baars & Gage, 2007).

Short-Term Memory

Each perceptual modality possesses its own short-term memory structures for temporarily retaining its situated forms and their associations. Visual short-term memory can hold the forms for only one scene at a time; changing scenes overwrites the previous scene. However, while iconic memory lasts for only a fraction of a second (excluding, perhaps, rare cases of eidetic imagery), visual short-term memory can last for many seconds—typically about ten seconds without rehearsal, but dependent on circumstances. The information content of visual short-term memory is probably limited to about 3-7 forms at any one time, along with their features. Short-term memory, therefore, is capable of holding a considerable amount of information about a scene or other perceptual occurrence over a period of several seconds.

Masking experiments (Potter, 1976) imply that it takes ½ to 1 second for perception to fully register the meaning (ie, content) of forms in a scene. This means the information in visual short-term memory probably remains stable for on the order of a second to enable this kind of processing to take place. Invoking the same overall structure as proposed previously for iconic memory, visual short-term memory may encompass a sequence of about ten maps (holding situated forms), which operate a similar bucket brigade, transferring content from one map to the next each second.

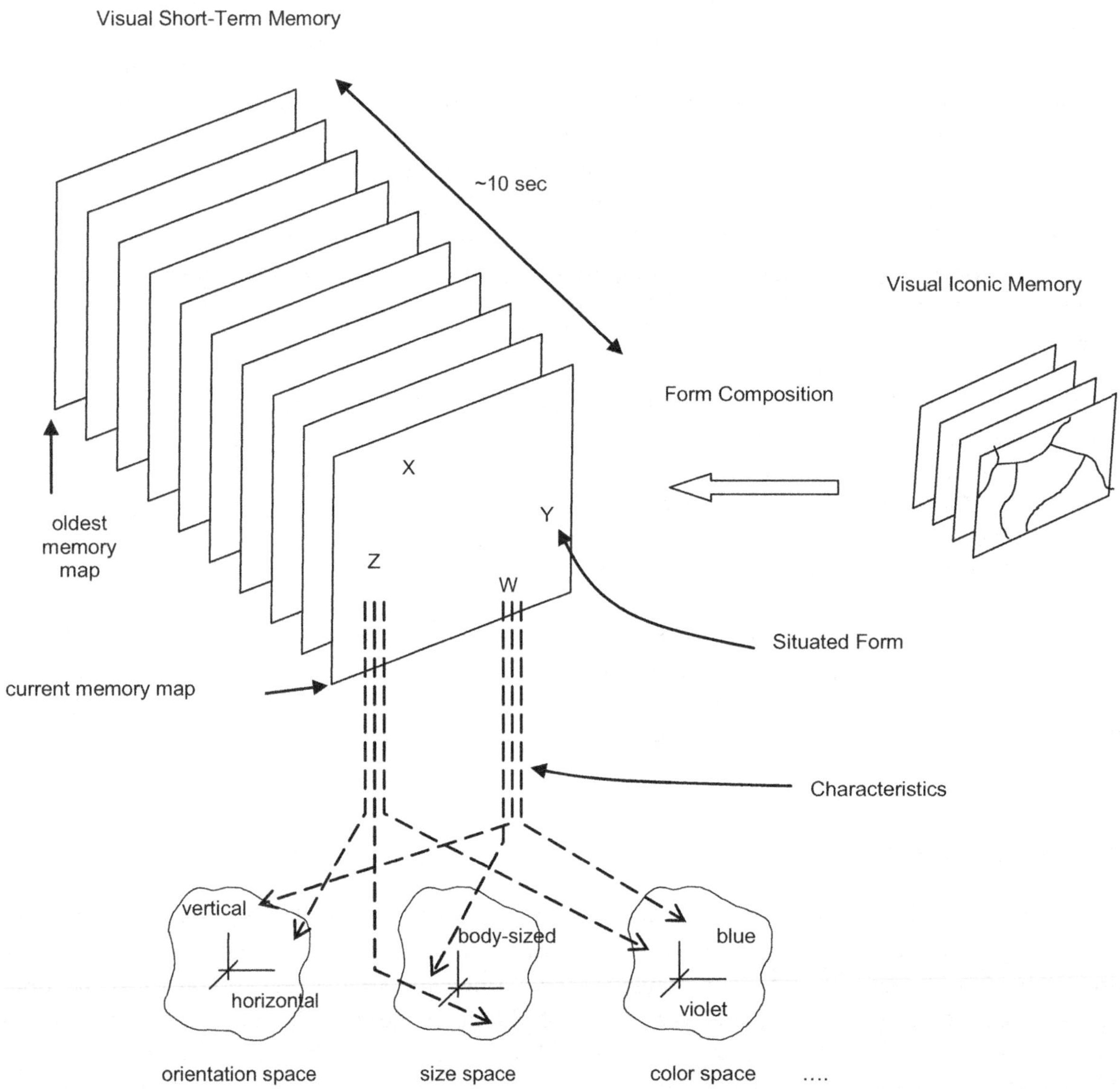

The feature associations linking situated forms with specific values of a feature in the diagram above are shown as explicit connections between short-term memory and the various feature spaces. However, it is highly unlikely that these associations reflect literal neural connections. For one thing, a traumatic bump on the head can completely erase the contents of short-term memory, indicating that the associations are fragile. Also, it takes some time to establish new synaptic connections like those involved in long-term memory. The short-term memory associations are more likely either encoded references to values in the distal feature spaces, or perhaps just simultaneous activations of neurons in different parts of the cortex brought about by the original stimulus. This latter kind of association—

proposed as a form of transitory binding (Crick & Koch, 2003)—requires the synchronized firing of loosely connected neurons in different parts of the brain, probably mediated by attention.

The relative locations of situated forms in the visual short-term memory maps may be the source of the layout features associated with each form. If form location is explicitly represented in a topographic map, then this places the visual, and possibly other, short-term memory maps in the "where" pathway of the posterior parietal cortex. Delayed response experiments with monkeys suggest the dorsal prefrontal cortex is also involved in location coding (Funahashi & Goldman-Rakic, 1989). That the short-term memory maps have a spatial layout something like the underlying iconic and attribute maps seems plausible, since visual situated forms are still essentially 2D representations (less for nonvisual forms). But once the jump to full 3D mental object representation is made, a 2D map is no longer adequate.

There is a case history (Zihl, Von Cramon, & Mai, 1983) of a woman who may have experienced the visual short-term memory maps individually due to a brain lesion. In one experiment she had trouble pouring liquid into a cup because she saw a succession of frozen images, much like a slowed down motion picture, that jumped from frame to frame as the cup filled—eventually to overflowing because she could not accurately predict the rate of filling. In order to produce the normal experience of smoothly flowing events, the information in the short-term memory maps possibly becomes integrated through activation of dynamic object *models* encoded in long-term memory. Most people seem able to accurately model environmental changes to objects, thereby phenomenally filling in temporal gaps, much like figural grouping fills in spatial gaps in occluded objects.

Temporal Analysis

Constructing perceptual figures, some of which evolve into situated forms appearing in short-term memory, involves spatial computations over the various perceptual fields. Once situated forms appear, they become available for temporal analysis, as well, extending over the sequence of short-term memory maps.

All sensory fields provide indirect information about the presence and qualities of stimuli over time. For example, a sound will typically arrive at one ear slightly before the other, allowing dynamic analysis in early perception to determine the direction from which the sound came. As previously discussed, such dynamic processes operate over very short time intervals (~100 ms or less), and do not involve organized memory. Short-term memory processes, however, can integrate the information associated with the situated forms in each map in order to extract temporal patterns exhibited in the map sequence over several seconds—temporal patterns like repetition, duration, change onset, average rate-of-change, etc. Such *temporal features* constitute an additional implicit description of the form over and above its static and dynamic features.

Altogether, a form's static, dynamic, and temporal features over several seconds amount to a *spatiotemporal* registration of the form in short-term memory extending over both space and time. Perceptual tracking normally functions such that a situated form remains the same form in spite of changes to its features between maps. Otherwise, each change would summon a new form. Assuming that perceptual tracking can distinguish whether each current situated form is either a new or continuing form, it can label the form appropriately. The spatiotemporal forms in short-term memory furnish the basis for subsequent representations of the enduring perceptual objects of common experience.

The temporal features of spatiotemporal forms are probably complex structures of multidimensional summary features, just as static and dynamic features are. For example, when an object activates retinal points in sequence, we typically perceive it as traversing a path from a source to a destination—as when one player tosses a ball to another. In this case, the ball's movement through space is one of the temporal features of its form—a feature which is probably structured into subfeatures like source, destination, and path. The source and destination subfeatures might point to the forms in short-term memory representing the first and second players, respectively. The path subfeature might point to a previously acquired perceptual value representation of a certain kind of (downward parabolic) arc—eg, distinct from the arc exhibited by a ball rolling off a table.

The registration of spatiotemporal forms takes time. The time between stimulus onset and a minimal level of awareness of an environmental object is usually estimated at 200-500 ms (excludes decision-making and subsequent motor response to the stimulus). The conclusion that we "live" in the recent past is inescapable; our experiences are delayed by up to ½ second from the time they actually occur. But then why do we not experience this delay, which would seem to make activities like driving a car even more dangerous than it is? A likely explanation is that we are able to use the short-term memory registrations of spatiotemporal forms to project their most recently observed features into the near future. We compensate for the delay by predicting what the realtime situation is based on the recent past. Projecting from past trends is subject to egregious error (we can't perceptually predict the sudden emergence of a novel form) and works only for relatively short time intervals (depending on how rapidly the situation is changing), but it probably represents a practical compromise between the need to analyze complex situations and the need to act. It has been calculated that a professional baseball player's swing must begin before the ball leaves the pitcher's hand, if he is to have any hope of hitting it. The batter may be attempting to predict where the ball will go from the nuanced movements of the pitcher's hand/arm before the release. The object models that we use to project and otherwise manipulate spatiotemporal forms in short-term memory are presumably encoded as procedures in implicit long-term memory (discussed further below under *Imagining Spatiotemporal Forms*) perhaps in the basal ganglia and cerebellar systems.

Once the situated forms in the most recent short-term memory map have coalesced and integrated with existing spatiotemporal forms spanning other maps, they are immediately and unconsciously projected. The conjecture here is that they project onto a privileged *realtime map*, structured very much like a short-term memory map. The content of the realtime map corresponds to our experience of "now". By retaining a short sequence of active short-term memory maps and a projected realtime map, perception opens a window (ie, an experience of "now" lasting for about a second, and an experience of the recent past lasting for about 10 seconds) on an internal world of enduring, more-or-less stable, spatiotemporal forms.

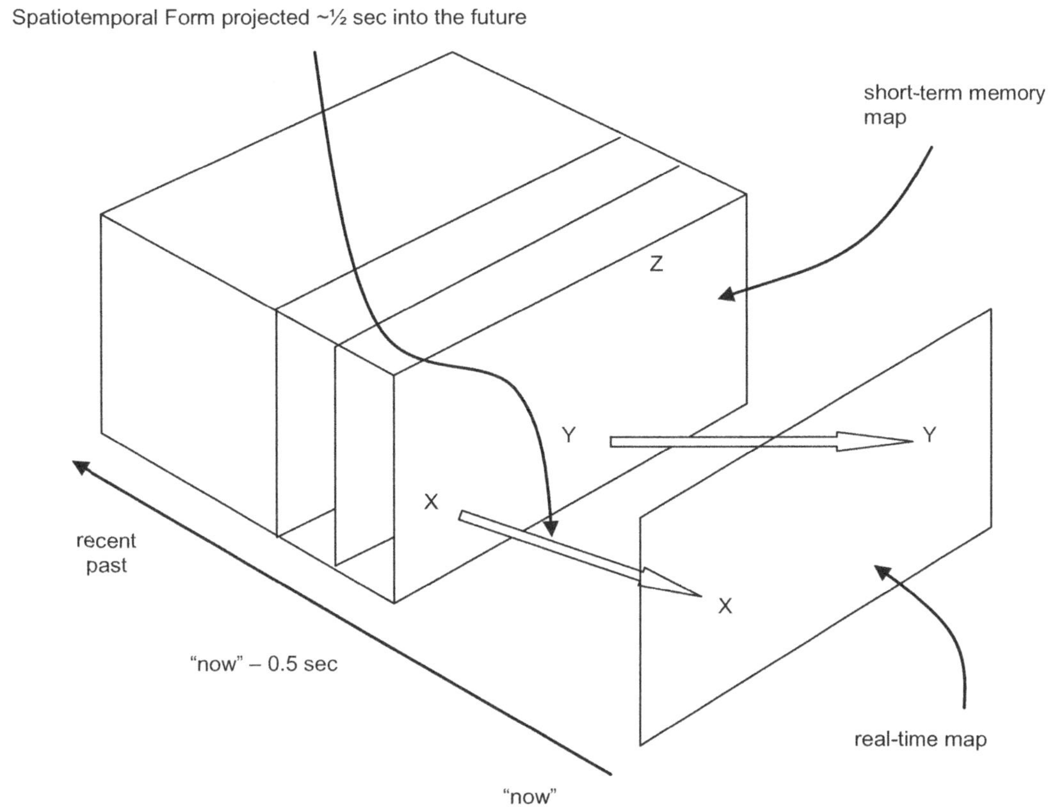

Primitive States and Events Associated with the Spatiotemporal Form

It is temporary, but persisting, spatiotemporal forms in short-term memory that provide the basis for follow-on perceptions of enduring environmental objects which have *states* and undergo changes, commonly known as *events*. If the experience of perceptual objects is grounded in spatiotemporal forms, the experience of perceptual states or events is grounded in changes to the features associated with those forms. Objects, states, and events therefore emerge as distinct, but correlated, experiences of a tracked perceptual spatiotemporal form. fMRI studies of brain areas MT and FEF show perception automatically segments the continuous stream of environmental stimuli into temporal units of meaningful states and events at both a course and a fine scale resolution—the rudiments of a hierarchical organization of perceived happenings (Hard, Lozano, & Tversky, 2006).

Looking at any single spatiotemporal form across the short-term memory maps shows the time sequence of its static and dynamic features (texture, size, orientation, shape, symmetry, position, layout, setting, affordance, spatial relationships) over time as the stimulus changes. For example, a rotating baton might suddenly stop rotating or change its direction of rotation. The change is marked by a value transition in one or more features between successive maps.

Temporal Sequence of a Characteristic's Values (for a situated form)

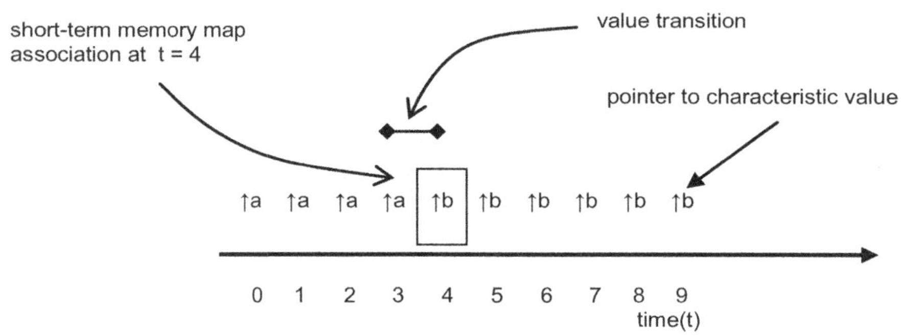

Such a transition marks a primitive event. Primitive events contrast with primitive states, like the continuing texture or shape of the baton, where the spatiotemporal form's features from map to map remain constant. (The term "state" refers to the stable features of a spatiotemporal form over several seconds in short-term memory. The related term "quality" is reserved to refer to a more-or-less permanent description applying to a perceptual object in long-term memory.) There are also other varieties of primitive event. In a transition event, the feature jumps from one value to another (eg, a light turning on), but the value may also immediately return to the base level (eg, a flash of light—one event, not two). We also seem to register a primitive event, possibly at the coarse level, if the initial transition lasts for a brief interval before returning to base level (eg, an apple falling from a tree, where the relevant feature is the apple's velocity), although such an event might also be viewed as two separate, but connected, transition events depending on the context. Primitive states and events, which we are normally capable of apprehending at a glance, are registered within the confines of short-term memory.

In both philosophy and physics it is common to give states and events ontological primacy over objects, such that the existence of real objects becomes theoretically debatable. Objects (ie, individuals) can be effectively excised from logic completely, as if they functioned only as props that merely simplify statements describing the world (Quine, 1953). Similarly, the field equations of physics make no mention of objects (except as convenient interpretations). Only the event-like changes and states of points in spacetime are ultimately relevant in describing reality. In perception however, the registration of objects, states, and events are unified through the associations of a spatiotemporal form with a set of features—its features. The experience of the enduring object having feature states and undergoing changes possibly reveals a subtle evolutionary consequence of a design that affords a "good enough" representation of the environment for our interests, without being exactly correct. In any case, primitive states and events constitute a vital perceptual construct, even if they do confer ontological status on objects by default.

Imagining Spatiotemporal Forms

Most people have the ability to imagine spatiotemporal forms, even after the direct experience has long faded. We can imagine our child's face, think the notes to *Row, Row, Row Your Boat*, imagine the smell of a rose, or visualize kissing a frog. Imagination, at least in the mind's eye, seems to take place in a 3D panoramic semi-geometric-topologic mental space modeled after physical space, and involves manipulating the spatiotemporal forms either recently perceived there or reconstructed there from long-term memory.[210] The integrated spatial maps associated with our experience of the environment may reside in the posterior parietal cortex and hippocampus, based on structuring information received from the prefrontal cortex (Kandel, 2006; Baars & Gage, 2007).

There is a good deal of controversy over the representational architecture of visual imagery. Some researchers like Kosslyn propose a "depictive representation" analogous to an array or network of locations with object representations linked to those locations, as with pins on a map (Kosslyn, Ball, & Reiser, 1978). Other researchers like Pylyshyn say the representation is more symbolic and knowledge based, and not like physical space (Pylyshyn, 2003). But whatever the underlying architecture, visual forms are evidently subject to a variety of imaginative transformations. We can imagine them growing larger or smaller, getting closer or farther away, moving around, rotating, changing features, articulating, interacting with each other, and even popping in and out of existence. We can also change our own point-of-view toward them, as when imagining getting in a car. It is as if we are watching or acting on a mental stage—ie, a visual buffer (Kosslyn, 1980) or Cartesian theater (Dennett, Consciousness Explained, 1991)—where we control the players and action, moment by moment.

The imaginary action may result from the application of procedures encoded in implicit long-term memory to the spatiotemporal forms represented in short-term memory. When the procedures run, they perform transformations on the forms very similar to the transformations of objects typically observed in the environment—akin to a simulation. The resulting transformed spatiotemporal forms are probably then experienced in the realtime projection map. Brain scan experiments (Le Bihan, et al., 1993) have further demonstrated that visual imagining consistently activates the same brain regions that are involved in visual perception, when the tasks are the same. Such results suggest imagination may use the same short-term memory structures as perception, and lend some support to the more global proposal that one of the main goals of perception is to build internal models which can be "run", as if they were real, to go beyond what is manifestly present or apparent.

---

[210] Our mental organization of physical space seems to have a gender bias. The male bias is for a more geometric layout, while for females the bias is toward the use of landmarks for reference (Lawton, 1994).

Features & Aspects

A spatiotemporal form in short-term memory constitutes a summary of the information garnered from its underlying figure over a number of seconds. Its situated features are presumed to include static, dynamic, and temporal representations of texture, size, symmetry, orientation, movement, shape, position, layout, setting, 2D parts, and perceptual affordances. Some of these forms, under the right conditions of long-term potentiation, will convert into long-term memory representations (probably in the temporal and prefrontal lobes, mediated by the hippocampus). Which forms contribute to long-term memories probably depends on their bottom-up salience in context, as well as on top-down expectations, and how much attention we pay to them. Also, they must have persisted in short-term memory long enough (without being overwritten) for the conversion process to operate.

A feature of a spatiotemporal form in short-term memory contributes to what will be called a corresponding *focal feature* in long-term memory. A focal feature combines small variations in values of a spatiotemporal feature over time into a single focal value. A focal feature, like a spatiotemporal feature, is an association; but rather than pointing to a transient particular situated value it points to a more permanent focal (average) value for the feature. It condenses recurrent values into a "typical" value.

Focal feature values build up over the duration of a single encounter and over repeated encounters with the same environmental object, or with numerically distinct but perceptually indistinguishable objects. They are also commonly forgotten over time in a piecemeal fashion (Hintzman & Ludlam, 1980). The very first encounter with a new environmental object necessarily just duplicates the feature values of its current spatiotemporal forms, creating an *aspect* of the new object corresponding to its underlying figure. The aspect binds together all of the concurrent features for a single spatiotemporal form that are remaining relatively constant in short-term memory (possibly at both a coarse and fine time scale). It represents the environmental object as a conjunction of focal values, perceived from one perspective in one sensory modality. A subsequent update of the focal features for the same environmental object will then either blend the current feature values with the existing focal values for the initial aspect by averaging (Strauss, 1979) or create a second aspect, binding alternative focal values. A significant value change in the features of a spatiotemporal form generates associations to new focal values, thereby generating a new aspect. Focal features which are repeatedly updated and activated at the same time probably establish long-term synaptic connections between the aspect and its focal values.

Repeated exposures to the same aspect of an environmental object also seem to encode *distribution measures* of its features. We generally have some idea of how different an aspect can be and still reasonably be regarded as the same aspect—ie, we know the allowable variance of its features. For example, the allowable size variation for a side view of my cat Angus (now that he is fully grown) is less than the allowable speed variation (during standing, walking, running) for the same related aspects. Together, a focal value and its variance describe a focal region for each feature of an aspect in long-term memory.

Throughout one encounter, simple object tracking provides a continuing reference for the updates. Over separate encounters, higher levels of object modeling and tracking (both perceptual and conceptual) will be required to reestablish reference to the same object. However, the relatively large number of static, dynamic, and temporal features, combined with high resolution value spaces, may generate sufficient distinctions for detecting recurrent aspects based on similarity in many circumstances. Even one glimpse of a stray dog may be enough to lay down a long-term memory perceptual aspect of the dog, which might be used to recognize it on the following day.

The term "aspect" is perhaps comparable to the term "exemplar" as used in cognitive science to describe a locus of information or an encounter concerning a single individual (Medin & Schaffer, 1978), except that an aspect contains no conceptual, only perceptual, information, and relates to only one point-of-view of the whole individual. Perceptual information specifically excludes kind or type information, so my aspectual representations of my cat, Angus, do not contain the information that he is a cat—or even male.

The focal values for the same focal feature of a single environmental object are mutually exclusive, since they depict it in different ways or from different viewpoints. A particular songbird's chirping calls, in contradistinction to its melodious song, exemplify two auditory focal values derived from the same individual. In the previous example of the cube, which can present either three, two, or one face depending on viewing angle, there are three corresponding focal values for its shape. A particular spatiotemporal form, constructed from any particular view of that cube, will usually have a shape feature value matching one of these three focal values. The visual spatiotemporal form will seem familiar, if its shape feature value is close to the focal value of some aspect of the cube. At this stage, the alternative focal shape features still describe the cube's shape without explicitly representing a 3D geometric model of it. The object modeling procedures in implicit long-term memory are presumed powerful enough to appropriately warp, rotate, and smoothly switch between focal values of its various aspects, as when imagining a rotating object.

All of the focal values for a perceptual aspect are equally valid registrations of that aspect, although some may be observed more frequently than others and on different occasions. While not all of the focal values for an aspect will be perceived on every occasion presenting that aspect, they all should be concurrently present. When my cat Angus' shape resembles a curled up ball (eg, when he is sleeping—one aspect of Angus), his movement and sounds are usually nil. When he is rubbing against my leg (a different aspect), he is usually meowing softly. Aspects, therefore, are the source of expectations about a spatiotemporal form, leading to heightened sensitivity for detecting missing features and, as a result, better object tracking across multiple encounters.

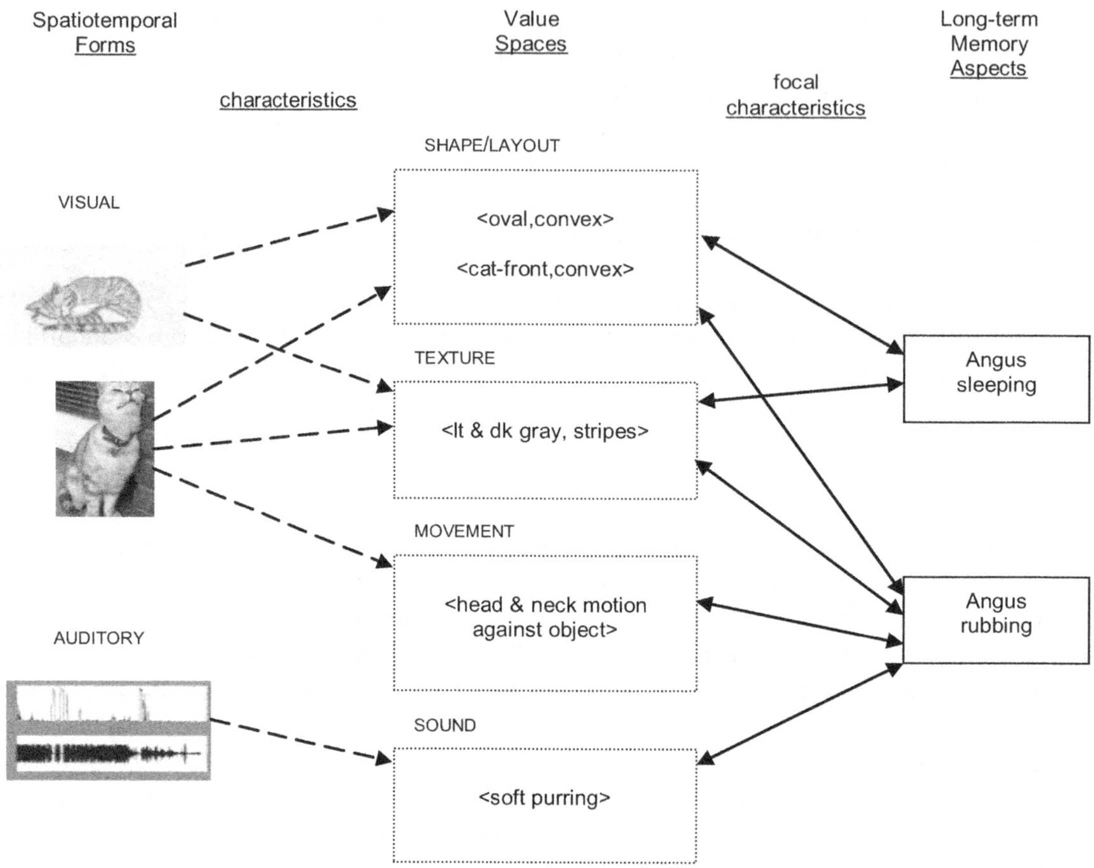

Perceptual assessments of the similarity between a spatiotemporal form and long-term memory aspects give rise to *prototype effects* (Rosch & Mervis, 1975) for an aspect. Prototype effects for an aspect imply that some spatiotemporal forms are better examples of an aspect than others. First, of two spatiotemporal forms, one may be judged as more similar to a given aspect, because its feature values more closely match the focal values of the aspect. Second, of two spatiotemporal forms that each precisely match an aspect's focal values, one may match more focal features than the other. Both factors probably contribute to deciding how to interpret the spatiotemporal form.

Aspectual Parts & Aspectual Features

The foregoing speculative discussion of long-term memory aspects and focal features is also oversimplified, because it assumes that aspects have no part structure. Yet most aspects do possess salient 2D parts, as derived from their underlying figural elements. In fact, salient parts are aspects in their own right and, when present, may receive representation too. They are described by focal features, just like a whole aspect, but at a lower level in long-term memory. The diagram below illustrates how the part structure of an aspect of the palm tree in my neighbor's yard might be represented—its two main parts being its tall trunk and the top foliage. The distinct part aspects are bound to the holistic aspect through mereological associations in long-term memory mirroring the transient mereological associations between figural elements and a whole figure in short-term memory. The layout of the parts, as specified in shape/layout space—ie, that the foliage is above the trunk—is only suggestive in the diagram below. The mental representation of layout may or may not have an explicit (spatial) map-like expression in neural arrays (Koch, 2004), although so-called place cells have been detected in the hippocampus of rats (O'Keefe & Nadel, 1978). Proximal locations among objects could also become represented by different types of neural connections.

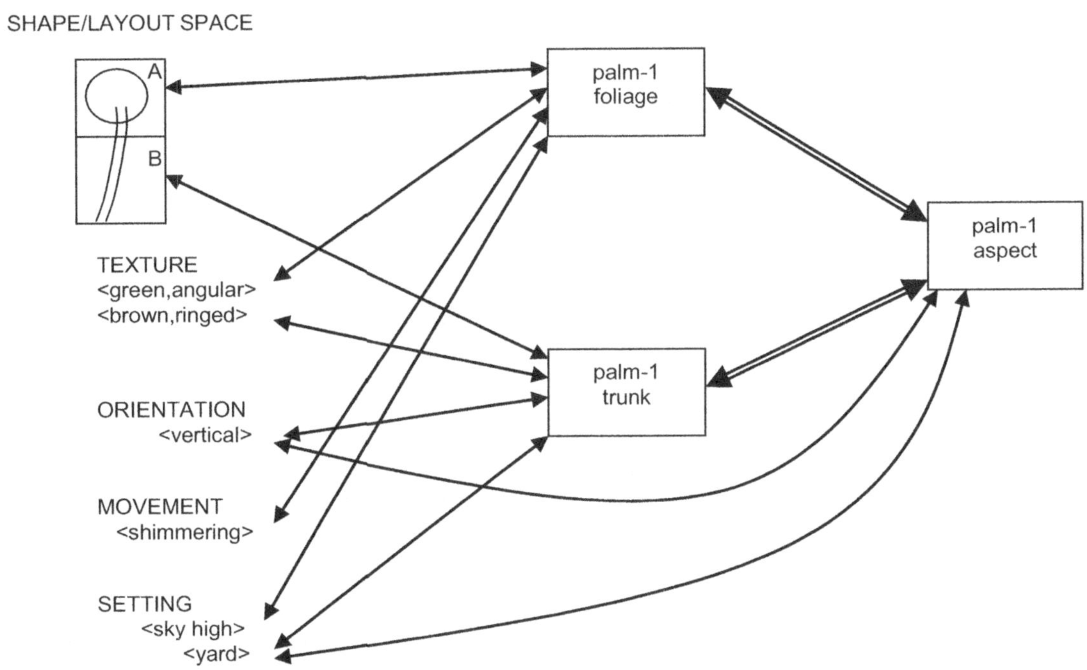

As you pay more attention to an environmental object, its perceptual parts become more elaborated in long-term memory. For the palm tree, noticing the splaying of the base of the trunk, or the fan-like shape of the branches at the top, establishes additional "parts of parts". Such additional parts take their place in a hierarchy of aspectual parts headed by the holistic aspect. The structure of the aspectual part hierarchy (in addition to focal features) is a powerful diagnostic tool for distinguishing a holistic aspect, since it will normally specify only one out of many potential combinations of parts.

We also seem to represent isolated focal features as dissociated aspects in long-term memory—ie, independently of a specific object. Kay and McDaniel have compiled evidence for the shared representation of focal values of certain colors spanning different cultures (Kay & McDaniel, 1978). Certain common shapes (eg, square or circular), textures (eg, checkered or polka dotted), settings, and other features probably achieve independent long-term memory representation, as well, based on genetic tendencies and perceptual experiences. An *aspectual feature*, then, is a minimal case of an aspect pointing to a single focal value. It functions just like a holistic aspect, but does not bind multiple focal values together. Such representations may facilitate the analysis of a new environmental object if insufficient data is available for an immediate match to an existing aspect; at least something about the encounter can be remembered.

The mix of holistic, part, and feature aspects in long-term memory, each associated with a unique selection of focal values, may serve as the foundation on which recognition rests. Successfully matching the features of a spatiotemporal form to any of these kinds of aspect results in recognizing something familiar about the spatiotemporal form. The matching process is assumed to use the same implicit procedures used by imagination to rotate or otherwise warp the relevant aspects to achieve a viable match. Recognition, then, is akin to hypothesis testing, rather than incremental construction, as was the case for early perception—broadly speaking, it is a top-down, rather than a bottom-up process. The matched aspect has become an interpretation of the underlying spatiotemporal form, grounded in what has been experienced before.

Aspects represent yet another step up the ladder of multidimensional value spaces. The consolidation of focal features for describing recurrent spatiotemporal forms in terms of aspects effectively produces an *aspect space* of high dimensionality, built out of the underlying feature spaces. An aspect of an environmental object is then a point in aspect space, such that similar aspects are represented by points that are close to each other. Assuming only useful, diagnostic focal features are retained for aspectual representation, the stage is set for perceptual clustering of aspects (Rogers & McClellan, 2008). In this way, prototypical daffodils, prototypical fingers, and prototypical shades of red become regarded as similar. Aspects may be the basic vocabulary perception uses to describe environmental objects, substances, and conditions of all sorts.

Murphy and Medin used the word "coherence" to encapsulate the idea that we do not perceive environmental objects as having random features, but rather recognize that an object's features spring from an underlying nature (Murphy & Medin, 1985). The world appears structured to us because certain feature values consistently occur together, and these correlations are useful guides for interacting with objects. Rosch's examples of creatures that have feathers, but not fur, are likely also to have wings apply more to conceptual than perceptual features, as the terms are used here, but the application to perception is straightforward. A perceptual example might note the correlation among having a green fine texture, being spread out over a large flat area, being soft/cushiony to the feel, etc, as an aspect of a grassy field—most grassy fields have these features. In this context, that the world is structured means that it is feasible to distinguish different aspects of it, such that the various aspects do not overlap too much. In other words, different aspects that we are interested in do not clump together (at some coarse or fine scale) in the high dimensional aspect space describing their focal features. Grassy fields are commonly distinguished from other similar things like green painted parking lots or even AstroTurf, because there are so many focal features available to make fine perceptual distinctions.

Moreover, because multiple focal values for an aspect tend to co-occur, they mutually reinforce each other as reliable indicators of different aspects.

The discussion of aspects has emphasized their development through tracking a single environmental object. But an aspect could hypothetically match any number of very similar objects. Some very similar objects are difficult to distinguish on purely perceptual grounds, and will therefore be lumped into the same aspect. For example, supposing a dollar bill has two aspects (a front and back side), it is probably impossible to tell one newly printed dollar bill's aspects from another without special equipment or noting serial numbers (a conceptual task). A coordinated set of perceptual aspects, therefore, is better seen as covering the perspectives on a class of indistinguishable individuals, as well as on a single individual. A degree of variance arises from the unavoidable limitations on perceptual discrimination. If the aspect truly corresponds to a single recurrent individual, it should eventually lead to a *uniqueness* determination, but this is a conceptual function. An aspect covers encounters with environmental objects without committing to the *identity* of those objects at this stage of processing. From a perceptual standpoint, we normally treat recurring indistinguishable individuals the same—oblivious to imposters and our own perceptual limitations. If a situation calls for enhanced discrimination, we pay more attention to latent focal features and parts, thereby refining our aspectual details or call on more sophisticated cognitive tracking techniques.

An aspect, then, can be understood as a segment of the environment derived from direct experience (expressed by its focal features and values), whose degree of specificity (ranging from specific to general depending on the number of focal features and 2D aspectual parts), is tailored to need. Highly specific aspects tend to be associated with unique (or indistinguishable) individuals, and highly general aspects with environmental features and conditions. Such generalized perceptual situations may be experienced by infants at two months, who probably do not yet experience their own mothers as unique individuals. Their perceptual aspects are probably assimilating successive encounters into nontyped perceptual aspects like "more Mama, more milk, more mouse" (later interpreted as a conceptual individual, substance, and kind, respectively, at 10-12 months), since their initial responses do not depend on *inference* which requires the abstract idea of uniqueness to function properly (Millikan, 2000). But the disparate unconnected aspects in infancy (eg, good mother, bad mother) need to become co-identified later if normal development is to proceed.

The Perceptual Object

As we perceptually track a rotating and/or slowly moving environmental object, it will often present different aspects in succession, which becomes cause for uniting them into an enduring object representation. Tracking can even operate over short time intervals, when the object is out of sight, but not out of mind. A coyote can track a rabbit even if it temporarily disappears behind a rock, or if the scent trail it has been following temporarily becomes confused. (However, after a relatively short interval, perceptual tracking must give way to conceptual tracking.) In general, the ability to perceptually track a moving object probably depends on the predictability of the changing position and values of its varying perceptual aspects. The sequence of aspects becomes unified into a single tracked object when the rate of change is not so great as to break the continuity of the single object experience.

A number of experiments seem to indicate that a tracked object is more than the simple sum of its features (or aspects, which are derived from features). While the nonspatial features processed in the "what" pathway are criterial for deciding how the object appears, the spatial location of the object, processed in the "where" pathway, seems to provide the hook on which features hang. The "where" pathway is probably ultimately responsible for our experience of the persistence of a single perceptual object in spite of its normally changing features. Even when all features are fully registered, there is no experience of an object until they become bound together at a single locale (Auyang, 1998).

The different aspects of a perceptually tracked environmental object, registered in different modalities, on different occasions, from different viewpoints, consolidate into a long-term memory *perceptual object* representation. For me, my cat Angus is a unified perceptual object linked to many perceptual aspects characterizing his appearances, movements, 2D part structure, home and yard settings, noises, smells, etc. I can call these long-term memory aspects of Angus to mind, individually, into short-term memory, if needed. However, the total perceptual object representation itself is probably not directly perceivable, since it encapsulates multiple *complementary* aspects in each modality. A perceptual object is like a catalog of distinct concrete aspects where one aspect is currently highlighted, and not like a Picasso painting where all the different aspects of the figure are integrated into a concrete whole. True aspect integration is more likely a conceptual phenomenon.

The perceptual object representations for vision seem more complex than for the other modalities. When we look at a typical environmental object, we have the impression of perceiving a complete 3D object, rather than an assortment of 2D aspects. We seem to have a sophisticated internal spatial model of how the aspects go together to make up the 3D object—unlike, say, the various sounds that a car makes (although sounds may be organized according to temporal sequence). The visual perceptual object probably includes a 3D object representation along the lines of Biederman's theory of geons (Biederman, 1987), where a geon is an elementary volumetric shape, as shown below.

The elementary shapes can be oriented, sized, and warped according to a few relevant parameters, and then arranged to make up a composite perceptual object. One potential simplification of the geon theory might be to reinterpret volumetric geons as pure surface "geons", which may either completely enclose a volume (like a brick) or not (like a hole in the ground). With this simplification, geons are a relatively straightforward idealization/completion of the 2D surfaces (oriented in 3D space) of lower-level visual aspects.

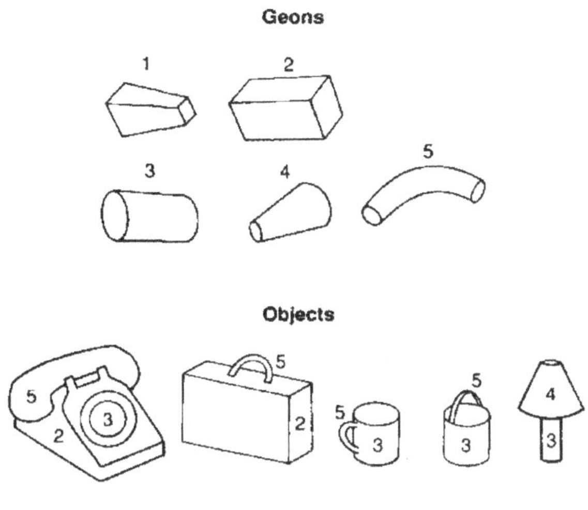

A perceptual object exhibits prototype effects, just like its aspects, but for a different reason. Some of an environmental object's aspects are better exemplars of that object than others, because one aspect may be more *diagnostic* than another. For example, the side view of a horse is more diagnostic than a top-down view (Palmer, Rosch, & Chase, 1981), because the side view conveys more information relevant to distinguishing it from other objects. The presence of prototype effects does not suggest that a perceptual object represents a prototypical or ideal or best example of an environmental object in long-term memory. Again, the perceptual object encompasses a cluster of diverse aspects which together cover its observations, as illustrated below. It is a composite of correlated aspects, where each aspect likely possesses its own diagnostic weight based on how successful it has been in the past for object detection.

Konorksi has put forward a theory called the gnostic (or grandmother) cell theory, in which a particular neuron comes to signal the presence of one particular individual (Konorski, 1967). Recent evidence for this idea was collected by Kreiman, who monitored a single neuron in the amygdala of a patient while showing him pictures (Kreiman, 2001). The neuron seemed to respond only to pictures containing Bill Clinton: a sketch of him, his portrait painting, and a group photo. Similar pictures of other male celebrities or presidents did not activate the neuron. Similar results were obtained for the hippocampus, with single neurons firing exclusively for famous buildings like the Tower of Piza (Quiroga, Reddy, Kreiman, Koch, & Fried, 2005). This degree of specificity for a single perceptual object under a wide range of observation conditions underlines the precision available through multiple aspects. However, such neurons should not be interpreted as indexing (or directly referring to or standing for) the particular individual. It is merely that repeated exposures to a (presumed singular) environmental object have fine-tuned its alternative aspects, so that in most situations, only Bill Clinton consistently matches them. A good impostor would still fool the detection process. Although a strict interpretation of the grandmother cell hypothesis (ie, a cell responding to only her aspects) does not appear to be a feasible way to represent perceptual objects (there are too many aspects for each individual vis-à-vis the number of brain cells), a modified approach where the same aspects are shared among perceptual objects seems more reasonable.

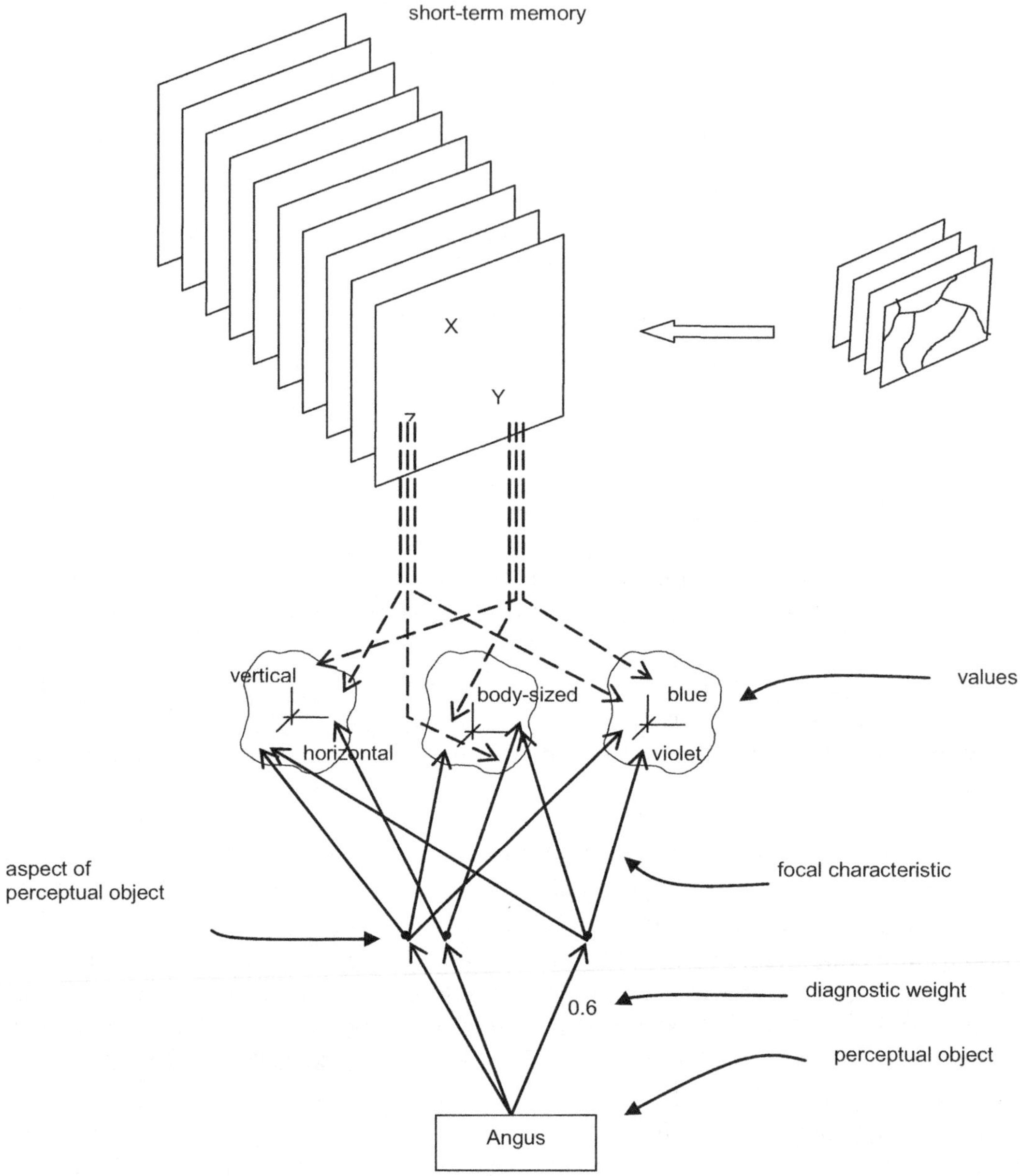

The assumption that a perceptual object is unique seems built into its functionality as a model of a persisting individual thing. We create two perceptual objects only if we perceive them as different objects. If we discover that two perceptual objects are really one and the same, they are readily merged into variant aspects of one perceptual object. Such a merger probably occurred when it was discovered that the morning star was the same object as the evening star (ie, Venus). The setting (morning or evening) then became the distinguishing feature for the variant aspects.

Some philosophical questions of identity relate to the idea of the perceptual object. Did the ship of Theseus remain the same ship even though every one of its planks was replaced over time? Perceptually speaking, since one could in theory track the ship as a perceptual object over time, the intuitive answer would be yes. Only one perceptual object is involved, and its various compositions at different times are simply different aspects. Metamorphosis is an even more extreme transition, but again, there is only one perceptual object, which happens to undergo a radical aspectual transformation. At a fundamental perceptual level, then, it is our ability to track an object that fixes identity, rather than its unique set of features. Perceptual features facilitate tracking, but effective continuity is criterial.

The construction of the perceptual object facilitates the socially adaptive shift from private to public experience. While a spatiotemporal form in short-term memory arises out of the subjective particularity of direct experience in the here-and-now, a perceptual object in long-term memory provides a basis for shared experiences. A perceptual object's relative permanence and stability allows different people to compare and contrast its aspects, and otherwise coordinate their representations, even when its spatiotemporal forms are no longer present. As long as we share enough aspects for an object to reliably recognize it in the current context (even if we are using different aspects), we can usually assume we are talking about the same object. The direct grounding provided by personal experience can get lost in public dialog, but we eventually come to trust that others possess perceptual object representations roughly similar to our own, based on the simple fact that the same object can be characterized in different ways.

Perceptual Schemas

The unabridged idea of a perceptual object, while useful as an expository construct, may be something of a misnomer. Only during continued perceptual tracking (including brief interludes) is it valid to assume that the perceptual object representation corresponds to a unique environmental object. Over isolated observations, the perceptual system cannot warrant that the same (identical) object is appearing again, as opposed to a similar object. Perception is based on how things appear, rather than what they are. For this reason a perceptual object representation, being composed of aspects, is probably better thought of as a grouping schema which encompasses a variety of alternative presentations. Highly clustered schemas—eg, ones exhibiting aspects with a large number of features and low variance—correspond to (indistinguishable) environmental objects, while those containing aspects with fewer features and high variance correspond to broader classes of environmental objects (and parts and features). A so-called perceptual object then is only one kind of tightly clustered perceptual schema.

In this approach a *perceptual schema* inherently represents a perceptual grouping of aspect-like experiences, where the limiting case is one aspect. Among other things a schema likely includes 1) pointers to the aspects which characterize its observations (probably including pointers to evaluative aspects in the limbic system for an object like a gun); 2) direct focal pointers to feature spaces that describe the range of aspects holistically in an overall sense (eg, overall shape, color, texture, orientation) independent of aspectual/part structure; 3) pointers to other schemas with which it has a consistent spatiotemporal or sensorimotor relation (eg, a dandelion can be in a lawn, a hammer is gripped in a certain way), and 3) a *3D perceptual model* integrating its 2D aspects (eg, a person's 3D body shape with legs, arms, head, and torso configured and articulated in a certain way). Perceptual schemas evidently can associate a remarkable amount of detail that is distributed widely across the cortex and limbic system.

For example, a doorknob schema would point to at least one aspect depicting the appearance of the doorknob from a typical down-looking vantage point, along with aspectual perceptual features like texture (smooth gold or chrome focal point in texture/color space), size (spatial extent or region with respect to some reference, probably the human hand), and shape. Its potential affordances would likely include being turnable, lockable/unlockable, pullable/pushable by hand. These kinds of behaviors would in turn be linked to motor schemas in the motor cortex. Its structural model would reference being part of a door, with subparts corresponding to a baseplate and knob, with screw holes and screws in the baseplate, and the lock button and keyhole as subparts of the knob (perhaps constructed from 3D geons). As already mentioned, the part structure of an environmental object may include an intricate structural hierarchy. However, the frame-like summary organization of a schema is distinct from the mereological organization of its 2D aspectual wholes, parts and subparts, which again is different from its 3D structural model representations.

As another example of two closely related but distinct perceptual schemas, consider the wire Necker cube. The two visual interpretations are conjectured to arise from two possible object schemas triggered by the stimulus. Assume a cube schema has six components, one corresponding to each surface, represented as (front, back, left, right, top, bottom). The first Necker schema then exhibits three of those surfaces corresponding to (1,0,1,0,1,0)—1 being visible, 0 being occluded; while the second schema's surfaces are (1,0,0,1,0,1). Without further perceptual clues, both are legitimate interpretations. In fact there is a small family of such component triples specifying other 3D views of a cube under various rotations. The remaining combinations (symbolizing views like back, left, bottom) are impossible, and wouldn't have schemas.

Perceptual Groups

Perceptual schemas are a versatile representational tool. They can unify appearances of unique recurring objects and parts (eg, Fido or just his tail), environmental conditions (eg, twilight), perceptually indistinguishable objects (eg, penny), and even single feature values (eg, pink things). But this is just the beginning. In yet another expression of this versatility, perceptual schemas support the representation of a group of objects perceived together.

Piaget's pioneering studies with children uncovered preferential strategies for grouping individual objects that differed from that of most adults. Adults (at least in Western societies) seemed to prefer grouping items according to taxonomic type (eg, a robin and a penguin go together because they are both birds) or origin (eg, humans and chimps go together because of a common evolutionary ancestry) or function (eg, a knife and scissors go together because they can cut), while young children prefer grouping according to appearance (eg, a penny and a washer go together because they are both round and about the same size) or theme (eg, a key and a lock go together because a key opens a lock). Today it is generally recognized that both children and adults use a wide variety of grouping strategies, depending on the task. However, there is still a fundamental divide between these strategies, based on whether a grouping employs conceptual or perceptual representations. Grouping strategies like type, origination, and function are conceptual, whereas theme and appearance-based groupings are perceptual.

The perceptual basis for grouping derives from the (focal) features and part structure that are associated with the aspects of perceptual objects. The appearance-based strategy uses (one or more) matching features like texture/color, shape, size, layout, or movement for grouping, while the theme-based strategy makes use of setting, affordances, and motor actions (ie, objects interacting/appearing together in the same place at the same time). This aspect-based explanation therefore anticipates perceptual groups based on spatial relationships (eg, objects located inside a box), 2D parts (eg, objects that have the same aspectual parts), affordances (eg, objects that are sitable), layout (eg, a constellation of stars depicting a familiar shape), and actions (eg, ways to throw a ball). All such perceptual grouping strategies rely on group schemas below the level of conception.

Take an example of a pair of pennies. Looking at two pennies on a table, we perceive them as distinct individuals with matching visual aspects, even though their various aspects may not be presented at the same time (eg, if one is heads and the other tails). In order to remember such a grouping, we probably do not activate two distinct perceptual schemas of *penny*, but rather build on the hierarchical structuring for parts, likening each penny to a perceptual "part" of a new two-penny group schema. The two-penny group schema can then point to a configuration in layout space (specifying their relative location), which in turn has two pointers to the same penny schema (or observed aspects thereof). The two-penny schema takes on aspectual features like the penny schema, but becomes interpreted as a group, as in the diagram below. The group interpretation is probably based on the relative connectedness of the object schemas pointed to by the layout configuration. Connected items come to signify parts, disconnected items come to signify group members.

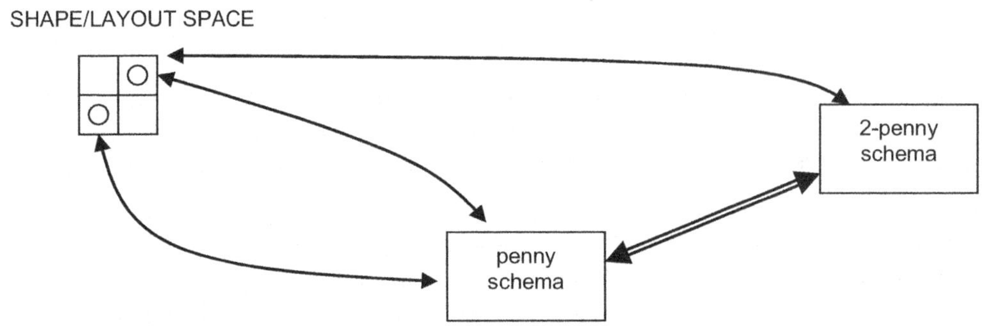

Experiments show small perceptual groups are *subitized*, rather than counted (Kaufman, Lord, Reese, & Volkmann, 1949). That is, they are grasped in a glance, rather than consciously enumerated. Subitized groups seem to be limited to 4-5 items, which is consistent with the previous spatiotemporal form limitations for visual short-term memory. Grasping the numerosity of subitized groups likely involves a kind of unconscious scanning process (akin to counting for larger groups), but the scan is very fast (about 50 ms per item), as compared with counting (about 250 ms per item). Grasping the approximate numerosity or layout of larger groups seems to involve global summaries of group configuration properties as opposed to individual item features.

Perceptual grouping processes operate over time as well as space, although we often interpret (rightly or wrongly) members of a similarity group encountered over time as a conceptual category. (A "group" refers to a class united by perceptual similarity—either similarity to each other or to an exemplar, whereas a "category" refers to a class united by shared criteria.) Over time, multiple encounters with similar individuals tend to generate open-ended *natural perceptual* groupings, whose features are the causal result of physical and biological principles. To a first approximation, the perceptual features of bears are distinct from those of wolves, although distinguishing timber wolves from gray wolves requires finer resolution of features. But perceptual grouping will also likely group porpoises and sharks together when viewed from a boat, requiring subsequent conceptual criteria or more refined perceptual criteria to distinguish them. Those groupings with the most distinguishing characteristics have been described as "basic level" (Rosch, 1978), because they give reliable medium level groupings based on many distinguishing characteristics. Groupings at a higher level (eg, animal, furniture) or at a lower level (eg, timber wolf, rocking chair) possess fewer distinguishing characteristics.

The development of high, basic, and low level perceptual groupings probably depends on an innate capacity for perceptual clustering. Clustering methods, as studied in machine learning and some neural network paradigms, take a set of object instances described by various feature values, and automatically derive a hierarchical organization scheme for the instances. The best organization is the one that minimizes the overlap between clusters of instances at each level of the hierarchy. In an analogous fashion, it appears that infants initially develop broad object groupings corresponding to animals, vehicles, humans, etc, and only later differentiate these perceptual groupings into basic level subgroups like dog and cat, car and airplane, and finally to lower-level groups like collie and terrier, sports car and van (Mandler, 2004). But these perceptual clusters should not be confused with conceptual object categories. In fact, using words like "vehicle", "car", and "dog" for these data-driven clusters is misleading since linguistic word meanings are more often tied to conceptual, rather than perceptual structures, and such words are normally used as category terms. Perceptual grouping often corresponds to conceptual categories, but not always (whales are mammals, not fish). Perceptual clusters are based only on perceptual features (a car has wheels, goes fast, etc), not on conceptual properties having to do with its use as a common means of ground transportation. Similarly, the animal cluster probably has something to do with subtle characterizations of self-initiated movement, irregular movement, contingent movement, and interaction at a distance (Mandler, 2004).

Prototype effects (Rosch, 1975) are also in force for groups, under the governance of a perceptual schema. The perceptual schema for a similarity group acts like a *prototype* for the group. A prototype, however, is not a best example, or even an ideal example of the group, along the lines of a spatiotemporal form possessing a particular constellation of feature values. Rather, a prototype represents the total (often infinite) extension of the group implicitly through the specification of acceptable variations in its features (via different aspects), thereby accounting for the customary family resemblance among members.

Perceptual Relations

Perception also builds relational schemas characterizing the elementary spatial and temporal relations between objects. While the schemas of perceptual objects and groups encapsulate recurring focal features, relational schemas link other schemas together. For example, perceiving that the moon is above a cloud involves a relational schema (above) linking two object schemas (moon and cloud).

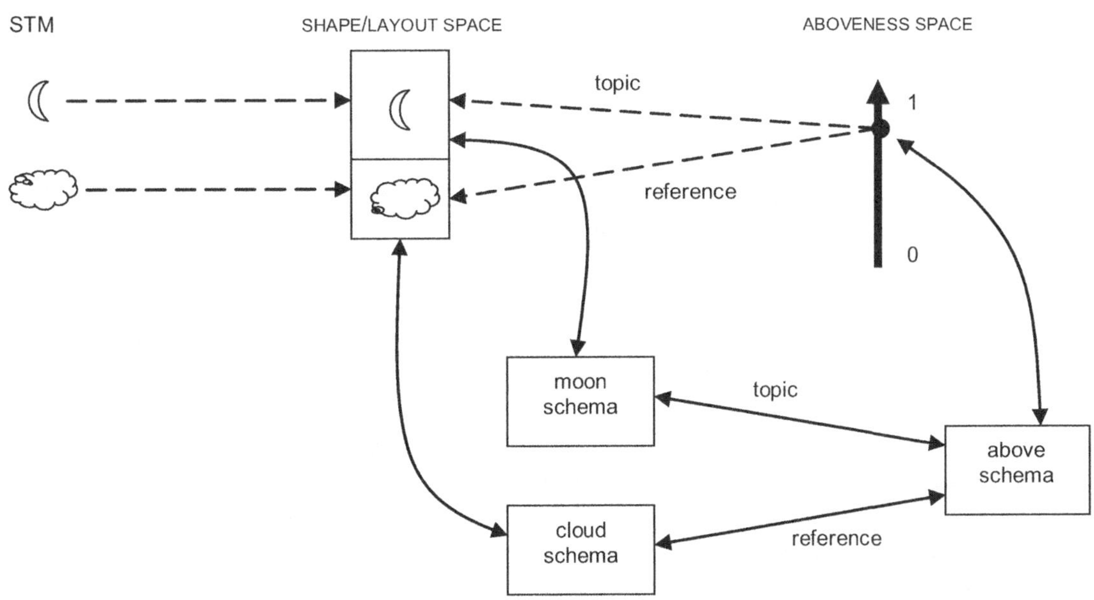

One way of organizing such relationships in long-term memory is via a feature space specifying the presence/absence of the particular relationship detected in shape/layout space (for vision). Accordingly a certain value of "aboveness" might become activated based on shape/layout space when distinct shapes are detected one above the other, just like a certain value might become activated in color space through a single colored shape. Assuming such transient object and relation associations become active together, they can then be transferred to long-term memory schemas. Long-term memory schema associations probably mirror the binary nature of the transient feature space associations of spatiotemporal forms in short-term memory, where one argument is a reference landmark and the other is the topic of the relation. Other possible perceptual relation schemas may include how objects can behave temporally (eg, a robin can transition from walking to flying), how objects can be contained, one in another (eg, coffee can be in a cup), or how objects may traverse a path in space. In general, the various kinds of relations encoded by perception are likely limited to spatial (eg, above/below), temporal (eg, before/after), and force dynamic (eg, push/pull) relations (Talmy, 2003). More abstract conceptual relations (eg, today's stocks closed above yesterday's high) are then analogical projections of basic perceptual relations (Lakoff, 1987).

Gentner has proposed that most relations are learned rather than being endowed genetically. The learning process is one of comparing repeated instances of the relation (eg, of one thing being above another), and aligning the roles of the objects participating in the relation (eg, the upper and lower roles)—the roles themselves are independent relations (Gentner, Holyoak, & Kokinov, 2001). In the current schema proposal, whereas a new object schema can be learned by consistent detection of its instance features, a relational schema might similarly be learned by detecting the correlation between its participants, and reifying the association. This explanation of the emergence of relations is sketchy, but it seems a plausible starting point.

The Recognition Process

If a spatiotemporal form in short-term memory can be correctly assigned to one or more preexisting perceptual schemas, we then have access to a great deal of useful, previously acquired information about it—information that may not be present in the occurrent spatiotemporal form. The added long-term memory information becomes the basis for a more informed analysis of the present, usually accompanied by a feeling of familiarity.

The rather specific feeling of familiarity that accompanies experiencing an object or situation we have experienced before derives from successfully matching focal features of an aspect. Aspects explicitly encode prior encounters and, when activated, directly stimulate limbic nodes responsible for evaluating those encounters. Perceptual schemas, on the other hand, are responsible for tying all of the various aspects of an object together, and therefore are required for correctly identifying the object perceptually. In this view recognition has two components: an aspectual component for matching a received spatiotemporal form, and a schematic one for interpreting it. These two components become dissociated when we can successfully match an occurrent object or situation, giving rise to the feeling of familiarity, but not have access to other perceptual information about it drawn from other aspects. Patients suffering from associative agnosia may experience a more permanent dissociation (McCarthy & Warrington, 1986).

Assuming aspects and schemas are the fundamental representational structures for perceptual long-term memory, full recognition would seem to involve both bottom-up triggering of relevant schemas, and top-down enhancements of early perceptual readings. Whatever salient cues are present in the spatiotemporal form act to trigger multiple schema hypotheses associated with those cues. Koch proposes an explicit *saliency map* (akin to an attribute map) for encoding how "conspicuous" a stimulus is in its immediate neighborhood (Koch, 2004). Those spatiotemporal form features associated with high salience become schema triggers (via their aspects). Once triggered, the schemas then actively enhance the sensitivity of the underlying forms and attribute maps to extract additional latent cues matching other schema aspects. This kind of unconscious bottom up attention to salience, distinct from conscious top-down focal attention, is constantly shifting from point to point, and difficult to suppress. Those schemas achieving better matches quickly inhibit the others until a single schema (or compatible set of schemas) emerges as the best interpretation of the occurrent spatiotemporal forms.

Occasionally the recognition process over enhances and sees things that are not there. In this case the downward projections from the schema are increasing the gain (ie, sensitivity) of the regions around the focal features of its aspects too much. Any small stimulus signal then becomes a confirmed hit for the schema, leading to unwarranted self-fulfillment of expectations or hallucinations in the extreme. Alternately, if the schema sets the gain too low, features that are present, but mixed with background noise, will not get processed. Effective recognition in varied contexts requires a delicate balance.

Reliable perceptual recognition takes time to develop. Young children seem to have difficulty making the fine discriminations that adults make, and it is all too easy to assume that children's' perceptual processes provide them with the same information as ours do. It takes several years of perceptual experience in a variety of contexts to form reliable schemas that include reliable features and exclude unreliable ones. For example, infants under one year typically confuse an object with a highly realistic picture of the object (DeLoache, Uttal, & Rosengren, 2004). A nine-month old may try to suck a picture of a bottle or put a foot into a picture of a shoe. These infants have developed a useful object schema, but it is overgeneralized. A competing schema of a "picture object" is not yet in place. Even up to three years children often make so-called scale errors, where they will try to sit on a doll's chair or slide down a tiny slide. Many four-year-olds likewise seem to have difficulty believing that turning a TV upside down showing a bowl of popcorn will not make a mess (Flavell, Flavell, Green, & Korfmacher, 1990). In the latter cases the children have apparently not yet tightened up the size and affordance feature value spaces of their schemas. There is further evidence that relational and group schemas develop later than object schemas, indicating that young children may not perceive some basic spatial and temporal relationships between objects and events that adults perceive readily.

Perceptual Schema Organization

As already mentioned, schemas naturally tend to become organized into mereological part/whole hierarchies as perception progressively constructs more global representations out of local ones. The long-term memory process composes perceptual objects out of aspects, and perceptual groups and relationships out of perceptual objects. Such hierarchies might more generally be termed *constituent hierarchies*, since the higher-level schemas include lower-level schemas as constituents.

Crosscutting the relatively deep constituent hierarchies are shallower *global feature hierarchies* involving the same schemas. A feature hierarchy is an organization of schemas according to shared global features, such that a lower-level schema shares (or inherits) its features with those higher-level schemas that are its ancestors. High-level schemas therefore possess relatively few features (eg, redness), while progressively lower-level descendants possess added features (redness + squareness, redness + squareness + checker-boardness, etc). Perceptual feature hierarchies enable a certain economy of representation by obviating repetition of features among similar schema.

Feature hierarchies are partitioned by schema type, since different features are relevant for characterizing different types of schema. Perceptual aspects have features relevant to their 2D surfaces, perceptual objects have features relevant to 3D shape, perceptual groups possess configuration/arrangement features, and relational schemas have features expressing the spatial or temporal roles that participating schemas play in the relation (eg, upper or lower role in the "above" relation). By and large, each feature hierarchy provides spatiotemporal descriptions of perceptual observations of a certain kind, organized into three discernible levels of generality, usually called the superordinate, entry, and subordinate levels (Jolicoeur, Gluck, & Kosslyn, 1984).

The middle *entry* level of the hierarchy contains the most readily available object schemas that we can assign observations to (eg, typical chair features, typical dog features). The *superordinate* level contains more general schemas (eg, possessing global features of furniture, animalism), while the *subordinate* level contains schemas carrying more specialized features (eg, rocking chair features, collie features). Experts in various fields seem to develop an additional level of schema organization that might be called the *sub-subordinate* level. For example, a dog breeder might develop schemas for recognizing specialized breeds of collie. Infants even become experts at the subordinate level in identifying their mothers, probably based on a variety of perceptual nuances, which in this case happen to characterize a single individual. In general, schemas seem to develop rapidly at all levels from infancy, although infants younger than four months may not yet have developed the fine perceptual discriminations required to support effective recognition at the subordinate level (Eimas, Quinn, & Cowan, 1994).

It is difficult to give an accurate verbal description of the meaning of such perceptual bundles of features. Schemas are different from the conceptual "categories" we talk about using language—use of the word "typical" above is an attempt to finesse the issue for the entry level. For example, the conceptual category "bird" includes all kinds of birds ranging from typical birds like robins and blue jays to atypical ones like penguins and ostriches. Typical birds have a more-or-less stable cluster of features (eg, size, shape, movement) not shared by atypical birds. Typical birds therefore are probably initially recognized via schemas at the entry level (in the absence of special circumstances like serious bird watching). Alternately, atypical birds are likely recognized initially at the subordinate level, since their features are distinct from typical birds, requiring a different, more specific, schematic representation. In general, any perceptual schema (used for recognition) is characterized strictly by its perceptual features, and exhibits a graded extension depending on how well an observation matches those features. Again, schemas deal with how concrete spatiotemporal things appear, rather than with what they are.

Superordinate and subordinate perceptual schemas also possess fewer features than entry level schemas. For example, the principle perceptual feature for a prototypical animal (from studies with infants) mainly seems to involve self-motion (Mandler, 2004), while for a prototypical collie it is long gold and white hair, and long snout. (In the current perceptual context "self-motion" means something like independent, feature, animal-like perambulations, rather than agentive attribution, which is conceptual.) Contrast these relatively few features with the more numerous highly correlated perceptual features of a typical dog: four leg-parts, tail-part, furry texture, bone-chewing, stick-chasing, barking, etc, etc.

Experiments by Gauthier, Anderson, et al, suggest that subordinate level processing involves the fusiform gyrus, possibly reflecting the increased spatial attention needed for fine discrimination (Gauthie, Tarr, Anderson, & Skudlarski, 1999). Other experiments (Jolicoeur, Gluck, & Kosslyn, 1984) suggest that semantic memory in the left inferior temporal gyrus is an important locus for superordinate level processing, but such semantic processing may relate more to conception than perception. Entry level processing, by comparison, is widespread, involving mostly regions of the left hemisphere (Laeng, Zarrinparb, & Kosslyn, 2003).

The Interface Between Perception & Conception

The current hypothesis under consideration is that perceptual schemas are the last of a long chain of perceptual representational structures extending back to the earliest sensory maps. Perceptual information processing progresses more-or-less automatically in stages marked by such intermediate representations. The proposed stages, depicted for a generic perceptual information processing channel, are outlined in the summary figure below.

The conceptual system is envisioned as a distinct functional component, developing separately from the perceptual system, but interacting with it. The conceptual system receives perceptual information from the perceptual schemas.[211] The content of schemas spans the spectrum of perceptually represented information, ranging from attributes like the local boundary between a shadow and a lighter background to full-fledged perceptual objects and perceptual relationships. Perceptual schemas are the grist for the

[211] The conceptual system also can affect perception, possibly through the local enhancement/bias of perceptual schemas.

conceptual mill in the sense that they are readily turned into conceptual representations by the conceptual system.[212]

Mediating between the perceptual and conceptual systems are the attentional processes. We are drawn to the information in the various perceptual representations either by its salience (bottom up) or by perceptual priming and conceptual goals (top down). The attentional processes select one from among the many perceptual representations to focus on at any one time (excluding perhaps the low-level sensory field maps).[213] Attention seems to shine a metaphorical spotlight on the contents of perceptual representations and attempts to validate only meaningful schema, as illustrated below. Some local perceptual adjustments (or distortions) may be needed to clarify the information, but we are generally pretty good at extracting even the faintest meaningful patterns. Attention is mostly serial rather than parallel, so the activation of schemas is sequential in character. Extracting the fuller meaning of a sequence of schemas is a conceptual task.

---

[212] But exactly how the continuous scalar characteristic values in a schema are turned into discrete propositional associations in a concept is still something of a mystery.

[213] This outline of how attention operates is significantly different from some other accounts. Here, attention is an independent global faculty, whereas others view attention as integrated into the perceptual processing stages, with feedback circuits from higher to lower stages controlling attention locally (Hawkins, 2004).

Appendix C: Late Perception

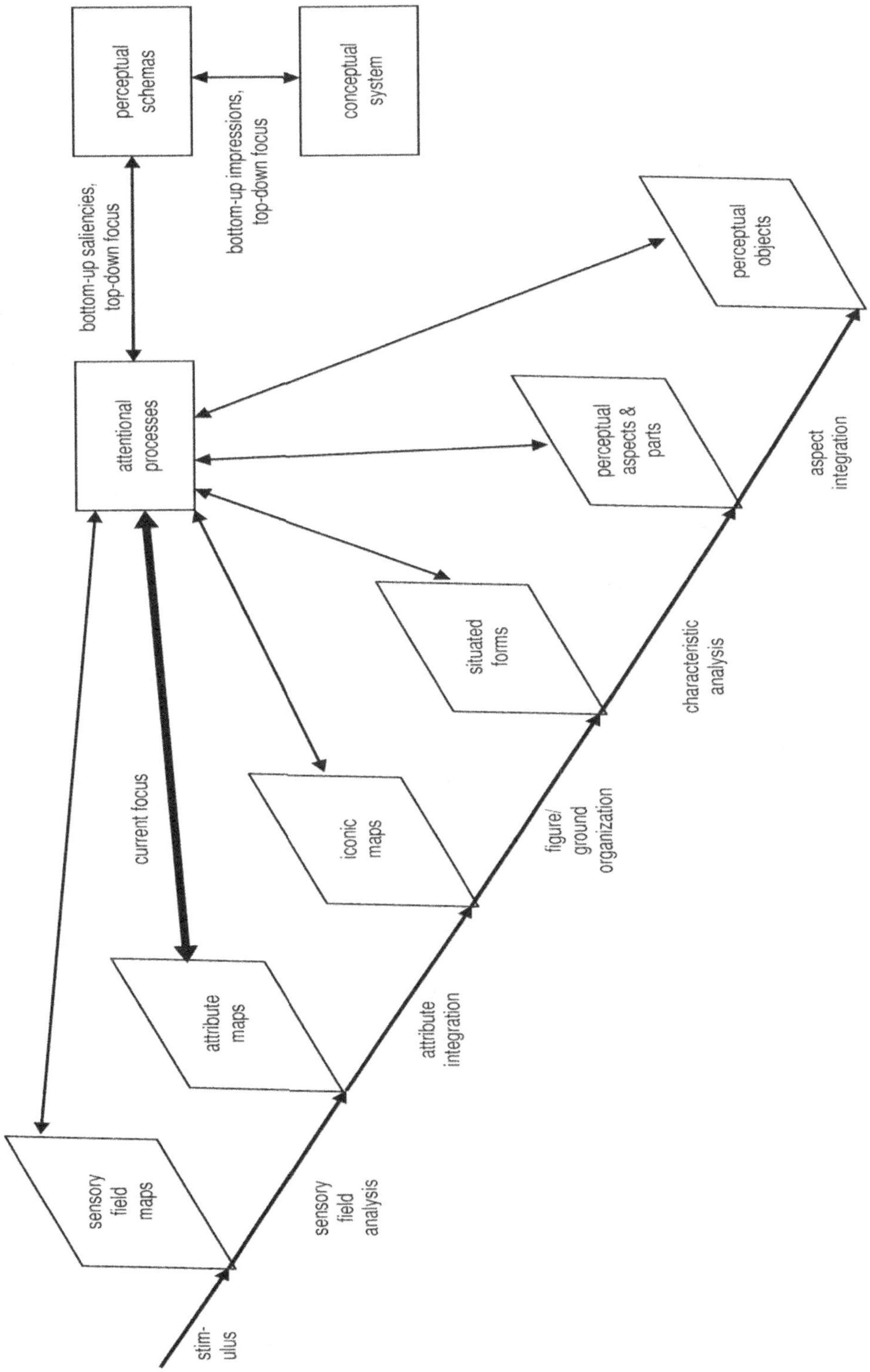

References

Aguiar, A., & Baillargeon, R. (2002). Development in young infants' reasoning about occuluded objects. *Cognitive Psychology 45(2)*, 267-336.

Andersen, H., Barker, P., & Chen, X. (2006). *The Cognitive Structure of Scientific Revolutions.* Cambridge, NY: Cambridge University Press.

Armstrong, D. M. (1989). *Universals: An Opinionated Introduction.* Boulder, CO: Westview Press.

Auyang, S. Y. (1998). *Foundations of Complex-System Theories: In Economics, Evolutionary Biology, and Statistical Physics.* Cambridge: Cambridge University Press.

Baars, B. J., & Gage, N. M. (2007). *Cognition, Brain, and Consciousness: Introduction to Cognitive Neuroscience.* London: Academic Press.

Baggott, J. (2013). *Farewell to Reality: How Modern Physics has Betrayed the Search for Scientific Truth.* New York, NY: Pegasus Books LLC.

Baker, C. I., Keysers, C., Jellema, T., Wicker, B., & Perrett, D. I. (2001). Neuronal representation of disappearing and hidden objects in temporal cortex of the macaque. *Experimental Brain Research 140*, 375-381.

Baldwin, D. A., Markman, E., & Melartin, R. (1993). Infants' Ability to Draw Inferences about Nonobvious Object Properties: Evidence from Exploratory Play. *Child Development, 64*, 711-728.

Ball, B. (2017, April 18). Alethic Pluralism and the Role of Reference in the Metaphysics of Truth. *The Southern Journal of Philosophy, 55*(1), 116-135.

Barsalou, L. W. (1992). Frames, concepts, and conceptual fields. In A. Lehrer, & E. Kittay, *Frames, Fields and Contrasts: New Essays in Semantical and Lexical Organization* (pp. 21-74). Hillsdale, NJ: Erlbaum.

Barsalou, L. W. (2003). Situated simulation in the human conceptual system. *Language and Cognitive Processes, 18*, 513-562.

Barsalou, L. W., Huttenlocher, J., & Lamberts, K. (1998). Basing categorization on individuals and events. *Cognitive Psychology, 36*, 203-272.

Bartoshuk, L. M., & Beauchamp, G. K. (1994). Chemical senses. *Annual review of psychology 45*, 419-449.

Biederman, I. (1987). Recognition-by-components: A theory of human image understanding. *Psychological Review 94*, 115-117.

Biederman, I., & Gerhardstein, P. C. (1993). Recognising depth-rotated objects: Evidence for three-dimensional viewpoint invariance. *Journal of Experimental Psychology: Human Perception and Performance, 19*, 1162-1182.

Billock, V. A., & Tsou, B. H. (2010). Seeing Forbidden Colors. *Scientific American, Feb*, 72-77.

Brown, H. I. (2007). *Conceptual Systems.* New York, NY: Routledge.

Burge, T. (1979). Individualism and the Mental. *Midwest Studies in Philosophy, 4*, 73-122.

Burge, T. (2010). *Origins of Objectivity.* Oxford, NY: Oxford University Press.

Caan, R. (1993). *Formal Semantics: An Introduction.* Cambridge: Cambridge University Press.

Campbell, K. (1997). The Metaphysic of Abstract Particulars. In D. H. Mellor, & A. Oliver, *Properties* (pp. 125-139). New York: Oxford University Press.

Carey, S. (2009). *The Origin of Concepts.* Oxford University Press.

Carey, S., & Xu, F. (2001). Infants' knowledge of objects: beyond object files and object tracking. *Cognition 80*, 179-213.

Casati, R., & Varzi, A. C. (1994). *Holes and Other Superficialities.* The MIT PRess.

Chaitin, G. (2005). *Meta Math!* New York: Pantheon Books.

Chalmers, D. (2006). The Foundations of Two-Dimensional Semantics. In M. Garcia-Carpintero, & J. Macia, *Two-Dimensional Semantics: Foundations and Applications* (pp. 55-140). Oxford: Oxford University Press.

Chi, M. T. (2008). Three Types of Conceptual Change: Belief Revision, Mental Model Transformation, and Categorical Shift. In S. Vosniadou, *Handbook of research on conceptual change* (pp. 61-82). Hillsdale, NJ: Erlbaum.

Chomsky, N. (2000). *New Horizons in the Study of Language and Mind.* Cambridge: Cambridge University Press.

Chun, M. M., & Potter, M. C. (1993). A two-stage model for multiple-target detection in rapid serial visual presentation. *Journal of Experimental Psychology: Human Perception & Performance 21*, 109-127.

Cimpian, A., & Salomon, E. (2014). The inherence heuristic: An intuitive means of making sense of the world, and a potential precursor to psychological essentialism. *BEHAVIORAL AND BRAIN SCIENCES, 37*, 461–527.

Cocchiarella, N. B. (2007). *Formal Ontology and Conceptual Realism.* Dordrecht, The Netherlands: Springer.

Collins, H. (2010). *Tacit and Explicit Knowledge.* Chicago: University of Chicago Press.

Coren, S., Ward, L. M., & Enns, J. T. (2004). *Sensation and Perception: Sixth Edition.* John Wiley & Sons, Inc.

Craik, F. I. (2007). Encoding: A Cognitive Perspective. In H. L. Roediger, Y. Dudai, & S. M. Fitzpatrick, *Science of Memory: Concepts* (pp. 129-135). New York: Oxford University Press.

Cree, G. S., & McRae, K. (2003). Analyzing the Factors Underlying the Structure and Computation of the Meaning of Chipmunk, Cherry, Chisel, Cheese, and Cello (and Many Other Such Concrete Nouns). *Journal of Experimental Psychology: General, Vol. 132, No. 2*, 163-201.

Crick, F., & Koch, C. (2003). A Framework for Consciousness. *Nature Neuroscience 6*, 119-126.

Croft, W., & Cruse, D. A. (2004). *Cognitive Linguistics.* Cambridge University Press.

Cruse, D. A. (1986). *Lexical Semantics.* Cambridge University Press.

Cruse, D. A. (2000). *Meaning in Language: An Introduction to Semantics and Pragmatics.* New York: Oxford University Press.

Daly, C. (1997). Tropes. In D. H. Mellor, & A. Oliver, *Properties* (pp. 140-159). New York: Oxford University Press.

Davidson, D. (1967). Truth and meaning. *Synthese, Volume 17, Number 1*, 304-323.

De Valois, R. L., Abramov, I., & Jacobs, G. H. (1966). Analysis of response patterns in LGN cells. *Journal of the Optical Society of America 56*, 966-977.

DeLoache, J. S., Uttal, D. H., & Rosengren, K. S. (2004). Scale errors offer evidence for a perception-action dissociation early in life. *Science 304*, 1027–1029.

Dennett, D. C. (1987). *The Intentional Stance.* Cambridge, MA: The MIT Press.

Dennett, D. C. (1991). *Consciousness Explained.* Little, Brown and Company.

Devitt, M. (2005). Rigid Application. *Philosophical Studies 125*, 139-165.

Dretske, F. (1981). *Knowledge and the Flow of Information.* Cambridge, MA: MIT Press.

Dretske, F. (1995). *Naturalizing the Mind.* The MIT Press.

Dummett, M. (1993). *Frege: philosophy of language, 2nd edition.* Harvard University Press.

Duncan, J. (2010). *How Intelligence Happens.* York: Yale University Press.

Dupré, J. (1993). *The Disorder of Things.* Cambridge, MA: Harvard University Press.

Edelman, S., & Duvdevani-Bar, S. (1997). A model of visual recognition and categorization. *Philosophical Transactions of the Royal Society of London, B352*, 1191-1202.

Eimas, P., Quinn, P., & Cowan, P. (1994). Development of exclusivity in perceptually-based categories of young infants. *Journal of Experimental Child Psychology 58*, 418-431.

Evans, V. (2009). *How Words Mean: Lexical Concepts, Cognitive Models, and Meaning Construction.* Oxford University Press.

Fauconnier, G. (1997). *Mappings in Thought and Language.* Cambridge University Press.

Flavell, J. H., Flavell, E. R., Green, F. L., & Korfmacher, J. E. (1990). Do young children think of television images as pictures or real objects? *Journal of Broadcasting and Electronic Media 34*, 399-419.

Floridi, L. (2003). Information. In L. (. Floridi, *The Blackwell Guide to the Philosophy of Computing and Information* (pp. 40-61). Oxford: Blackwell.

Floridi, L. (2005). Is Information Meaningful Data? *Philosophy and Phenomenologicfal Research: 70(2),* 351-370.

Fodor, J. A. (1980). On the impossibility of acquiring "more powerful" sutrctures: Fixation of belief and concept acquisition. In M. (. Piatelli-Palmerini, *Language and Learning* (pp. 142-162). Cambridge: Harvard University Press.

Fodor, J. A. (1994). Fodor's Guide to Mental Representation: The Intelligent Auntie's Vade-Mecum. In S. P. Stich, & T. A. Warfield, *Mental Representation: A Reader* (pp. 9-33). Cambridge, MA: Blackwell Publishers, Ltd.

Fodor, J. A. (2008). *LOT 2.* Oxford: Clarendon Press.

Freeman, W. J., Burke, B. C., & Holmes, M. D. (2003). Aperiodic phase re-setting in scalp EEG of beta-gamma oscillations by state transitions at alpha-theta rates. *Human Brain Mapping 19*, 248-272.

Funahashi, C. J., & Goldman-Rakic, P. S. (1989). Mnemonic coding of visual space in the monkey's dorsolateral prefrontal cortex. *Journal of Neurophysiology 61*, 331-349.

Gardenfors, P. (2000). *Conceptual Spaces: The Geometry of Thought.* The MIT Press.

Gauthie, I., Tarr, M. J., Anderson, A. W., & Skudlarski, P. (1999). Activation of the middle fusiform 'face area' increases with expertise in recognizing novel objects. *Nature Neuroscience 2*, 568-573.

Gazzaniga, M. S. (2002, Updated from the July 1998 issue). The Split Brain Revisited. *Scientific American.*

Gentner, D. (1988). Metaphor as structure mapping: The relational shift. *Child Development 59*, 47-59.

Gentner, D., Holyoak, K. J., & Kokinov, B. K. (2001). *The Analogical Mind.* The MIT Press.

Gershkoff-Stowe, L., & Rakison, D. H. (2005). *Building Object Categories in Developmental Time.* Mahwah, NJ: Lawrence Erlbaum Associates.

Gibson, J. J. (1966). *The Senses Considered as Perceptual Systems.* Boston: Houghton Mifflin.

Gibson, J. J. (1979). *The Ecological Approach to Visual Perception.* Boston: Houghton Mifflin.

Goodale, M. A., & Milner, A. D. (1991). A neurological dissociation between perceiving objects and grasping them. *Nature 349*, 154-156.

Gopnik, A., Glymour, C., Sobel, D., Schulz, L., Kushnir, T., & Danks, D. (2004). A theory of causal learning in children: Causal maps and Bayes nets. *Psychological Review, 111*, 1-31.

Gould, S. J. (2002). *The Structure of Evolutionary Theory.* Belknap Press.

Guarino, N., & Welty, C. (2000). Ontological Analysis of Taxonomic Relationships. *Proceedings of ER-2000: The International Conference on Conceptual Modeling.* Springer-Verlag.

Hale, B., & Wright, C. (1997). *A Companion to the Philosophy of Language.* Oxford: Blackwell Publishers.

Hard, B. M., Lozano, S. C., & Tversky, B. (2006). Hierarchical encoding: Translating perception into action. *Journal of Experimental Psychology: General 135*, 588-608.

Hawkins, J. (2004). *On Intelligence.* Owl Books.

Helbig, H. (2006). *Knowledge Representation and the Semantics of Natural Language.* Berlin Heidelberg: Springer-Verlag.

Hintzman, D. L., & Ludlam, G. (1980). Differential forgetting of prototypes and old instances: simulation by an exemplar-based classification model. *Memory & Cognition 8*, 378-382.

Horridge, M. (2009). *A Practical Guide To Building OWL Ontologies Using Protege 4, Edition 1.2.* Manchester, England: The University Of Manchester.

Irwin, D. E. (1992). Memory for position and identity across eye movements. *Journal of Experimental Psychology: Learning, Memory, & Cognition18*, 307-317.

Jackendoff, R. (2002). *Foundations of Language.* New York: Oxford University Press.

Johnson, K. O. (2001). The roles and functions of cutaneous mechanoreceptors. *Current Opinion in Neurobiology, 11*, 455-461.

Jolicoeur, P., Gluck, M., & Kosslyn, S. M. (1984). Pictures and names: Making the connexion. *Cognitive Psychology 19*, 31-53.

Kahneman, D., & Treisman, A. (1984). Changing views of attention and automaticity. In R. Parasuraman, & D. R. Davies, *Varieties of Attention* (pp. 29-62). New York: Academic Press.

Kandel, E. R. (2006). *In Search of Memory.* New York: W. W. Norton & Company.

Kaplan, D. (1977 & 1989). Demonstratives. In J. Almog, J. Perry, & H. Wettstein, *Themes from Kaplan* (pp. 481-563). Oxford: Oxford University Press.

Kaufman, E. L., Lord, M. W., Reese, T. W., & Volkmann, J. (1949). The discrimination of visual number. *American Journal of Psychology 62*, 498–525.

Kay, P., & McDaniel, C. K. (1978). The linguistic significance of the meanings of basic color terms. *Language 54*, 610-646.

Kellman, P. J., & Shipley, T. F. (1991). A theory of visual interpolation in object perception. *Cognitive Psychology 23(2)*, 141-221.

Koch, C. (2004). *The Quest for Consciousness: A Neurobiological Approach.* Englewood: Roberts and Company Publishers.

Koenderink, J. J., van Doorn, A. J., & Kappers, A. M. (1996). Pictorial surface attitude and local depth comparisons. *Perception & Psychophysics 58(2)*, 163-173.

Konorski, J. (1967). *Integrative Activity of the Brain: An Interdisciplinary Approach.* Chicago: University of Chicago Press.

Kosslyn, S. M. (1980). *Image and Mind.* Cambridge: Harvard University Press.

Kosslyn, S. M., Ball, T. M., & Reiser, B. J. (1978). Visual images preserve metric spatial information: Evidence from studies of image scanning. *Journal of Experimental Psychology: Human Perception & Performance 4*, 47-60.

Kreeft, P. (2010). *Socratic Logic.* St. Augustine's Press.

Kreiman, G. (2001). On the neuronal activity in the human brain during visual recognition, imagery and binocular rivalry. *PhD Thesis, California Institute of Technology*.

Kuhn, T. S. (2000). *The Road Since Structure.* Chicago, IL: The University of Chicago Press.

Kumaran, D., Summerfield, J., Hassabis, D., & Maguire, E. (2009). Tracking the Emergence of Conceptual Knowledge during Human Decision Making. *Neuron, Volume 63, Issue 6*, 889-901.

Laeng, B., Zarrinparb, A., & Kosslyn, S. M. (2003). Do separate processes identify objects as exemplars versus members of basic-level categories? Evidence from hemispheric specialization. *Brain and Cognition 53*, 15-27.

Lakoff, G. (1987). *Women, Fire, and Dangerous Things: What Categories Reveal about the Mind.* The University of Chicago Press.

Lakoff, G., & Johnson, M. (1999). *Philosophy in the Flesh.* New York, NY: Basic Books.

Lawton, C. A. (1994). Gender differences in way-finding strategies: Relationship to spatial ability and spatial anxiety. *Sex Roles, 30*, 765-779.

Le Bihan, D., Turner, R., Zeffiro, T. A., Cuenod, C. A., Jezzard, P., & Bonnerot, V. (1993). Activation of human primary visual cortex during visual recall: A magnetic resonance imaging study. *Proceedings of the National Academy of Sciences USA 90*, (pp. 11802-11805).

Lepore, E., & Ludwig, K. (2005). *Donald Davidson: Meaning, Truth, Language, and Reality.* Oxford University Press.

Leslie, S.-J. (2008). Generics: Cognition and Acquisition. *Philosophical Review, v117 n1*, 1-47.

Livingstone, M. S., & Hubel, D. (1988). Segregation of form, color, movement, and depth: Anatomy, physiology, and perception. *Science 240 (4853)*, 740-749.

Logan, W. B. (1989). *The Smithsonian Guide to Historic America: The Pacific States.* Harry N Abrams, Inc.

Lowe, E. J. (1997). Objects and criteria of identity. In B. Hale, & C. Wright, *A Companion to the Philosophy of Language* (pp. 613-633). Malden, MA: Blackwell Publishers Ltd.

Lowe, E. J. (1999). *The Possibility of Metaphysics: Substance, Identity, and Time.* Oxford: Oxford University Press.

Lowe, E. J. (2006). *The Four-Category Ontology: A Metaphysical Foundation for Natural Science.* Oxford: Oxford University Press.

Machery, E. (2009). *Doing Without Concepts.* Oxford University Press.

Mack, A. (1978). Three modes of visual perception. In M. H. Pick, *Modes of Perceiving and Information Processing* (pp. 171-186). Hillsdale, NJ: Erlbaum.

MacLean, P. D. (1990). *The Triune Brain in Evolution: Role in Paleocerebral Function.* New York: Plenum Press.

Macnamara, J. (1986). *Border Dispute: The Place of Logic in Psychology.* Cambridge: MIT Press.

Mandler, J. M. (2004). *The Foundations of Mind: Origins of Conceptual Thought.* New York: Oxford University Press.

Mareschal, D., Quinn, P., & Lea, S. (2010). *The Making of Human Concepts.* New York: Oxford University Press.

Marr, D., & Nishihara, H. K. (1978). Representation and recognition of the spatial organisation of three-dimensional shapes. *Proceedings of the Royal Society of London, B200*, 269-294.

Marslen-Wilson, W. (1989). *Lexical Representation and Process.* MIT Press.

McCarthy, R. A., & Warrington, E. K. (1986). Visual associative agnosia: a clinico-anatomical study of a single case. *Journal of Neurological Neurosurgery and Psychiatry 49*, 1233–1240.

Medin, D. L., & Schaffer, M. M. (1978). Context theory of classification learning. *Psychological Review 85*, 207-238.

Mellor, D. H., & Oliver, A. (1997). *Properties.* New York: Oxford University Press.

Millikan, R. G. (1999). A Common Structure for Concepts of Individuals, Stuffs, and Real Kinds. In E. Margolis, & S. Laurence, *Concepts: Core Readings* (pp. 525-547). Cambridge, MA: MIT Press.

Millikan, R. G. (2000). *On Clear and Confused Ideas: An Essay about Substance Concepts.* Cambridge: Cambridge University Press.

Millikan, R. G. (2004). *Varieties of Meaning.* The MIT Press.

Mintz, T. B., & Gleitman, L. R. (2002). Adjectives really do modify nouns: The incremental and restricted nature of early adjective acquisition. *Cognition, 84*, 267-293.

Mitchell, M. (2009). *Complexity: A Guided Tour.* Oxford University Press.

Montmayeur, J. P., & Matsunami, H. (2002). Receptors for bitter and sweet taste. *Current Opinion in Neurobioloy 12*, 366-371.

Moscovici, S., & Personnaz, B. (1991). Studies in social influence: VI. Is Lenin orange or red? Imagery and social influence. *European Journal of Social Psychology 21*, 101-118.

Murphy, G. L. (2002). *The Big Book of Concepts.* The MIT Press.

Murphy, G. L., & Medin, D. L. (1985). The role of theories in conceptual coherence. *Psychological Review 92*, 289-316.

Nadal, L. (2007). Consolidation: The demise of the fixed trace. In H. L. Roediger III, Y. Dudai, & S. M. Fitzpatrick, *Science of Memory: Concepts* (pp. 177-181). New York: Oxford University Press.

Nersessian, N. J. (2008). *Creating Scientific Concepts.* Cambridge, MA: The MIT Press.

Nirenburg, S., & Raskin, V. (2004). *Ontological Semantics.* Cambridge: The MIT Press.

Nosofsky, R. M. (1986). Attention, similarity, and the identification–classification relationship. *Journal of Experimental Psychology: General, 115*, 39-57.

Nosofsky, R. M. (1988). Similarity, frequency, and category representations. *Journal of Experimental Psychology: Learning, Memory, and Cognition, 14*, 54-65.

O'Keefe, J., & Nadel, L. (1978). *The Hippocampus as a Cognitive Map.* Oxford: Clarendon Press.

Palmer, S. E. (1999). *Vision Science: Photons to Phenomenology.* The MIT Press.

Palmer, S. E., & Rock, I. (1994). Rethinking Perceptual Organization: The Role of Uniform Connectedness. *Psychonomic Bulletin and Review 1*, 29-55.

Palmer, S. E., Rosch, E., & Chase, P. (1981). Cannonical perspective and the perception of objects. In J. Long, & A. Baddeley, *Attention and Performance (vol 9)* (pp. 135-151). Hillsdale: Erlbaum.

Parsons, T. (1990). *Events in the Semantics of English: A Study in Sobatomic Semantics.* The MIT Press.

Pauling, L. (1970). *General Chemistry.* New York: Dover Publications, Inc.

Peterson, M. A., & Gibson, B. S. (1994). Must figure-ground organization precede object recognition? An assumption in peril. *Psychological Science 5*, 253-259.

Pollack, I. (1952). The information of elementary auditory displays. *Journal of the Acoustical Society of America 24*, 745-749.

Potter, M. C. (1976). Short-term conceptual memory for pictures. *Journal of Experimental Psychology: Human Learning and Memory 2*, 509-522.

Putnam, H. W. (1975). The Meaning of 'Meaning'. *Minnesota Studies in the Philosophy of Science 7*, 131-193.

Putnam, H. W. (1981). *Reason, Truth, and History.* Cambridge: Cambridge University Press.

Pylyshyn, Z. W. (2003). *Seeing and Visualizing: It's Not What You Think.* Massachusetts Institute of Technology.

Quine, W. V. (1951). Two Dogmas of Empiricism. *The Philosophical Review 60*, 20-43.

Quine, W. V. (1953). *From a Logical Point of View.* New York: Harper & Row.

Quine, W. V. (1960). *Word & Object.* The MIT Press.

Quine, W. V. (1969). *Ontological Relativity & Other Essays.* New York: Columbia University Press.

Quine, W. V. (1987). *Quiddities: An Intermittently Philosophical Dictionary.* President and Fellows of Harvard College.

Quine, W. V., & Ullian, J. S. (1978). *The Web of Belief.* McGraw-Hill.

Quirk, R., Greenbaum, S., Leech, G., & Svartvik, J. (1985). *A Comprehensive Grammar of the English Language.* Longman.

Quiroga, Q. R., Reddy, L., Kreiman, G., Koch, C., & Fried, I. (2005). Invariant visual representation by single neurons in the human brain. *Nature 435*, 1102-1107.

Rescorla, R. A. (1988). Pavlovian Conditioning: It's not what you think it is. *American Psychologist, 43(3)*, 151-160.

Robertson, L. C., Lamb, M. R., & Knight, R. T. (1988). Effects of lesions of the temporal-parietal junction on perceptual and attentional processing in humans. *Journal of Neuroscience 8*, 3757-3769.

Rock, I. (1975). *An Introduction to Perception.* New York: Macmillan.

Rock, I. (1983). *The Logic of Perception.* Cambridge: The MIT Press.

Rogers, T. T., & McClellan, J. L. (2008). Precis of Semantic Cognition: A Parallel Distributed Processing Approach. *Behavioral and Brain Sciences 31*, 618-632.

Rosch, E. (1975). Cognitive Representation of Semantic Categories. *Journal of Experimental Psychology (General) 104*, 192-233.

Rosch, E. (1978). Principles of Categorization. In E. Rosch, & B. B. Lloyd, *Cognition and Categorization* (pp. 27-48). Hillsdale: Lawrence Erlbaum Associates.

Rosch, E., & Mervis, C. B. (1975). Family resemblances: Studies in the internal structure of categories. *Cognitive Psychology 7*, 573-605.

Rosch, E., Mervisa, C., Gray, W., Johnson, D., & Boyes-Braem, P. (1976). Basic objects in natural categories. *Cognitive Psychology*, 382-439.

Ross, B., Perkins, S., & Tenpenny, P. (1990). Reminding-based category learning. *Cognitive Psychology 22*, 460-492.

Savage-Rumbaugh, E. S., Murphy, J., & al. (1993). Language comprehension in ape and child. *Monographs of the society for Research in Child Development, 58 (3-4, sn 333).*

Schneider, S. (2011). *The Language of Thought: A New Philosophical Direction.* Cambridge: MIT Press.

Schyns, P. G., Goldstone, R. L., & Thibaut, J. (1998). The development of features in object concepts. *Behavioral and Brain Sciences 21*, 1-54.

Searle, J. R. (2004). *Mind: A Brief Introduction.* New York: Oxford University Press.

Shannon, C. E. (1948). A Mathematical Theory of Communication. *Bell System Technical Journal, 27,* 379–423 & 623–656.

Shepp, B. (1978). From perceived similarity to dimensional structure: A new hypothesis about perspective development. In E. Rosch, & B. Lloyd, *Cognition and Categorization* (pp. 135-167). Hillsdale, NJ: Erlbaum.

Shipley, T. F. (2008). An Invitation to an Evet. In T. F. Shipley, & J. M. Zacks, *Understanding Events* (pp. 3-30). New York: Oxford University Press, Inc.

Siegler, R. S. (1983). How Knowledge Influences Learning. *American Scientist, 71*, 631-638.

Simons, P. (1987). *Parts: A Study in Ontology.* Oxford: Oxford University Press.

Smith, E. E., Osherson, D. N., Rips, L. J., & Keane, M. (1988). Combining concepts: A selective modification model. *Cognitive Science, 12*, 485-527.

Smith, L. B., Gasser, M., & Sandhofer, C. M. (1997). Learning to Talk about the Properties of Objects: A Network Model of the Development of Dimensions. In Elsevier, *Psychology of Learning and Motivation* (pp. 219–255). Academic Press, Inc.

Soames, S. (2010). *What Is Meaning?* Princeton: Princeton University Press.

Sowa, J. F. (2000). *Knowledge Representation: Logical, Philosophical, and Computational Foundations.* Brooks/Cole.

Spelke, E. S. (1990). Principles of Object Perception. *Cognitive Science 14*, 29-56.

Spelke, E. S., & Kinzler, K. D. (2007). Core Knowledge. *Developmental Science 10*, 89–96.

Squire, L. R. (1993). The organization of declarative and nondeclarative memory. In T. Ono, L. R. Squire, M. E. Raichle, D. I. Perrett, & M. Fukuda, *Brain Mechanisms of Perception and Memory: From Neuron to behavior* (pp. 219-227). New York: Oxford University Press.

Sternberg, R. J. (2003). *Cognitive Psychology: Third Edition.* Belmont: Wadsworth.

Strauss, M. S. (1979). The abstraction of prototypical information by adults and l0-monthold. *Journal of Experimental Psychology: Human Learning and Memory 50*, 618-632.

Talmy, L. (2003). *Toward a Cognitive Semantics, Volume 1: Concept Structuring Systems.* The MIT Press.

Tarr, M. J., & Bulthoff, H. H. (1995). Is human object recognition better described by geon structural descriptions or by multiple views? Comment on Biederman and Gerhardstein 1993. *Journal of Experimental Psychology: Human Perception and Performance, 21*, 1494-1505.

Thagard, P. (2012). *The Cognitive Science of Science.* Cambridge: The MIT Press.

Thornton, C. (2000). *Truth from Trash: How Learning Makes Sense.* Cambridge: The MIT Press.

Toyabe, S., Sagawa, T. U., Muneyuki, E., & Sano, M. (2010). Experimental demonstration of information-to-energy conversion and validation of the generalized Jarzynski equality. *Nature Physics, doi:10.1038/nphys1821*.

Ullman, S. (1998). Three-dimensional object recognition based on the combination of views. *Cognition, 67*, 21-44.

van Inwagen, P. (2001). *Ontology, Identity, and Modality: Essays in Metaphysics.* Cambridge University Press.

van Tonder, G. J., & Ejima, Y. (2000). From image segmentation to anti-textons. *Perception*, 1231-1247.

von Baeyer, H. C. (2003). *Information: the new language of science.* Great Britain: Weidenfeld & Nicolson.

Wegner, D. M. (2002). *The Illusion of Conscious Will.* Cambridge: The MIT Press.

Wellman, H. M., & Gelman, S. A. (1992). Cognitive development: Foundational theories of core domains. *Annual Review of Psychology 43*, 337-375.

Wheeler, J. A. (1990). Information, Physics, Quantum: The Search for Links. In W. H. Zurek, *Complexity, Entropy and the Physics of Information.* Santa Fe Institute: Westview Press.

Wiener, N. (1965). *Cybernetics or Control and Communication in the Animal and the Machine, 2nd ed.* Cambridge: The MIT Press.

Wierzbicka, A. (2015, March 21). *Semantic explications.* Retrieved from Griffith University: http://www.griffith.edu.au/humanities-languages/school-languages-linguistics/research/natural-semantic-metalanguage-homepage/semantic-explications

Williams, D. C. (1997). On the Elements of Being: I. In D. H. Mellor, & A. Oliver, *Properties* (pp. 112-124). New York: Oxford University Press.

Wittgenstein, L. (1963). *Philosophical Investigations: Section 43.* (G. Anscombe, Trans.) Oxford: Basil Blackwell.

Wynn, K. (1992). Addition and subtraction by human infants. *Nature 358*, 749-750.

Xu, F., & Carey, S. (1996). Infants' metaphysics: the case of numerical identity. *Cognitive Psychology 30*, 111-153.

Zalta, E. N. (1988). *Intensional Logic and the Metaphysics of Intentionality.* Cambridge, MA and London: The MIT Press.

Zalta, E. N. (1999). *Principia Metaphysica.* http://mally.stanford.edu/principia.pdf.

Zihl, J., Von Cramon, D., & Mai, N. (1983). Selective disturbance of movement vision after bilateral brain damage. *Brain 106*, 313-340.

SUBJECT INDEX

D

E

F

G

H

I

K

L

M

N

O

P

Q

R

S

T

U

V

W

Made in the USA
Las Vegas, NV
15 June 2026

48876731R10195